ASPECTS OF HOMOGENEOUS CATALYSIS

Volume 3

ASPECTS OF HOMOGENEOUS CATALYSIS

A Series of Advances

EDITED BY
RENATO UGO
ISTITUTO DI CHIMICA GENERALE ED INORGANICA
MILAN UNIVERSITY

VOLUME 3

D. REIDEL PUBLISHING COMPANY
DORDRECHT-HOLLAND / BOSTON-U.S.A.

The Library of Congress Cataloged the First Issue of this Work as Follows:

Ugo, Renato (ed.)

Aspects of homogeneous catalysis. vol. 1.

1970–

Milano, C. Manfredi.

v. illus. 25 cm. annual.

'A Series of Advances'.

Editor: 1970– R. Ugo.

1. Catalysis–Periodicals. I. Ugo, Renato (ed.).

QD501.A83 541'.395 72–623953

ISBN 90–277–0786–3

Published by D. Reidel Publishing Company,
P.O. Box 17, Dordrecht, Holland

Sold and distributed in the U.S.A., Canada and Mexico
by D. Reidel Publishing Company, Inc.
Lincoln Building, 160 Old Derby Street, Hingham,
Mass. 02043, U.S.A.

Printed in The Netherlands

Editorial Board

Contents of Volume 3

Transition Metal Complexes as Catalysts for the Addition of Oxygen to Reactive Organic Substrates

J. E. Lyons

Structure and Electronic Relations Between Molecular Clusters and Small Particles: An Essay to the Understanding of Very Dispersed Metals
J. M. Basset and R. Ugo

Asymmetric Hydrosilylation by Means of Homogeneous Catalysts with Chiral Ligands
Iwao Ojima, Keiji Yamamoto, and Makoto Kumada

ASPECTS OF HOMOGENEOUS CATALYSIS

Volume 3

Chapter 1

Transition Metal Complexes as Catalysts for the Addition of Oxygen to Reactive Organic Substrates

JAMES E. LYONS

Sun Oil Company, P.O. Box 1135, Marcus Hook, Pa. 19061, U.S.A.

1. INTRODUCTION

In recent years, the liquid phase oxidation of organic substrates using transition metal compounds as catalysts has become a profitable means of obtaining industrially important chemicals. Millions of tons of valuable petrochemicals are produced in this manner annually [1]. Typical examples of such processes are the production of vinyl acetate or acetaldehyde *via* the Wacker process, equations (1) and (2); the Mid-Century process for the oxidation of methyl aromatics, such as *p*-xylene to terephthalic acid, equation (3); and the production of propylene oxide from propylene using alkyl hydroperoxides, equation (4).

$$CH_2 = CH_2 + 1/2\ O_2 \xrightarrow[-H_2O]{PdCl_2,\ CuCl_2} CH_3CHO \tag{1}$$

$$CH_2 = CH_2 + 1/2\ O_2 + HOAc \xrightarrow[CuCl_2]{PdCl_2} CH_2 = CHOAc + H_2O \tag{2}$$

$$p\text{-}CH_3C_6H_4CH_3 + O_2 \xrightarrow{Co(OAc)_2} p\text{-}HOOCC_6H_4COOH \tag{3}$$

$$CH_3CH = CH_2 + ROOH \xrightarrow{\text{Mo compounds}} CH_3\underset{\diagdown O \diagup}{CH-CH_2} + ROH \tag{4}$$

The vast majority of liquid phase transition metal catalyzed oxidations of organic compounds fall into these three broad categories: (*a*) free radical autoxidation reactions, (*b*) reactions involving nucleophilic attack on coordinated substrate such as the Wacker process, or (*c*) metal catalyzed reactions of organic substrates with hydroperoxides. Of these three classes of oxidations only the first represents the actual interaction of dioxygen with an organic substrate. The function of oxygen in the Wacker process is simply to re-oxidize the catalyst after each cycle [2].

Although some autoxidation reactions can be controlled in a useful way, organic substrates more often tend to oxidize in an unselective manner. Oxygen is, after all, a highly reactive molecule and many reaction pathways are open to it. It is imperative, however, that a reaction be selective if it is to have utility either as an economically attractive process or a convenient laboratory synthesis. Despite the rapid advances made during the past decade in the area of liquid phase oxidation, much is still to be learned concerning the efficient control of reactions of molecular oxygen.

Stimulated by recent success in the catalytic activation of other small molecules, an intensive effort has been made to activate molecular oxygen by coordination to a metal center with the ultimate goal being catalytic transfer of oxygen to a reactive substrate in a selective manner. It is with this approach to the chemistry of metal catalyzed oxidation that this review will be concerned. Therefore, the subject matter will deal primarily with catalytic oxidation reactions in which at least one step in the catalytic cycle is likely to involve coordination of oxygen and/or the reacting substrate. For this reason, only the oxidation of those organic substrates which are known to be easily incorporated into the coordination sphere of metal complexes will be considered. The oxidation of aliphatic hydrocarbons, for example, will not be discussed but rather emphasis will be placed on the oxidation of those organic compounds which possess lone pairs of *p*-electrons such as oxygen, nitrogen, sulfur and phosphorous compounds and those having π-systems susceptible to coordination such as olefins, acetylenes and some aromatic hydrocarbons. In addition, detailed consideration will be given only to those reactions which actually incorporate oxygen into the reactive substrate. The large number of formal oxidations of organic substrates which do not result in oxygenation such as conversion of alcohols to carbonyl compounds, amines to nitriles, radical eliminations, and oxidative coupling reactions, will not be discussed in depth. In addition the catalysts to be covered will be limited to complexes of the first row transition metals as well as second and third row metals of groups VI, VII, and VIII. Lanthanide or actinide complexes will not be treated in this chapter. Furthermore, emphasis will be placed on the literature of the past decade since earlier literature has been adequately summarized in numerous accounts.

Prior to the discussion of catalytic oxidation reactions it will be instructive to consider the various ways in which oxygen can interact with metal complexes. Since this subject has been ably and comprehensively reviewed recently [3–7], only selected aspects of oxygen coordination of particular relevance to oxidation will be treated.

2. STRUCTURE AND BONDING IN DIOXYGEN COMPLEXES

Ground state molecular oxygen is paramagnetic with the two unpaired electrons occupying degenerate π^* antibonding orbitals. The chemistry of ground state oxygen may be described as a series of electron transfers to and from the neutral molecule [3]. The effect of adding electrons into the partially vacant antibonding orbitals of the ground state oxygen molecule is to decrease the O–O bond dissociation energy

Table 1

PROPERTIES OF SOME DIOXYGEN SPECIES[a]

Species	Bond order	O–O bond length, Å	O–O vibrational frequency, cm^{-1}	O–O bond dissoc'n energy, k cal/mole
O_2^+	2.5	1.12	1905	154
O_2	2.0	1.21	1580	118
O_2^-	1.5	1.33	1113	93.8
$O_2^=$	1.0	1.49	802	49.5

[a] Data from Table 1, reference [3] unless otherwise noted.

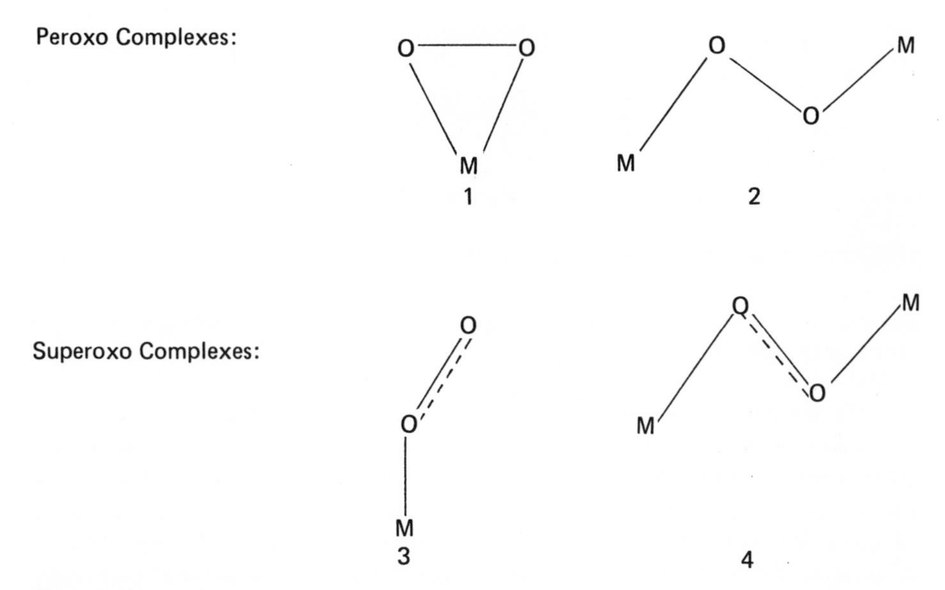

Figure 1. Geometries of peroxo and superoxo complexes.

and the O–O vibrational frequency and to increase the O–O interatomic distance [3]. Table 1 lists some properties of dioxygen species of relevance to this discussion.

Dioxygen is covalently bound to transition metal centers as opposed to the ionic superoxide (O_2^-) or peroxide (O_2^{2-}) species listed in Table 1. Nonetheless, covalently bound dioxygen is similar in some respects to either of these ionic species (i.e. O–O bond length and vibrational frequency). Transition metal dioxygen complexes may be categorized into two classes: peroxo and superoxo which can be thought of as the formal (although not chemical) covalent analogs of the ionic peroxide or superoxide. In superoxo complexes dioxygen is a univalent anionic ligand whereas in peroxo complexes oxygen is formally divalent. Vaska represents the types of bonding which occur in these complexes as shown in Figure 1 above [3].
Examples of both peroxo and superoxo complexes have been identified in which oxygen is coordinated either to one metal center (1, 3) or to two metal centers (2, 4) [3]. Examples of peroxo complexes are far more numerous than are those of transition metal superoxo complexes. Of the four different types of bonding shown in Figure 1, peroxo complexes of type 1 are the most common while superoxo complexes of type 4 are exhibited only in the case of some Co(III) complexes. In fact, cobalt is the only metal for which examples of all four bonding modes have been observed to date [3].

Peroxo complexes of type 1, Figure 1, often referred to as side-bonded or π-bonded dioxygen complexes, are generally diamagnetic 1:1 complexes of dioxygen. Complexes of the following metals have been shown by X-ray crystallographic structure determination to exhibit this type of bonding: Ti, V, Nb, Cr, Mo, W, U, Co, Rh, Ir and Pt [3]. Several paramagnetic peroxo chromium complexes have also been pre-

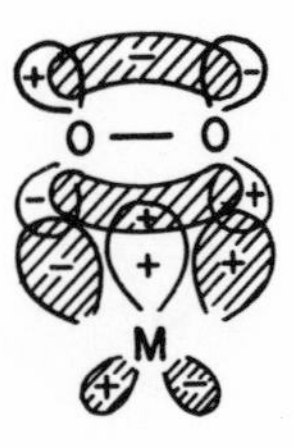

Figure 2. Bonding orbitals in a mononuclear transition metal peroxo complex.

pared [8]. In complexes of type 1, Figure 1, the O–O bond remains intact but is longer than that of ground state oxygen and each of the metal–oxygen bond lengths are the same.

The bonding in mononuclear transition metal peroxo complexes (1, Figure 1) has been described as involving two π bonds formed by transfer of electrons from filled metal d-orbitals into empty π^* orbitals of ligated dioxygen, and a σ bond formed by overlap of a filled dioxygen π bonding orbital with an unfilled d-orbital of the metal, Figure 2 [3]. As sufficient electron density is transferred to the dioxygen π^* orbitals, the O–O bond approaches a single bond with considerable lengthening relative to free O_2. The length and stability of the O–O bond in metal complexes may be of great importance to the use of these complexes as catalysts for oxidation reactions. Table 2 lists representative examples of some mononuclear peroxo complexes.

Transition metal peroxo complexes having two metal centers bound to a dioxygen (2, Figure 1) are also diamagnetic and have been reported for the metals: Mn, Fe, Co, Rh and Mo. An example of this type of complex is the diamagnetic μ-peroxobis(pentaaminecobalt)(4+). The bonding of cobalt to dioxygen in this complex has been proposed to involve a four center molecular orbital. A π bond is formed between the empty antibonding π^* orbitals of the oxygen, and two doubly occupied d-orbitals of the two cobalts [4], Figure 3. This type of bonding would bring spin density into the antibonding π^* orbital of the ligated dioxygen and thereby weaken the O–O bond.

Table 2 lists structural data for some representative peroxo complexes. It is interesting to note that for the majority of these complexes there is very little variation in O–O bond lengths despite differences in metal, ligands, valence state and structure. Fourteen of the twenty-two complexes listed in Table 2 have O–O bond lengths between 1.40 and 1.49 Å. In fact, the O–O bond length is not greatly affected by whether the dioxygen is coordinated to one or two metal centers [3]. The most

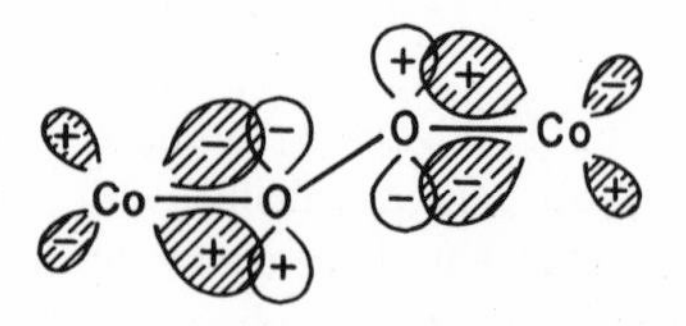

Figure 3. Bonding orbitals in a dinuclear cobalt μ-peroxo complex.

Table 2

ESTABLISHED STRUCTURES OF SOME TRANSITION METAL PEROXO COMPLEXES

Structure	X		O–O, Å	M–O, Å	Ref
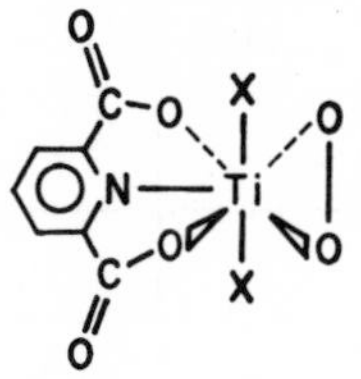 [structure: Ti pyridinedicarboxylate peroxo complex with axial X]	H_2O		1.46	1.85	11
	F		1.46	1.85	11
	M				
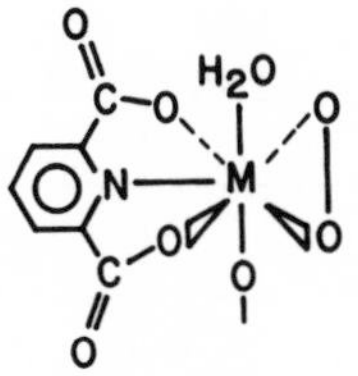 [structure: M pyridinedicarboxylate peroxo complex with H_2O]	Ti		1.45	1.89	12
	V		1.44	1.87	13
Structure	M	L–L			
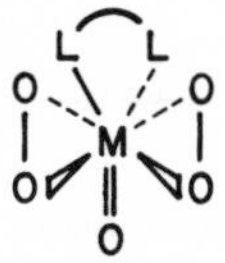 [structure: M(O)(O₂)₂(L–L)]	Cr	Bipy	1.40	1.85	14
	Mo	$C_2O_4^=$	1.55	1.97	15
	X	L			
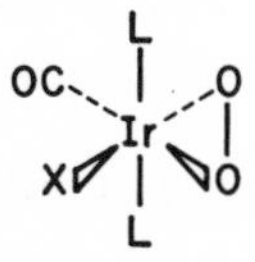 [structure: Ir(CO)X(O₂)L₂]	Cl	Ph_3P	1.30	2.07	16, 17
	Br	Ph_3P	1.36	2.00	18
	I	Ph_3P	1.51	2.06	19, 20
	Cl	Ph_2EtP	1.47	2.06	21
	M	L–L			
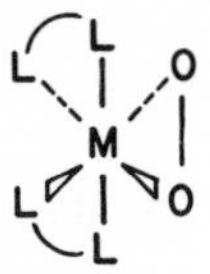 [structure: M(O₂)(L–L)₂]	Rh	$Ph_2PCH_2CH_2PPh_2$	1.42	2.03	9
	Ir	$Ph_2PCH_2CH_2PPh_2$	1.62	1.98	10, 116
			1.52	2.05	
	Co	$Ph_2PCH{=}CHPPh_2$	1.42	1.89	22

Structure	xS	O–O, Å	M–O, Å	Ref
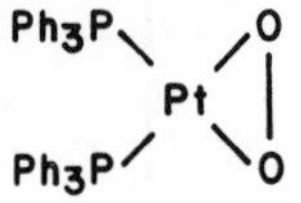	2.0 $CHCl_3$	1.51	2.01	24
	1.5 C_6H_6	1.45	2.01	25
	1.0 $C_6H_5CH_3$	1.26		23
	n L's			
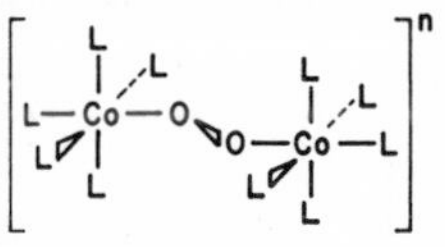	+4 10(NH_3)'s	1.47	1.88	26
	+4 2(en)'s, 2(dien)'s	1.49	1.90	27
	0 2($salen^=$)'s, 2(DMF)'s	1.34	1.91	28
	0 2($N_3C_{20}H_{23}^=$)'s	1.45	1.93	29
	–6 10(CN^-)'s	1.45	1.98	30
	L			
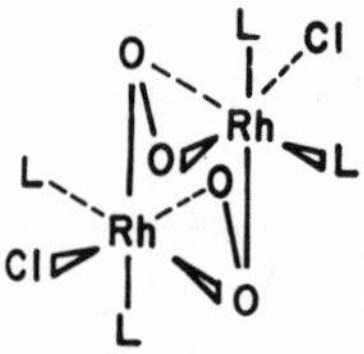	Ph_3P	1.44	2.08	31

notable exception to this constancy of bond lengths is a series of complexes of iridium. Ibers and co-workers first noted this rather remarkable trend in the lengthening of O–O bond distance as the ancillary ligands become more electron releasing (dp > PPh_2Et > PPh_3 and I > Br > Cl) [9, 10]. From the viewpoint of catalysis, an understanding of those factors which promote O–O bond weakening are most important, however, the trends observed in iridum complexes have no clear analogs as yet in other metal complexes.

Superoxo complexes (3, 4, Figure 1) are paramagnetic complexes in which either one or two metal centers may be coordinated to dioxygen. Table 3 lists structures and bond lengths for some representative superoxo complexes whose crystal structure has been determined. The majority of superoxo complexes whose crystal structure has been determined are complexes of cobalt. The bonding in mononuclear cobalt superoxo complexes having one cobalt–oxygen bond has been interpreted as involving a coordinative σ bond formed between one of the oxygen lone pairs and the d_z^2 orbital of cobalt [4]. A certain amount of back donation from one of the doubly occupied *d*-orbitals of the cobalt to the empty antibonding π^* orbitals of ligated dioxygen may result in O–O bond lengthening. The unpaired electron resides in a molecular orbital made up of a 3*d* orbital of Co and the empty antibonding π^*

Table 3

ESTABLISHED STRUCTURES OF SOME TRANSITION METAL SUPEROXO COMPLEXES

Structure		O–O, Å	M–O, Å	Ref
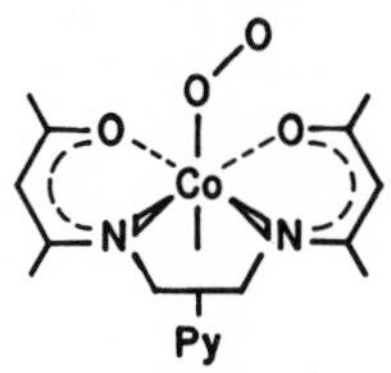 O O O O Co N N Py		1.26	1.86	32
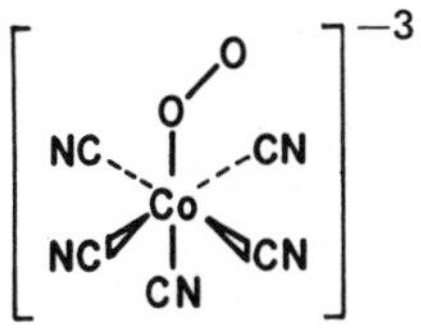 O O NC CN Co NC CN CN −3		1.2–1.3	1.93	33
	X's			
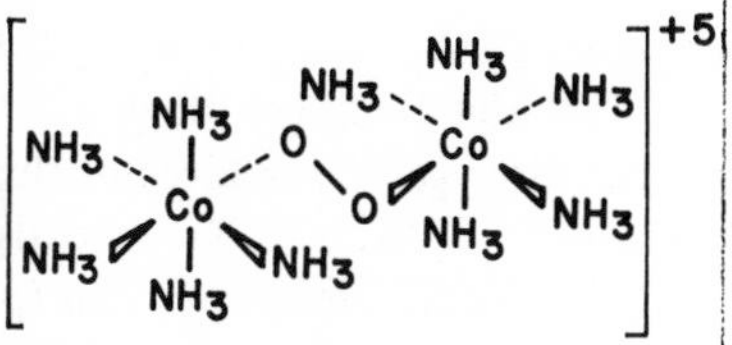 NH3 NH3 NH3 NH3 NH3 O Co NH3 NH3 NH3 O Co NH3 NH3 NH3 NH3 NH3 +5	$5(NO_3^-)$'s $1(SO_4^=)3(HSO_3^-)$'s	1.31 1.32	1.89 1.92	34 35
	L's			
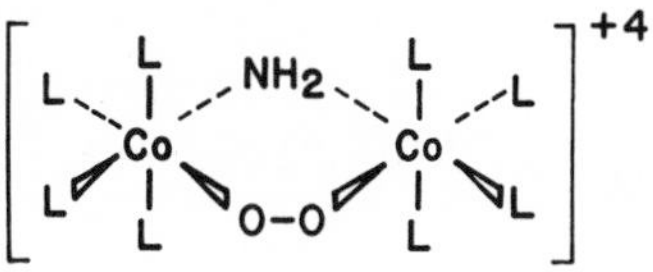 L L NH2 L L Co Co L L O–O L L L +4	$8(NH_3)$'s 4(en)'s	1.32 1.36	1.87 1.88	36 37

orbital of coordinated dioxygen [4]. Its principle residence, however, may be considered to be on the dioxygen ligand [3]. The ESR spectra of the 1:1 superoxo complexes are in accord with the Co(III)–O_2^- group and not with that of Co(II)–O_2^0 [5].

3. PROPERTIES OF COORDINATED DIOXYGEN

The most widely measured property of metal dioxygen complexes has been the vibrational spectrum of coordinated dioxygen. The infrared spectral frequency in the

Table 4

DIOXYGEN VIBRATIONAL FREQUENCIES IN METAL COMPLEXES[a]

Type[b]	Metal	No.	ν_{O_2}, cm^{-1}	
			Range	Average
1	Fe Co	9	1103–1195	1134
2	Co	5	1075–1122	1110
	Total, superoxo complexes:	14	1075–1195	1125
3	Ti V Cr Co Ni	33	818–932	881
	Zr Nb Mo Ru Rh Pd	66	800–929	872
	Ta W Os Ir Pt U	75	807–911	850
4	Fe Co Cu	8	742–844	799
	Total, peroxo complexes:	182	742–932	860

[a] This table taken from reference [3].
[b] From Figure 1.

800–900 cm^{-1} range (assigned to the O–O stretching motion) is diagnostic for the presence of a side-bonded peroxo group (1, Figure 1) [3]. Metal–oxygen single bond vibrations give rise to bands in the 500 cm^{-1} region for dioxygen complexes of this type (1, Figure 1) [3]. The infrared absorption band near 1100 cm^{-1} has recently become characteristic for superoxo complexes having one metal–dioxygen bond (3, Figure 1) and is assigned to the O–O stretching vibration of the superoxo ligand [3]. Far fewer infrared results are available for dioxygen bridged complexes than for non-bridged species.

Electronic spectral data corroborate structural and infrared results which show that on oxygenation of a metal complex, the metal is oxidized and the dioxygen is reduced [3]. Visible spectral evidence is available for a Co(III) dioxygen complex from oxygenation of Co(I).

Most peroxo complexes are diamagnetic, however, a number of exceptions exist. For example, while $[(O_2)_2Cr^{VI}(O)C_5H_5N]$, $(S = O)$, d^0, C.N. = 6, is diamagnetic, $K_3[(O_2)_2Cr^{IV}(CN)_3]$, $(S = 1)$, d^2, C.N. = 7 and $K_3[(O_2)_4Cr^V]$, $(S = 1/2)$, d^1, C.N. = 8, are paramagnetic [8]. In the case of superoxo complexes, paramagnetism $(S = 1/2)$ is displayed when the total number of electrons is odd (and diamagnetism when this number is even) [3].

4. FORMATION AND STABILITY OF DIOXYGEN COMPLEXES

The dioxygen complexes which have been discussed may be prepared either by reaction of a metal complex with molecular oxygen or with peroxidic species. Both of these reactants are usually present during catalytic oxidation so a consideration of dioxygen complexes in general, regardless of the nature of their origin, is relevant to this topic. In this section, we will consider first those complexes whose principle mode of formation is *via* direct reaction with molecular oxygen and then those complexes which are generally formed by reactions with hydroperoxides. The first group of complexes consists mainly of the group VIII metal compounds, manganese, copper, and gold; whereas, formation of dioxygen complexes of metals of groups IVB, VB, and VIB is usually accomplished with hydroperoxides.

4.1. Oxygenation reactions

The chemistry of cobalt dioxygen complexes is replete with nearly every type of reaction of relevance to catalysis. It is the only metal for which all four types of metal dioxygen species have been unambiguously identified [3]. In addition, peroxo and superoxo complexes of cobalt have been prepared both by direct oxygenation *and* by reactions of cobalt species with peroxide.

Cobalt dioxygen complexes having side-bonded dioxygen (1, Figure 1) may be prepared by direct oxygenation of cobalt(I). When the planar complex, $[Co(Ph_2PCH=CHPPh_2)]^+ BF_4^-$ is exposed to atmospheric oxygen, the corresponding peroxo complex is formed, equation (5), [22].

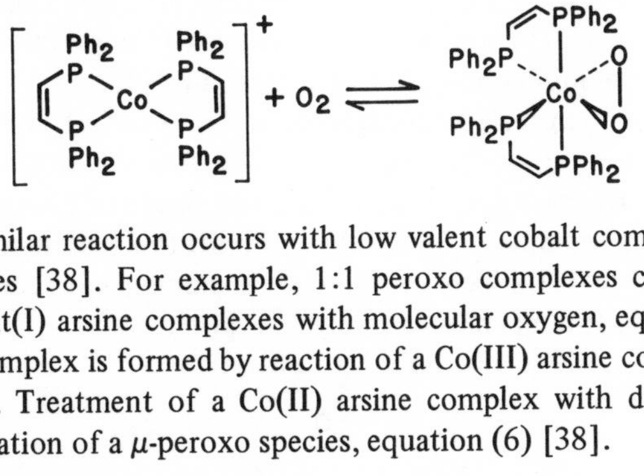

[Ph2 P Co P Ph2 ...]+ + O2 ⇌ Ph2P, PPh2, Co, O, O (5)

A similar reaction occurs with low valent cobalt complexes with diarsines and tetraarsines [38]. For example, 1:1 peroxo complexes can be prepared by reaction of cobalt(I) arsine complexes with molecular oxygen, equation (6) [38]. The same type of complex is formed by reaction of a Co(III) arsine complex with hydrogen peroxide [38]. Treatment of a Co(II) arsine complex with dioxygen, however, leads to the formation of a μ-peroxo species, equation (6) [38].

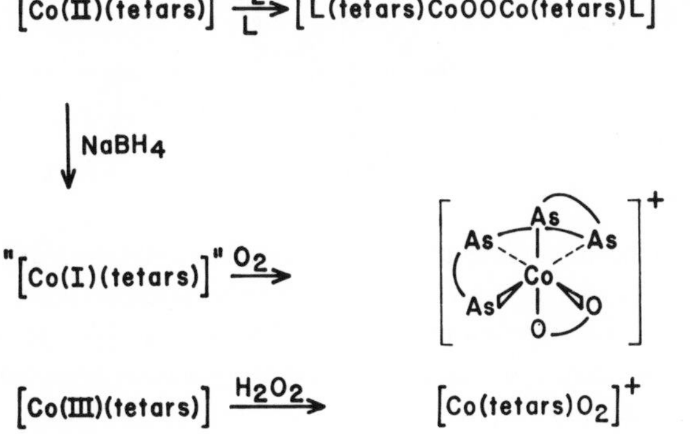

$$[Co(II)(tetars)] \xrightarrow[L]{O_2} [L(tetars)CoOOCo(tetars)L]^{4+}$$

$\downarrow NaBH_4$

$$\text{"}[Co(I)(tetars)]\text{"} \xrightarrow{O_2} [\text{As}_4\text{Co}(O_2)]^+ \qquad (6)$$

$$[Co(III)(tetars)] \xrightarrow{H_2O_2} [Co(tetars)O_2]^+$$

μ-Peroxo complexes are readily formed by oxygenation of $[Co(NH_3)_6]^{2+}$ in aqueous ammonia in a reversible manner, equation (7) [4, 5]. Chelating aliphatic amines such as ethylenediamine (en) and ethylenetriamine (dien) can be used instead of ammonia and similar μ-peroxo complexes are also formed, Table 1.

$$2[Co(NH_3)_6]^{2+} + O_2 \rightleftharpoons \left[(NH_3)_5Co{-}O{-}O{-}Co(NH_3)_5\right]^{4+} \qquad (7)$$

Stimulated largely by the existence of important biochemical systems, a number of cobalt complexes which are considered to be models for the natural oxygen carriers have been prepared and studied. One of the most extensively investigated complexes is [2,2′-ethylenbis(nitrilomethylidene)diphenolato] cobalt(II), Co(salen), I, and its

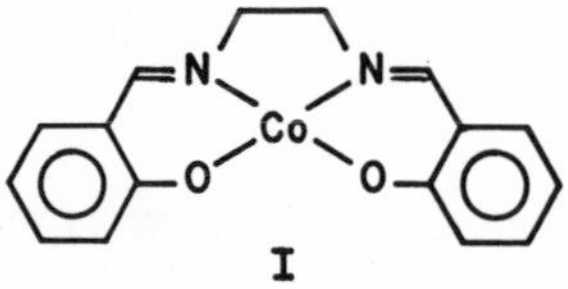

I

derivatives. Reversible oxygenation of I occurs in both the solid [39] and in solution in the presence of certain organic bases [40–42]. The solid takes up oxygen in a 2:1 ($Co:O_2$) ratio and oxygen may be removed simply by heating the dioxygen complex. The efficiency of this process is such (50% active even after 3×10^3 cycles) that this system was once considered for the production of oxygen from air [43].

In solution, either 2:1 ($Co:O_2$) μ-peroxo complexes or 1:1 ($Co:O_2$) superoxo complexes have been formed from I or its derivatives. The product which was obtained was dependent on the temperature, solvent, chelating ligand or base which was used [40–42]. A similar complex, [4,4′-ethylenedinitrilo-2-pentanonato] cobalt(II), Co(acacen), II, forms a 1:1 superoxo complex in non-aqueous solvents containing pyridine, equation (8), at temperatures below 0 °C [44].

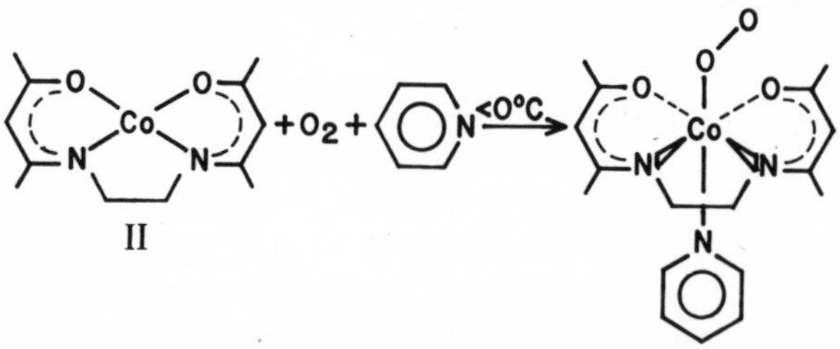

$$\text{II} + O_2 + \text{py} \xrightarrow{<0^{\circ}C} \text{Co(acacen)(py)}O_2 \qquad (8)$$

Superoxo 1:1 ($Co:O_2$) complexes are usually unstable relative to formation of bridged μ-peroxo complexes. In fact the mechanism of formation of most μ-peroxo complexes involves intermediacy of the 1:1 superoxo complex [45], Wilkins, *et al.* [46–48] have presented kinetic evidence for such a mechanism for the oxygenation of cobalt(II) amine complexes in water, equations (9), (10).

$$[(NH_3)_5Co(H_2O)]^{2+} \rightleftharpoons \left[(NH_3)_5Co{-}O{-}O\right]^{2+} + H_2O \tag{9}$$

$$\left[(NH_3)_5Co{-}O{-}O\right]^{2+} + [Co(NH_3)_5(H_2O)]^{2+}$$

$$\longrightarrow \left[(NH_3)_5Co{-}O{-}O{-}Co(NH_3)_5\right]^{4+} + H_2O \tag{10}$$

The binuclear μ-peroxo complex can be oxidized further by a number of reagents to yield the corresponding μ-superoxo complex [45]. An efficient procedure for the preparation of the decamine μ-superoxo complex from the μ-peroxo species utilizes ammonium persulfate as the oxidant, equation (11).

$$\left[(NH_3)_5Co{-}O{-}O{-}Co(NH_3)_5\right]^{4+}$$

$$\xrightarrow{(NH_4)_2S_2O_8} \left[(NH_3)_5Co{-}O{-}O{-}Co(NH_3)_5\right]^{5+} \tag{11}$$

In the case of $K_6[(CN)_5Co(O_2)Co(CN)_5]$ oxygen itself seems capable of oxidizing the 6- to the 5-ion [49], equation (12).

$$\left[(CN)_5Co{-}O{-}O{-}Co(CN)_5\right]^{6-} \xrightarrow{-e^-} \left[(CN)_5Co{-}O{-}O{-}Co(CN)_5\right]^{5-} \tag{12}$$

Thus the sequence represented in equation (13) illustrates the relationships which exist between 1:1 superoxo, μ-peroxo and μ-superoxo complexes.

$$Co(II) \xrightarrow{O_2} (O_2^-)Co(III) \xrightarrow{Co(II)} Co(III)(O_2^{2-})Co(III) \xrightarrow{-e^-} Co(III)(O_2^-)Co(III) \tag{13}$$

The only presently known route to μ-superoxo complexes, in fact is *via* oxidation of a μ-peroxo complex [3].

As in the case of 1:1 ($Co:O_2$) peroxo complexes of Co, a 1:1 ($Co:O_2$) superoxo complex can be formed either by direct oxygenation or *via* a peroxide, equation (14). For example, the Co(III) aquo cobalamin complex reacts with solutions of superoxide in DMF (prepared by electrochemical reduction of dissolved O_2) [50] to form a superoxo complex.

$$Co(II) + O_2 \longrightarrow Co(III)(O_2^-) \longleftarrow O_2^- + Co(III) \tag{14}$$

At this point it might be well to review those factors which are important in determining which type of dioxygen compound is formed from cobalt complexes in solution. Clearly low valent cobalt complexes, (Co(I)), stabilized by good π-acceptor ligands form 1:1 side-bonded peroxo complexes with dioxygen (1, Figure 1) whereas amine complexes of Co(II) give end-bonded (3, Figure 1) 1:1 superoxo complexes. In most instances rapid reaction of the 1:1 superoxo complexes form μ-peroxo species (2, Figure 1) which can be converted to μ-superoxo complexes (4, Figure 1) with a variety of oxidants.

The rather unique capability of Co(II) to undergo oxygenation is not as readily shared by most other first row transition metal species. This uniqueness has been accounted for on both thermodynamic and kinetic grounds [51]. From thermodynamic considerations it has been determined that most of the first row transition metal complexes ($Ti^{2,3+}$aq., V^{3+}aq., Cr^{2+}aq., $Mn(CN)_6^{4-}$, $Co(CN)_5^{4-}$, Cu^+/NH_3) could form 2:1 μ-peroxo complexes though in most cases oxygenation would be irreversible [51]. From this standpoint, Co(II) and Fe(II) complexes are probable candidates for reversible formation of 1:1 superoxo complexes. Some Ti(II), Cr(II) and Mn(II) complexes are also listed as possibilities [51]. However, not all of the thermodynamically feasible dioxygen complexes are kinetically stable enough to be reversible. For instance O_2^- is replaceable in aqueous medium resulting in the transformation: (M^{n+}aq. $\rightarrow$ M^{n+1}aq.) for μ-superoxo complexes of Ti(III), V(III) Mn(III), Fe(III), and Cu(II) [51]. On the other hand, Co(III)(O_2^-) complexes would be substitution inert resulting in oxygenation to a kinetically stable dioxygen complex. Thus the uniqueness of the oxygenation of Co(II) complexes is mainly the result of their substitution inertness [51]. In non-aqueous medium, however, many more first row dioxygen species are possible.

Oxygen adds readily to solutions of a large number of iron complexes. Much of the work in this area has been related to the nature of oxygen binding in natural oxygen carriers. The naturally occurring oxygen carriers, hemoglobin and myoglobin form FeO_2 complexes whereas hemerythrin is considered to form FeO_2Fe compounds [52–55]. Although the nature of coordinated dioxygen in biological systems has been the subject of some controversy historically, detailed consideration of these systems is beyond the scope of this article. Synthetic oxygen carriers of the 1:1 superoxo and μ-peroxo type have been prepared, however, the existence of side bonded 1:1 peroxo (1, Figure 1) or μ-superoxo iron complexes has not thus far been established [3].

A number of iron porphyrin complexes reversibly take up oxygen in the presence of a basic organic ligand such as pyridine. In contrast to cobalt systems, many iron complexes of this sort have been found to irreversibly oxidize to a μ-oxo dimer, [54, 56], equation (15).

$$\mathrm{BFe^{II}} + \mathrm{O_2} \longrightarrow \mathrm{BFeO_2} \xrightarrow{\mathrm{FeB}} \mathrm{BFeOOFeB} \longrightarrow \mathrm{BFe^{III}OFe^{III}B} \tag{15}$$

Structural modifications such as highly hindered ligands [53, 55] or a neighboring imidazole group [57] prevent dimerization and stabilize the iron dioxygen complex. More recently it has been found that, just as in the case of cobalt dioxygen complexes [58], solvent polarity and low temperature favor iron oxygen adduct stability [59–61]. Thus reversible oxygen uptake activity has been demonstrated for some unhindered iron porphyrin complexes at low temperatures in polar aprotic solvents such as methylene chloride and DMF [59–61], equation (16).

$$Fe(py)_2TPP + O_2 \underset{}{\overset{CHCl_3 - 78\,°C}{\rightleftharpoons}} Fe(O_2)py\ TPP \tag{16}$$

Iron dioxygen complexes having ligands other than porphyrins have also been prepared. As in the case of cobalt, equation (17), DMG complexes of iron reversibly take up oxygen, equation (18) [62–64]. These complexes, however, have not been completely characterized.

$$Co(DMG)_2 \xrightarrow{O_2,\ pyridine} Co(DMG)_2(py)_2O_2 \tag{17}$$

$$Fe(DMG)_2 \xrightarrow{O_2,\ pyridine} Fe(DMG)_2(py)_2O_2 \tag{18}$$

Evidence has also accumulated for the existence of dioxygen complexes of most of the other first row transition metals although the precise nature of their structure and bonding is often not known. For example, the reaction of N,N′-ethylenebis-(salicylaldiminato)(vanadium(III)), $V(salen)^+$, and oxygen in pyridine indicates that a $V(salen)^+$-oxygen adduct is formed [65]. Oxygen also reacts rapidly with *tris* (diisopropylaminato)chromium(III) to give an unstable chromium dioxygen complex [66]. Solutions of copper complexes have been shown to take up oxygen in a reversible manner [67]. A dioxygen adduct of a manganese porphyrin has recently been formed reversibly when *meso*-tetraphenylporphyrin(pyridine)manganese(II) is oxygenated in toluene at $-79\,°C$ [68]. These and other examples serve to illustrate that most first row metals in certain valence states are capable of interacting with molecular oxygen providing they have suitable ligand systems under appropriate reaction conditions.

Second and third row group VIII metal dioxygen complexes have been more extensively studied than have first row complexes with the exception of cobalt. Many dioxygen complexes are readily formed merely by bubbling oxygen into solutions of the parent complex. Iridium dioxygen complexes have been the most thoroughly studied of this group. Vaska first discovered that $[IrCl(CO)(PPh_3)_2]$ in benzene solution reversibly takes up molecular oxygen to form a 1:1 side-bonded peroxo complex (1, Figure 1), equation (19) [16, 69].

$$[IrCl(CO)(PPh_3)_2] + O_2 \underset{k_{-1}}{\overset{k_2}{\rightleftharpoons}} [IrCl(CO)(O_2)(PPh_3)_2] \tag{19}$$

Since this discovery, analogous iridium dioxygen complexes have been reported by a number of authors (listed in references 3–7). In all cases the forward reaction was found to follow a bimolecular rate law, equation 20 (see references 5 and 7).

$$-\frac{d[ML_n]}{dt} = k_2[ML_n][O_2] \tag{20}$$

Vaska and Chen have found that the rates of dioxygen addition to complexes of the formula: $[IrX(CO)L_2]$ increase with the basicity of the neutral ligand L (Table 5). In general, reactivity is also enhanced by more polarizable ligands and is lowered by strong π-acceptor ligands such as CO. Highly electronegative ligands or a positive charge on the metal also lower its reactivity toward oxygen. In short, an increase in electron density at the metal usually improves oxygen uptake properties. For example, the relative rates of reaction of a series of iridium complexes follows the order:

$$[IrCl(PPh_3)_3] > [IrCl(CO)(PPh_3)_2] > [IrF(CO)(PPh_3)_2] \gg [IrCl(NO)(PPh_3)_2]$$

The geometry of the complex and of the individual ligands also has a profound effect on the reactivity of iridium complexes with oxygen. For example $[Ir(Ph_2PCH_2CH_2PPh_2)_2]^+$ forms a dioxygen complex whereas the tetrahedrally-distorted complex, $[Ir(PPh_2CH_3)_4]^+$ does not [72]. In addition, although $[IrCl(CO)(PPh_3)_2]$ and $[IrCl(CO)((p\text{-}CH_3C_6H_5)_3P)_2]$ readily react with oxygen to form dioxygen complexes, $[IrCl(CO)((o\text{-}CH_3C_6H_5)_3P)_2]$ does not – presumably because of steric hindrance created by the *o*-methyl substituent on the triarylphosphine ligands [71].

Data from the first nine entries in Table 5 may be expressed in a linear free energy relationship, equation (21), where k_2 is the second order rate constant for oxygenation of the metal complex and $\sigma(X)$ is the Hammett constant for the *para* substituent in the triarylphosphine ligand [7].

$$\log k_2 = 0.194 - 14.1\sigma(X) \qquad (21)$$

Rhodium(I) complexes add molecular oxygen, equation (22), however, their reactivity toward dioxygen is considerably less than the analogous iridium complexes.

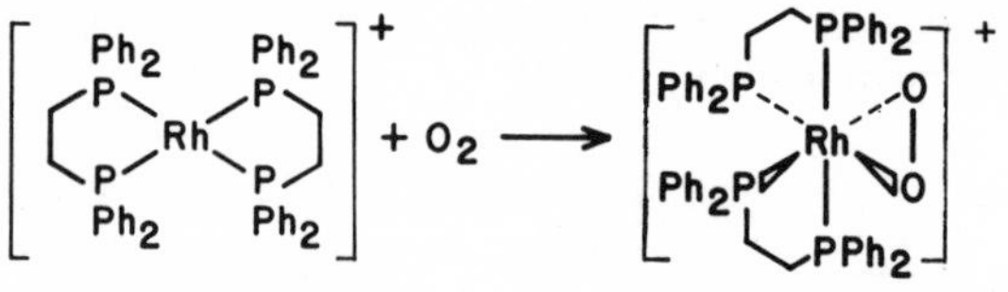

Ph2 Ph2 P P Rh P P Ph2 Ph2 $]^+$ + O_2 ⟶ PPh2 Ph2P O Rh Ph2P O PPh2 $]^+$ (22)

In general, reactivity toward molecular oxygen is lower for complexes of the second row transition metals than for third row complexes [3–7]. It has been found, however, that a cobalt(I) phosphine complex is far more reactive toward O_2 than either the rhodium(I) or iridium(I) compounds having the same ligand system [70]. This leads, in at least one instance, to the rather unexpected reactivity sequence: Co $\gg$ Ir $>$ Rh (Table 6). As in the iridum case, rhodium(I) forms side-bonded peroxo complexes with dioxygen. Reaction products, however, can be quite different as in the oxygenation of *tris*(triphenylphosphine)chlororhodium(I), equation (23) [31].

Table 5

KINETIC AND THERMODYNAMIC PARAMETERS FOR THE FORMATION OF SOME IRIDIUM DIOXYGEN COMPLEXES[a]

$$IrL_2(CO)X + O_2 \underset{k_{-1}}{\overset{k_2}{\rightleftharpoons}} IrL_2(CO)X(O_2)$$

chlorobenzene, 40°

X	L	$10^2 k_2$, M^{-1} sec^{-1}	$10^6 k_{-1}$ sec^{-1}	$10^{-3} K_{eq}$ M^{-1}	ΔH_2^* kcal/mol	ΔH_{-1}^* kcal/mol	ΔS_2^* eu	ΔS_{-1}^* eu
F	PPh_3	1.48	50.9	0.291	13.6	23.7	−24	−0.5
NCO	PPh_3	2.99	90.2	0.331	11.1	21.0	−30	−10
$OClO_3$	PPh_3	6.96			4.01		−50	
N_3	PPh_3	7.33	13.1	5.60	9.57	26.2	−33	3
Cl	PPh_3	10.1	13.8	7.32	9.50	26.5	−33	4
Br	PPh_3	20.6	3.32	62.0	8.42	28.8	−35	8
ONO_2	PPh_3	36.3			5.50		−43	
I	PPh_3	72.3	0.84	857	5.76	29.0	−41	6
NO_2	PPh_3	172			3.48		−46	
Cl	$P(C_6H_{11})_3$	0.127	5.85	0.217	9.5		−42	
Cl	$P(p\text{-}C_6H_4Cl)_3$	3.10	8.67	3.58	10.8		−31	
Cl	$AsPh_3$	14.2	8.25	17.2	8.1		−37	
Cl	$PPh_2C_2H_5$	14.2	6.75	21.0	8.4		−36	
Cl	$P(p\text{-}C_6H_4CH_3)_3$	21.6	12.2	17.7	9.3		−32	
Cl	$P(n\text{-}C_4H_9)_3$	26.1	3.90	66.9	9.0		−33	
Cl	$P(C_2H_5)_3$	33.8	1.57	215	8.5		−34	
Cl	$P(p\text{-}C_6H_4OCH_3)$	48.4	12.5	38.7	8.5		−33	

[a] Table taken from reference 5. Data taken from references 70 and 71.

Cl, PPh3, Rh, Ph3P, PPh3 $+ O_2 \longrightarrow$ L, O, Cl, Rh, L, O, O, L, Rh, Cl, O, L (23)

Binuclear rhodium complexes having bridging di-oxygen of both the superoxo $[Cl(L)_4Rh(O_2)Rh(L)_4Cl]^{+3}$, and of the peroxo $[Cl(L_4)Rh(O_2)Rh(L_4)Cl]^{+2}$, (L = py, pic) have also been reported [74]. James and Ng have found that DMA solutions of $[RhCl(C_8H_{14})]_2$ readily absorb oxygen to give a 1:1 dioxygen complex having an infrared absorption at 895 cm^{-1} characteristic of side-bonded dioxygen (1, Figure 1). These authors reported, however, the solution showed an ESR signal thought to be due to the presence of a rhodium(II) superoxide species, $Rh(II)\text{–}O_2^-$ [75].

Ruthenium complexes also add dioxygen, usually with concurrent expulsion of a neutral ligand. For example, a series of dioxygen complexes having the formula, $[RuX(NO)(Ph_3P)_2(O_2)]$ where X = OH, Cl, Br, I, NCS, NCO, N_3 or CN have been prepared by reactions of low valent ruthenium complexes with molecular oxygen.

Table 6

COMPARISON OF KINETIC AND ACTIVATION PARAMETERS FOR THE FORMATION OF GROUP VIII COMPLEXES (Co, Rh, Ir)[a]

1.[b] $[M(Ph_2PCH{=}CHPPh_2)_2]^+ + O_2 \xrightarrow[\text{chlorobenzene}]{k_2} [M(Ph_2PCH{=}CHPPh_2)_2(O_2)]^+$

M	ΔH* kcal/mol	ΔS* eu	ΔG* kcal/mol	k_2(25°), M^{-1} sec^{-1}
Co	3.4	– 28	10.3	1.7×10^4
Rh	11.6	– 24	18.8	0.12
Ir	6.5	– 38	17.8	0.47

2.[c] $[M(Ph_2PCH_2CH_2PPh_2)_2]^+ + O_2 \xrightarrow[\text{chlorobenzene}]{k_2} [M(Ph_2PCH_2CH_2PPh_2)_2(O_2)]^+$

M	ΔH* kcal/mol	ΔS* eu	k_2(30°), M^{-1} sec^{-1}
Rh	7.9	– 35	1.1
Ir	3.6	– 44	3.3

[a] Taken from reference 5. [b] Data from reference 70. [c] Data from reference 73.

The products exhibit infrared absorption bands between 800 and 900 cm^{-1} and have been formulated as side-bonded ruthenium peroxo complexes [76–78]. Typical of these reactions is the addition of oxygen to $[RuCl(CO)(NO)(Ph_3P_2)]$ to give the dioxygen complex (IR 876 cm^{-1}), equation (24).

PPh3, CO, ON—Ru, Cl, PPh3 + O_2 ⟶ PPh3, ON, O, Ru, Cl, O, PPh3 (24)

X-ray crystallographic studies confirm this geometry [79]. Similar reactions have been observed with other ruthenium complexes [78, 80, 81]. In these cases the neutral ligands which are lost are either triphenylphosphine, equations (25) and (26), or triphenylarsine, equation (27).

$$Ru(CO)(CNR)(PPh_3)_3 + O_2 \xrightarrow{\text{benzene}} Ru(CO)(CNR)(PPh_3)_2O_2 \quad (25)$$

$$Ru(CO)_2(PPh_3)_3 + O_2 \xrightarrow{\text{benzene}} Ru(CO)_2(PPh_3)_2O_2 \quad (26)$$

$$RuCl_2(AsPh_3)_3 + O_2 \xrightarrow{\text{benzene}} RuCl_2(AsPh_3)_2O_2 \quad (27)$$

The dioxygen complexes are quite stable, however, oxygen can be removed from the products of equations (25) and (26) by warming in ethanol with excess PPh_3. It is interesting to note that in contrast to iridium and rhodium peroxo complexes, $[RuCl_2(AsPh_3)_2(O_2)]$ is paramagnetic [81].

Relatively few osmium dioxygen complexes have been reported. The first report of an osmium dioxygen complex appears to have been made in 1972 by Cavit, Grundy and Roper [80]. It was prepared by the slow reaction of $[Os(CO)_2(PPh_3)_3]$ with oxygen, equation (28). An infrared absorption band at 820 cm^{-1} [7] indicates a side-bonded peroxo species analogous to dioxygen complexes of Ir, Rh and Ru.

$$Os(CO)_2(PPh_3)_3 + O_2 \longrightarrow Os(CO)_2(PPh_3)_2O_2 \tag{28}$$

Group VIII d^{10} complexes (Ni(O), Pd(O), Pt(O)) generally react with molecular oxygen to give 1:1 peroxo complexes with ligand displacement, equation (29), (30) and (31).

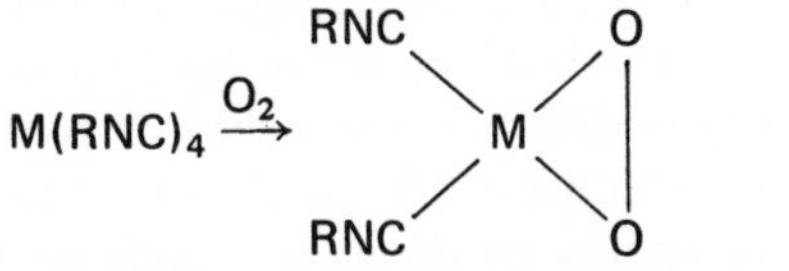

$$M(RNC)_4 \xrightarrow{O_2} (RNC)_2M(O_2) \tag{29}$$

M = Ni, Pd; R = *t*-Bu, cyclo-C_6H_{12}

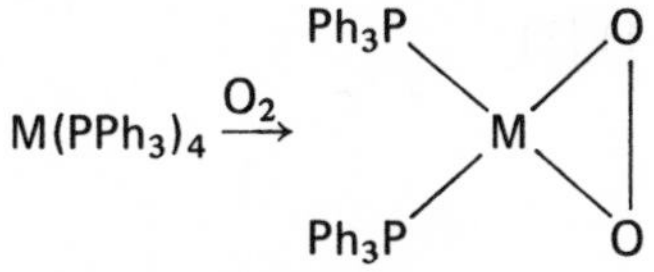

$$M(PPh_3)_4 \xrightarrow{O_2} (Ph_3P)_2M(O_2) \tag{30}$$

M = Ni, Pd, Pt

Compounds formed *via* reaction (29) are diamagnetic and thermally unstable [82, 83]. Infrared, NMR, electronic spectra and magnetic data have been used to deduce the square planar structure. Reaction (30) is essentially irreversible and forms stable, square planar dioxygen complexes having infrared bands between 800 and 900 cm^{-1}. Oxygen reacts rapidly and irreversibly with a benzene solution of $[Pt(Ph_3P)_4]$ to yield $[Pt(Ph_3P)_2(O_2)]$, equation (31). The platinum dioxygen complex is a well characterized square planar peroxo complex (1, Figure 1) having IR absorptions at 815 and 824 cm^{-1} [84–88].

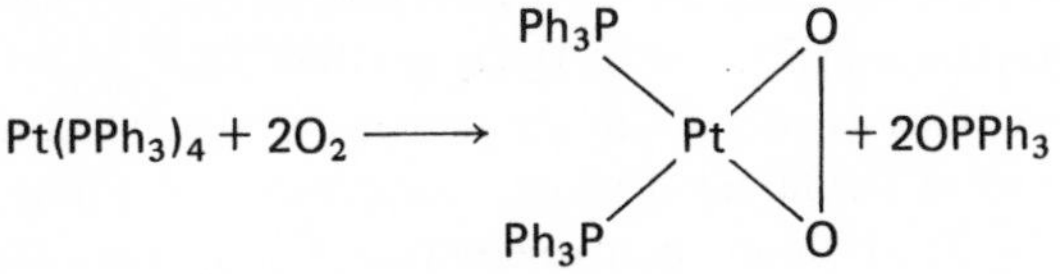

$$Pt(PPh_3)_4 + 2O_2 \longrightarrow (Ph_3P)_2Pt(O_2) + 2OPPh_3 \tag{31}$$

In summary, d^7 cobalt(II) complexes reaction with molecular oxygen to give 1:1 superoxo complexes (3, Figure 1) which form binuclear μ-peroxo complexes (2, Figure 1) with Co(II) at varying rates. μ-Superoxo complexes (4, Figure 1) are formed from μ-peroxo complexes in suitable oxidizing media. Cobalt(I) complexes, on the other hand, tend to form 1:1 peroxo complexes on reaction with dioxygen. They do so much faster than do other group VIII d^8 complexes having comparable ligand systems. The reactivity of other group VIII d^8 complexes toward O_2 is dependent on a number of factors. The metal center is important: reactivity increases in going from second to third row complexes (Os(O) > Ru(O), Ir(I) > Rh(I)). In general, reactivity is increased by strong σ-donor ligands and by highly polarizable ligands [5]. Electron withdrawal from the metal whether by very strong π-acceptor ligands, positive charge on the metal complex or other factors, lessens reactivity toward molecular oxygen [5]. Group VIII d^{10} complexes, Ni(O), Pd(O), Pt(O) also react readily with O_2 to give stable square planar dioxygen complexes. Neutral ligands are lost, and in some cases are oxidized in the process.

In all cases oxygen appears to initially form a 1:1 complex with the metal. This interaction will result in an increase in oxidation and coordination numbers of the metal center by either one unit to give a superoxo complex (3, Figure 1) or by two units to give a peroxo complex (1, Figure 1). Which oxidation reaction occurs depends on the number, arrangement and types of ligands, the d-electron configuration and the oxidation state and oxidation potential of the metal in the complex [3]. It appears that the principle determining factors are the metal's oxidation state and the relative stabilities of its higher valences. Thus, low valent metal complexes tend to produce peroxo species (1, Figure 1) whereas metals in their normal oxidation states tend to give superoxo complexes (3, Figure 1) [3].

4.2. Reactions with hydrogen peroxide

Peroxo complexes of groups V and VI are usually formed by reaction of the parent metal complex with hydrogen peroxide. Hydroperoxides are often present during catalytic oxidation of organic substrates. Since certain of these peroxo complexes are capable of selectively oxidizing unsaturated organic compounds, a brief discussion of methods of formation of group V and VI metal peroxo complexes is in order.

Highly colored aqueous solutions of titanium, vanadium, chromium, molybdenum and tungsten peroxo complexes have been known for many years [90]. They may be prepared by treating the appropriate metal ion with H_2O_2 in either acidic or neutral medium [91, 92]. Recently, reactions carried out in the presence of organic bases or chelates have led to the synthesis of stable metal peroxo complexes containing one or more organic ligands [11–15]. The titanium peroxo complex, III, [11] is a stable, orange-red crystalline solid which is formed when an acidic aqueous solution of the titanium peroxo complex is treated with NaOH and dipicolinic acid. The chromium complex, IV, [14] may be prepared by treatment of the peroxo complex formed from aqueous Cr(VI) and H_2O_2, with dipyridyl. The molybdenum dioxygen complex, V, [15], is also known. All three structures: III, IV, V, have been established by X-ray crystallography [11, 14, 15].

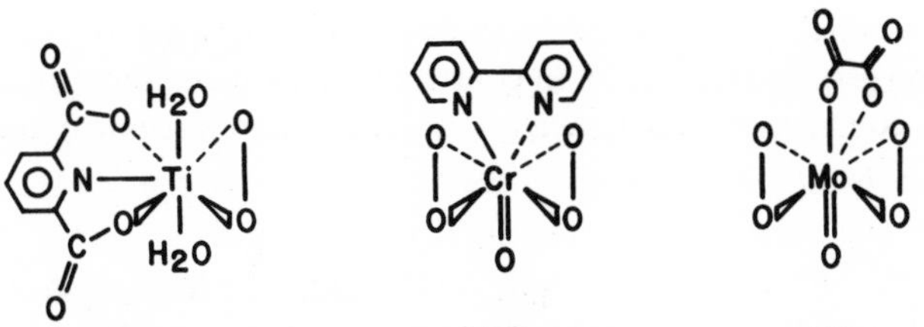

III IV V

Both the peroxo chromium etherate, VI, formed from reaction of H_2O_2 with acidic aqueous solution of chromates and ether, as well as the molybdenum complexes, VII, formed from reactions of H_2O_2 with MoO_3 in the presence of hexamethylphosphoramide(HMPA), selectively transfer oxygen to unsaturated organic molecules [93–95].

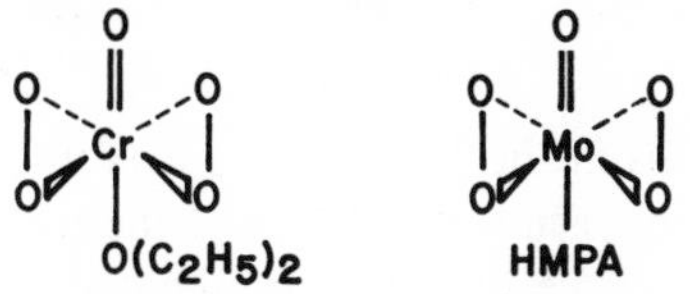

$O(C_2H_5)_2$ HMPA

VI VII

5. TRANSFER OF COORDINATED DIOXYGEN TO REACTIVE SUBSTRATES

During the past decade a sizable number of reactions have been reported which involve the oxidation of substrates by dioxygen in the coordination sphere of the metal. For example, the cobalt μ-superoxo complexes, $[L_4Co(NH_2)(O_2)CoL_4]$, react with SO_2 and SeO_2 to cleave the O–O bond of coordinated dioxygen and form sulfato and selenato linkages, equation (32) [96].

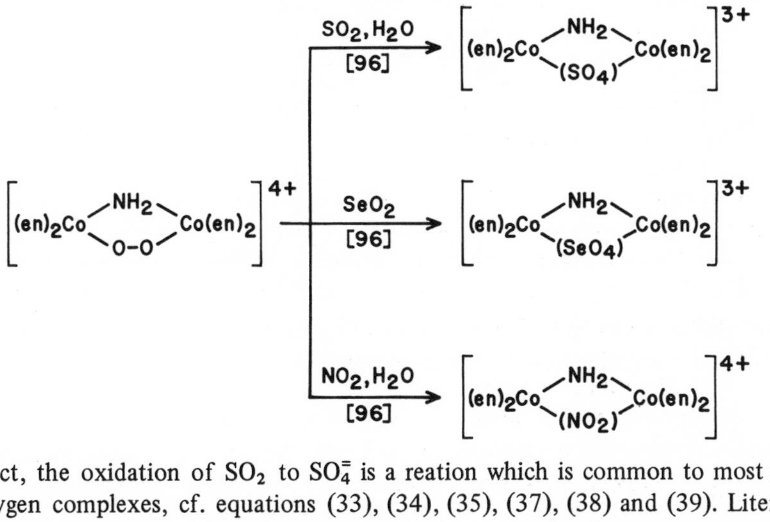

$$\left[(en)_2Co\begin{smallmatrix}NH_2\\O{-}O\end{smallmatrix}Co(en)_2\right]^{4+} \begin{cases} \xrightarrow[{[96]}]{SO_2,\,H_2O} \left[(en)_2Co\begin{smallmatrix}NH_2\\(SO_4)\end{smallmatrix}Co(en)_2\right]^{3+} \\ \xrightarrow[{[96]}]{SeO_2} \left[(en)_2Co\begin{smallmatrix}NH_2\\(SeO_4)\end{smallmatrix}Co(en)_2\right]^{3+} \\ \xrightarrow[{[96]}]{NO_2,\,H_2O} \left[(en)_2Co\begin{smallmatrix}NH_2\\(NO_2)\end{smallmatrix}Co(en)_2\right]^{4+} \end{cases} \quad (32)$$

In fact, the oxidation of SO_2 to $SO_4^=$ is a reation which is common to most metal dioxygen complexes, cf. equations (33), (34), (35), (37), (38) and (39). Literature references are given in square brackets under the arrows for the respective reactions.

Another reaction which has been widely studied is the oxidation of NO_2 to NO_3^- by coordinated dioxygen, equation (33). In contrast to most metal dioxygen complexes, the μ-superoxo complex in effect merely undergoes a displacement reaction with NO_2, equation (32).

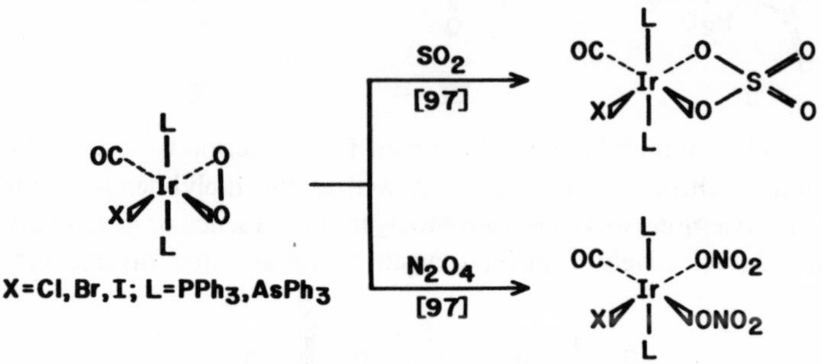

L, OC, Ir, O, O, X, L

X=Cl, Br, I; L=PPh_3, $AsPh_3$

SO_2 [97] → L, OC, Ir, O, S, O, O, O, X, L

N_2O_4 [97] → L, OC, Ir, ONO_2, ONO_2, X, L

(33)

Under mild conditions in the absence of a catalyst, oxygen will not react with SO_2. Consequently, the dioxygen molecule can be thought of as undergoing activation toward this substrate by virtue of being coordinated to the metal. In like manner, many other small molecules are oxidized far more readily in the coordination sphere of a metal dioxygen complex than with molecular oxygen alone.

$$RhCl(PhP[(CH_2)_3PPh_2]_2)(O_2) \xrightarrow[\text{[98]}]{SO_2} RhCl(PhP[(CH_2)_3PPh_2]_2)(SO_4) \qquad (34)$$

The ruthenium complex, $[RuCl(NO)(Ph_3P)_2(O_2)]$, reacts readily with both SO_2 and CO, and interestingly in the latter case, coordinated NO is oxidized rather than CO, equation (35).

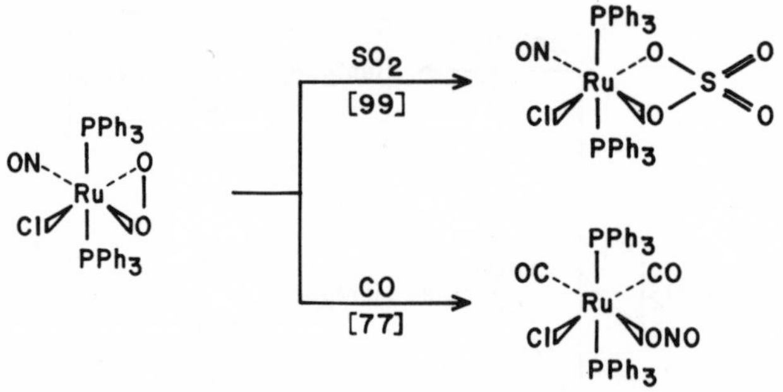

PPh_3, ON, Ru, O, O, Cl, PPh_3

SO_2 [99] → PPh_3, ON, Ru, O, S, O, O, O, Cl, PPh_3

CO [77] → PPh_3, OC, Ru, CO, Cl, ONO, PPh_3

(35)

Another curious oxidation reaction involves heating a ruthenium dioxygen complex containing an isocyanide ligand; in ethanol, equation (36).

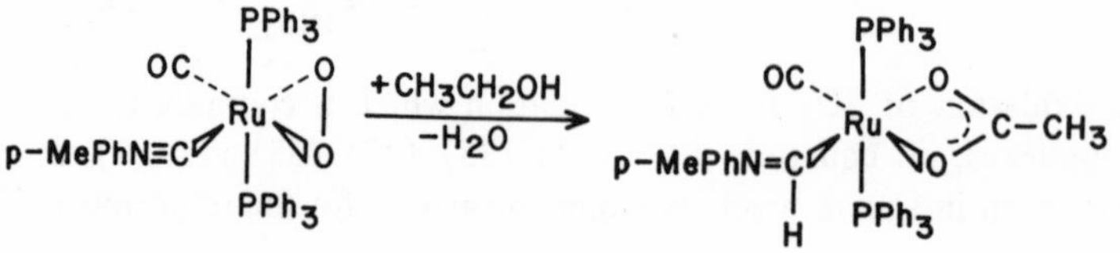

PPh_3, OC, Ru, O, O, p−MePhN≡C, PPh_3 $\xrightarrow[-H_2O]{+CH_3CH_2OH}$ PPh_3, OC, Ru, O, C−CH_3, O, p−MePhN=C, H, PPh_3 (36)

The alcohol is oxidized to coordinated acetate with hydride transfer to the isocyanate [100].

Dioxygen complexes of nickel, palladium and platinum undergo similar series of reactions with reactive substrates, equations (37), (38) and (39). All react with SO_2, N_2O_4, NO, CO_2, RNC, and PPh_3, however, there are some differences in the products which are formed. For example, although Pd, Pt and Ni dioxygen complexes all react with N_2O_4, nickel forms a *trans*-nitrato complex, equation (37*b*) whereas Pd and Pt form *cis*-complexes, equation (38*c*) and (39*c*).

$(RNC)_2Ni(O_2)$ [82,83,86,89,97] R = t-Bu; X = Cl, Br

a. SO_2 → $(RNC)_2Ni(O_2SO_2)$

b. N_2O_4 → *trans*-$(RNC)_2Ni(ONO_2)_2$

c. 2NO, RNC → $[Ni(NO)(RNC)_3]^+ NO_3^-$

d. 4CO → $(RNC)_2Ni(CO)_2 + CO_2$

e. CO_2, RNC → $(RNC)_2Ni(O_2C{=}O)$ + RNCO

f. 4RNC → $(RNC)_2Ni(CNR)_2$ + 2RNCO

g. $4Ph_3P$ → $(RNC)_2Ni(PPh_3)_2 + 2OPPh_3$

h. TCNE → $(RNC)_2Ni(C(CN)_2{-}C(CN)_2) + O_2$

i. 2PhCOX → $(RCN)_2NiX_2$ + PhC(=O)OC(=O)Ph

j. $2Ph_3CBr$ → $(RNC)_2NiX_2 + Ph_3COOCPh_3$

(37)

$$(Ph_3P)_2Pd(O_2) \; [86,97] \longrightarrow \begin{cases} \xrightarrow[\text{a.}]{SO_2} (Ph_3P)_2Pd(O_2SO_2) \\ \xrightarrow[\text{b.}]{NO} (Ph_3P)_2Pd(ONO)_2 \\ \xrightarrow[\text{c.}]{N_2O_4} (Ph_3P)_2Pd(ONO_2)_2 \end{cases} \tag{38}$$

$$(Ph_3P)_2Pt(O_2) \; [86,87,88,97,101,102,103,104,105] \longrightarrow \begin{cases} \xrightarrow[\text{a.}]{SO_2} (Ph_3P)_2Pt(O_2SO_2) \\ \xrightarrow[\text{b.}]{2NO} (Ph_3P)_2Pt(ONO)_2 \\ \xrightarrow[\text{c.}]{N_2O_4} (Ph_3P)_2Pt(ONO_2)_2 \\ \xrightarrow[\text{d.}]{CO} (Ph_3P)_2Pt(O_2C{=}O) \\ \xrightarrow[\text{e.}]{CO_2} (Ph_3P)_2Pt(O{-}O{-}C({=}O){-}O) \\ \xrightarrow[\text{f.}]{RCHO} (Ph_3P)_2Pt(O{-}O{-}CHR{-}O) \\ \xrightarrow[\text{g.}]{R_2CO} (Ph_3P)_2Pt(O{-}O{-}CR_2{-}O) \\ \xrightarrow[\text{h.}]{CS_2} (Ph_3P)_2Pt(OCS_2)_2 \\ \xrightarrow[\text{i.}]{3PPh_3} (Ph_3P)_2Pt{-}PPh_3 + 2OPPh_3 \\ \xrightarrow[\text{j.}]{TCNE} (Ph_3P)_2Pt(O{-}O{-}C(CN)_2{-}C(CN)_2) \end{cases} \tag{39}$$

All three metal complexes also react with NO, however, palladium and platinum dioxygen complexes give *cis*-di-nitrito complexes (38*b*) and (39*b*) while nickel has been reported to give a tetrahedral nitrosylnitrato complex, equation (37*c*). While platinum dioxygen has been reported to form a cyclic oxygen containing complex with tetracyanoethylene, equation (39*j*), a nickel dioxygen complex forms an olefin complex with expulsion of O_2. Platinum complexes react with CO to form a cyclic carbonate, equation (39*d*), whereas a nickel peroxo complex gives CO_2 and a zero valent dicarbonyl complex, equation (37*d*). Nickel and platinum, however, yield similar complexes when reacted with SO_2, CO_2, and PPh_3, equations (37) and (39).

Group VI peroxo complexes also interact with reactive substrates. The molybdenum peroxo complex [MoO_5(HMPA)] has been shown to oxidize cyclohexene to cyclohexene oxide, equation (40) [95]. It was postulated that a cyclic peroxy intermediate was formed [95, 106].

(HMPA) ⇌ + HMPA

→ [ ] → MoO_4^+ (40)

Oxygen reacts readily with *tris*(di-isopropylaminato)chromium(III). Cold pentane solutions yield [$Cr(O_2)(NPr^i_2)_3$] while under different conditions ESR evidence has been found for a chromium(IV) complex of di-isopropylnitroxide, [$CrO(NPr^i_2)_2(ONPr^i_2)$] [66]. The chromium peroxo complex VI, selectively oxidizes tetraphenylcyclopentadienone, however, some disagreement exists concerning the nature of the organic product obtained [93, 94].

A number of logical explanations have been advanced for the enhancement of the reactivity of dioxygen upon coordination, yet each has its exceptions or drawbacks [3–7]. First, since coordinated dioxygen is usually diamagnetic, reactions with diamagnetic reactants to give diamagnetic products fulfill spin conservation requirements. However, the paramagnetic dioxygen complex [$RuAs(PPh_3)_2Cl_2(O_2)$], in which coordinated dioxygen apparently retains triplet character, reacts readily with SO_2 to give a sulfate complex. Secondly, since coordinated dioxygen is partially reduced, the increased electron density on the oxygen may result in a more active species. Electron transfer to coordinated dioxygen should lengthen the O–O bond and enhance O–O bond breaking. Nevertheless, there appears to be no correlation between reactivity of dioxygen complexes and O–O bond lengths. Thirdly, the orientation of the substrate and dioxygen may be such that both are in *cis*-positions in the coordination sphere thus lowering the activation energy for substrate–oxygen interactions. In several instances, however, there is some question of whether the substrate is coordinated to the metal prior to oxidation. Fourthly, proposals have also been made that activation of coordinated dioxygen results in promotion of the oxygen to one of its excited states. Nonetheless this author is aware of no examples of reactions of coordinated dioxygen which parallel the characteristic diagnostic reactions for singlet oxygen. It would seem therefore that these and other reasons for enhanced oxygen activity may have varying degrees of validity in individual instances but a unified picture which adequately rationalizes oxygen activation in all instances has not yet emerged.

6. OXIDATION OF COORDINATED LIGANDS

In the foregoing section coordinated dioxygen was reacted with various substrates resulting in either oxidation and incorporation into the coordination sphere or oxidation and expulsion from the coordination sphere. Clearly the latter process is the most desirable one to extend to catalytic reactions. In this section reactions of coordinated ligands with molecular oxygen are considered. In many cases, the results are the same as those reviewed in the previous section suggesting the possibility of common intermediates. This is particularly true in the reactions of SO_2 complexes, equations (41)–(46). Literature references again appear under the arrow for these reactions.

$$Ir(PPh_3)_2(CO)Cl(SO_2) + O_2 \xrightarrow[\text{[99, 107, 108]}]{} Ir(PPh_3)_2(CO)Cl(SO_4) \quad (41)$$

$$Rh(PPh_3)_2(CO)Cl(SO_2) + O_2 \xrightarrow[\text{[97]}]{} Rh(PPh_3)_2(CO)Cl(SO_4) \quad (42)$$

$$Ru(PPh_3)_2(NO)Cl(SO_2) + O_2 \xrightarrow[\text{[97]}]{} Ru(PPh_3)_2(NO)Cl(SO_4) \quad (43)$$

$$Ni(t\text{-}BuNC)_3(SO_2) + O_2 \xrightarrow[\text{[89]}]{} Ni(t\text{-}BuNC)_2(SO_4) \quad (44)$$

$$Pd(Ph_3P)_3(SO_2) + O_2 \xrightarrow[\text{[86, 97]}]{} Pd(Ph_3P)_2(SO_4) \quad (45)$$

$$Pt(Ph_3P)_3(SO_2) + O_2 \xrightarrow[\text{[86, 97]}]{} Pt(Ph_3P)_2(SO_4) \quad (46)$$

It should be pointed out, however, that although many of the dioxygen complexes react readily with SO_2 to give sulfato complexes, many of the analogous SO_2 complexes react more slowly, or not at all with molecular oxygen. Reactions of coordinated CO also give rise to carbonate complexes except in the case of rhodium complexes which tend to form coordinated or free CO_2, equations (47)–(50).

$$Ir(TDPME)(CO)Cl + O_2 \xrightarrow[{[109]}]{} Ir(TDPME)Cl(CO_3) \quad (47)$$

$$Rh_2(PPh_3)_4(CO)_4 + O_2 \xrightarrow[{[110]}]{} Rh_2(PPh_3)_3(CO)_2(CO_2) \quad (48)$$

$$Ru(PPh_3)_2(CNR)(CO)_2 + O_2 \xrightarrow[{[78]}]{} Ru(PPh_3)_2CNR(CO)(CO_3) \quad (49)$$

$$Os(PPh_3)_2(CO)(NO)Cl + O_2 \xrightarrow[{[77]}]{} Os(PPh_3)_2(NO)Cl(CO_3) \quad (50)$$

Fewer examples of the oxidation of coordinated NO exist, however, some ruthenium complexes are among those which undergo this reaction, equation (51).

$$Ru(PPh_3)_2(NO)_2 + O_2 \xrightarrow[{[111]}]{} Ru(PPh_3)_2NO(NO_3)O_2 \quad (51)$$

Inspection of equations (41)–(51) reveals some interesting correlations concerning the relative reactivity of several ligands in the coordination sphere of various metals. For instance, one might conclude from equations (43), (49) and (51) that the order of reactivity of ligands toward O_2 in ruthenium complexes is $SO_2 > NO, CO > CNR$, Ph_3P. Coordinated phosphines, however, do react with oxygen under the proper conditions and, in fact several ruthenium complexes are catalysts for phosphine oxidation. In most instances when coordinated phosphines are oxidized they are readily expelled from the coordination sphere of the metal, however, in the case of Co(II), the phosphine oxide is retained, equation (52).

$$Co(P(C_2H_5)_3)_2Cl_2 + O_2 \xrightarrow[{[112]}]{} Co(OP(C_2H_5)_3)_2Cl_2 \quad (52)$$

Since metal hydrides and metal alkyls are often intermediates of catalytic reactions of unsaturated hydrocarbons, reactions of these species with O_2 are of interest. Cationic hydrido complexes of Ir, Rh, Ru and Os react with molecular oxygen to insert oxygen between the metal and the hydrido ligand, equations (53)–(56).

$$[Ir(PPh_3)_3(NO)H]^+ + O_2 \xrightarrow[{[111]}]{} [Ir(PPh_3)_2(NO)(OH)]^+ + OPPh_3 \quad (53)$$

$$[Rh(NH_3)_5H]^{2+} + O_2 \xrightarrow[{[113]}]{} [Rh(NH_3)_4(OH)(OOH)]^+ \quad (54)$$

$$[Ru(PPh_3)_2(NO)_2H]^+ + O_2 \xrightarrow[{[111]}]{} [Ru(PPh_3)_2(NO)_2OH]^+ \quad (55)$$

$$[Os(PPh_3)_2(NO)_2H]^+ + O_2 \xrightarrow[{[111]}]{} [Os(PPh_3)_2(NO)_2OH]^+ \quad (56)$$

Oxygen has also been shown to insert between the cobalt–alkyl bond in cobaloximes, equation (57) [114].

$$\text{oxime Co(III)—R} \xrightarrow[h\nu]{O_2} \text{oxime Co(III)—OOR} \quad (57)$$

Iridium alkylperoxide complexes have also been reported [115], but they were formed by reaction of iridium(I) complexes with alkylhydroperoxides, equation (57*a*).

$$\text{trans-Ir(CO)LXL} + \text{ROOH} \longrightarrow \text{Ir(OOR)(OOR)(L)(L)(Cl)(CO)} \qquad (57a)$$

$L = PPh_3, AsPh_3, PPh_2Me$; $X = Cl, Br$; $R = t\text{-Bu}, PhMe_2$.

7. CATALYTIC OXIDATION REACTIONS

At this point we have considered the interaction of oxygen with metal complexes in solution and have presented ways in which coordinated dioxygen is transferred to substrates which may themselves be coordinated to the metal. In several instances we have seen that the oxidized substrate leaves the coordination sphere. For catalysis to occur this is a necessary step. Thus, those substrates which on oxidation become more weakly held ligands, can be displaced by more unoxidized substrate and a catalytic cycle, equation (58), is possible. If, on the other hand, the oxidized substrate is more strongly held than the original substrate, catalysis does not occur.

$$L_2M(O_2) \longrightarrow M(LO)_2 \xrightarrow[-2LO]{+2L,\ O_2} L_2M(O_2) \qquad (58)$$

Catalysis of oxidation is seldom as straightforward as the simplified model, equation (58), portrays. Although some examples which are not too different from the simple case will be presented, complex patterns of reactivity are the rule in most instances. The reason for this is the high reactivity of dioxygen with the organic substrate, the metal center, ancillary ligands and with the various organometallic intermediates which may be formed. Reaction pathways leading to the generation of organic free radicals are often possible and organic peroxides may form. The metal catalyzed decomposition of organic hydroperoxides to give organic radicals and the subsequent radical initiated autoxidation of the substrate often compete with coordination catalysis. Metal complexes vary widely in their ability to decompose hydroperoxides, and to some extent in the manner in which they decompose them. Selective catalytic conversion of an organic hydroperoxide to stable oxidation products in the presence of certain metal complexes affords convenient routes to useful chemicals. Unselective decomposition to radicals which initiate further unselective reactions often occurs, however.

We will consider both selective and unselective reactions in the next few sections of this chapter. The only criterion will be the nature of the substrate which is catalytically oxidized. The substrates considered will have systems of p or π electrons by which they are known to react readily with many of the metal complexes used as catalysts. Those instances which are clear examples of coordination catalysis will serve to illustrate principles of catalytic oxygen activation. Those instances in which coordination catalysis plays a less important role may serve to illustrate areas in which future work could provide improvements by means of selective activation and transfer of dioxygen.

8. CATALYTIC OXIDATION OF PHOSPHINES TO PHOSPHINE OXIDES

8.1. Complexes of Ni, Pd and Pt as catalysts

A large number of transition metal complexes have been found to catalyze the oxidation of organic phosphines to phosphine oxides. Wilke, Schott and Heimbach [85] showed that complexes of the type: $[M(PPh_3)_4]$ catalyze the oxidation of triphenylphosphine to triphenylphosphine oxide in toluene solution, equation (59).

$$2PPh_3 + O_2 \xrightarrow{[M(PPh_3)_4]} 2OPPh_3 \qquad (59)$$

When M is Ni reaction occurs at temperatures above $-35\ ^\circ C$ whereas the corresponding Pd and Pt complexes show catalytic activity when warmed at temperatures above $90\ ^\circ C$. Using $[Ni(PPh_3)_4]$, about 50 moles of phosphine were oxidized per mole of nickel complex while for each mole of the palladium complex used, over 500 moles of triphenylphosphine were converted to triphenylphosphine oxide.

The oxidation of triphenylphosphine catalyzed by $[Pt(PPh_3)_4]$ has been extensively studied [87, 88, 117, 118] by Halpern and co-workers who have elucidated many aspects of the reaction mechanism. By the use of optical absorption spectroscopy it has been shown that the predominant species present in a benzene solution of $[Pt(PPh_3)_4]$ is $[Pt(PPh_3)_3]$, equation (60).

$$Pt(PPh_3)_4 \longrightarrow Pt(PPh_3)_3 + PPh_3 \qquad (60)$$

Subsequent dissociation to $[Pt(PPh_3)_2]$ and triphenylphosphine or recombination with triphenylphosphine occur to only a minor extent over a rather wide range of phosphine concentrations.

The complex, $[Pt(PPh_3)_3]$, has been shown to add molecular oxygen with expulsion of triphenylphosphine to give a dioxygen complex, equation (61).

$$Pt(PPh_3)_3 + O_2 \xrightarrow{k_1} Pt(PPh_3)_2(O_2) + PPh_3 \qquad (61)$$

$$\frac{-d[Pt(PPh_3)_3]}{dt} = k_1[Pt(PPh_3)_3][O_2]$$

$$k_1 = 2.61\,\text{mol}^{-1}\,\text{s}^{-1} \text{ at } 25\,^\circ\text{C}$$

The platinum dioxygen complex reacts with more triphenylphosphine, equation (62), to form triphenylphosphine oxide and regenerate the original species, $[Pt(PPh_3)_3]$, thus creating a catalytic cycle.

$$Pt(PPh_3)_2(O_2) + PPh_3 \xrightarrow{k_2} Pt(PPh_3)_3(O_2) \xrightarrow[\text{fast}]{2PPh_3} Pt(PPh_3)_3 + 2OPPh_3 \qquad (62)$$

$$\frac{+d[Pt(PPh_3)_3]}{dt} = k_2[Pt(PPh_3)_2(O_2)][PPh_3]$$

$$k_2 = 0.151\,\text{mol}^{-1}\,\text{s}^{-1} \text{ at } 25\,^\circ\text{C}$$

This reaction sequence was established by investigating each reaction separately and monitoring the concentration of $[Pt(PPh_3)_3]$ spectrometrically. The reaction of $[Pt(PPh_3)_3]$ with dioxygen, equation (61), was studied using an excess of the metal complex relative to dioxygen to insure minimum interference from the subsequent reaction with triphenylphosphine. Reaction of the dioxygen complex, $[Pt(PPh_3)_3(O_2)]$, with triphenylphosphine, equation (62) was studied using excess phosphine in benzene. The resulting rate laws appear under each reaction, equations (61) and (62).

The validity of this reaction sequence was also checked while both reactions were occurring simultaneously using a large excess of triphenylphosphine with constant oxygen pressure and measuring the rate of oxygen consumption, equation (63).

$$\frac{-d[O_2]}{dt} = k_1[Pt(PPh_3)_3][O_2] \qquad (63)$$

If stationary state conditions are assumed, equation (64) results. Making use of the conservation relation, equation (65), the rate law for oxygen consumption becomes, equation (66). The excellent agreement between the measured rate of oxygen consumption and values calculated from equation (66), with the rate constants for equations (61) and (62), provides strong support for this reaction sequence. This agreement between measured and calculated values of rate of oxygen consumption also argues against equation (67) as a possible pathway by which initial reaction between $[Pt(PPh_3)_3]$ and O_2 might proceed. A significant contribution from a step with stoichiometry corresponding to equation (67) would result in a higher value in rate than that calculated from equations (61) and (62).

$$\frac{d[Pt(PPh_3)_3]}{dt} = k_1[Pt(PPh_3)_3][O_2] - k_2[Pt(PPh_3)_2(O_2)][PPh_3] = 0 \qquad (64)$$

$$[Pt(PPh_3)_3] + [Pt(PPh_3)_2(O_2)] = [Pt]_{total} \qquad (65)$$

$$\frac{-d[O_2]}{dt} = \frac{k_1k_2[Pt]_{total}[PPh_3][O_2]}{k_1[O_2] + k_2[PPh_3]} \qquad (66)$$

$$Pt(PPh_3)_3 + 1.5O_2 \longrightarrow Pt(PPh_3)_2(O_2) + Ph_3PO \qquad (67)$$

Having established the likelihood of the reactions (61) and (62), for the catalytic oxidation of triphenylphosphine, Halpern has postulated a mechanism involving a dissociative oxygen insertion step as the means of oxygen transfer to the phosphorous, equation (68).

$$(Ph_3P)_2Pt(O_2) \xrightarrow[\text{slow}]{PPh_3} Ph_3P{-}Pt(O{\cdots}PPh_3)_2 \longrightarrow Ph_3P{-}Pt(OPPh_3)_2 \xrightarrow[\text{fast}]{+2PPh_3,\ -2OPPh_3} Pt(PPh_3)_3 \xrightarrow{O_2} (Ph_3P)_2Pt(O_2) \qquad (68)$$

Stern [118*a*] has suggested a modification of this mechanistic scheme, equation (68*a*). This is in accord with the observations that the predominant species in solutions of $[Pt(PPh_3)_4]$ is $[Pt(PPh_3)_3]$, [118*b*], and that PPh_3 reduces platinum and palladium oxides to zero valent metal complexes [118*c*, *d*].

$$Pt(PPh_3)_2(O_2) \xrightarrow[\text{slow}]{PPh_3} (PPh_3)_2Pt(O_2{\cdots}PPh_3) \longrightarrow (Ph_3P)_2Pt{-}OOPPh_3 \longrightarrow (PPh_3)_2PtO + OPPh_3 \xrightarrow[\text{fast}]{PPh_3} Pt(PPh_3)_2 + OPPh_3 \xrightarrow{O_2} Pt(PPh_3)_2(O_2) \qquad (68a)$$

8.2. Ruthenium complexes

Ruthenium complexes also catalyze the oxidation of triphenylphosphine to triphenylphosphine oxide [76, 79, 81, 119]. Among the catalytically active ruthenium complexes are $[Ru(NCS)(CO)(NO)(PPh_3)_2]$ and $[Ru(NCS)(NO)(PPh_3)_2(O_2)]$ which have been found to catalyze this reaction at 50–80 °C in xylene solution [76, 79]. These two complexes are related through the equilibrium reaction shown by equation (69).

$$O_2 + Ru(NCS)(CO)(NO)(PPh_3)_2 \overset{K_1}{\rightleftharpoons} Ru(NCS)(NO)(PPh_3)_2(O_2) + CO \qquad (69)$$

The actual catalytic cycle is reported to involve reaction of the dioxygen complex with triphenylphosphine followed by a slow oxygen atom transfer reaction to give a triphenylphosphine oxide complex. This complex undergoes ligand displacement to give a ruthenium phosphine complex liberating triphenylphosphine oxide. The ruthenium phosphine complex readily reacts with oxygen to form the dioxygen complex and the catalytic cycle begins again, equation (70).

$$\begin{array}{ccc} Ru(NCS)(NO)(PPh_3)_2(O_2) & \underset{}{\overset{K_2PPh_3}{\rightleftharpoons}} & Ru(NCS)(NO)(PPh_3)_3(O_2) \\ \uparrow O_2\text{, fast} & & k_2 \downarrow \text{slow} \\ Ru(NCS)(NO)(PPh_3)_2 & \underset{-2OPPh_3}{\overset{+PPh_3}{\longleftarrow}} & Ru(NCS)(NO)(PPh_3)(OPPh_3)_2 \\ & \text{fast} & \end{array} \tag{70}$$

The individual steps in the mechanism proposed for the ruthenium catalyzed oxidation of triphenylphosphine are similar to those of the platinum catalyzed reaction. However, the stoichiometry of one step and the relative rates of several other steps appear to be somewhat different. The observed rate laws are quite different from those of the platinum complex. The rate equations for the proposed mechanism are shown in equation (71) for the dioxygen complex and equation (72) for the carbonyl complex ($RuO_2 = [Ru(NCS)(NO)(PPh_3)_2(O_2)]$; $RuCO = [Ru(NCS)(CO)(NO)(PPh_3)_2]$).

$$\frac{-d[PPh_3]}{dt} = \frac{k_1 K_2[RuO_2][PPh_3]}{1 + K_2[PPh_3]} \tag{71}$$

$$\frac{-d[PPh_3]}{dt} = \frac{k_1 K_1 K_2[RuCO][O_2][PPh_3]}{[CO] + K_1[O_2] + K_1 K_2[PPh_3][O_2]} \tag{72}$$

Analysis of the data on the basis of these equations yields $k_1 = 1.26 \times 10^{-2}\ sec^{-1}$ and $K_2 = 1.63 \times 10^{-2}\ l\,mol^{-1}$. In contrast to the observations of Halpern and co-workers who found that the rate of triphenylphosphine oxidation catalyzed by $[Pt(O_2)(PPh_3)_2]$ depended linearly on $[PPh_3]$, the above two rate equations require that the reciprocal of the rate depends on $[Ph_3P]^{-1}$.

The catalytic cycle is initiated by entry of triphenylphosphine into the coordination sphere of the ruthenium dioxygen complex. Graham *et al.*, formulate the intermediate, $[Ru(NCS)(NO)(PPh_3)_3(O_2)]$ as a ruthenium complex in which the nitrosyl ligand is bound as NO^- forming a bent Ru–N–O linkage. These authors note that the ability of the nitrosyl group to adapt its ligand properties from a linear three electron donor to a bent one electron donor, seems to be playing a special role in allowing the third phosphine to coordinate.

The rate determining step in the catalytic cycle, equation (70), is the oxygen-atom-transfer process to give a triphenylphosphine oxide complex. Since such species should be relatively unstable it readily undergoes ligand exchange to produce a coordinatively unsaturated complex which readily takes up oxygen to reform the catalyst, $[Ru(NCS)(NO)(PPh_3)_2(O_2)]$.

Cenini, Fusi and Capparella [119] found that $[RuCl_2(PPh_3)_3]$ is an effective catalyst for oxidizing triphenylphosphine to triphenylphosphine oxide at room temperature and atmospheric pressure. For example, at 20 °C in benzene (20 ml), $[RuCl_2(PPh_3)_3]$ (0.15 mM) oxidized PPh_3 (1.5 mM) nearly quantitatively in about 15 min under a pure atmosphere of dioxygen. These authors have observed considerable dissociation of the ruthenium complex in solution, equation (73) followed by oxygen uptake, equation (74), but have not isolated the dioxygen adduct. The mechanism of catalytic oxidation is expected to be similar to those cases previously discussed.

$$RuCl_2(PPh_3)_3 \rightleftharpoons RuCl_2(PPh_3)_2 + PPh_3 \quad (73)$$

$$Ru(PPh_3)_2Cl_2 + O_2 \longrightarrow RuCl_2(PPh_3)_2(O_2) \quad (74)$$

Another dioxygen complex, $[RuCl_2(AsPPh_3)_3]$, also appears to catalyze the oxidation of triphenylphosphine to triphenylphosphine oxide [81], however, a mechanism study has not yet been carried out.

8.3. Complexes of rhodium and iridium

Rhodium complexes are also active catalysts for phosphine oxidation. For example, Poddar and Agarwal [120] report that when air was bubbled through a solution of $[Rh(PPh_3)_3Cl]$ (0.05 g) and triphenylphosphine (0.50 g) in refluxing toluene for 4 hr, a quantitative yield of triphenylphosphine oxide was obtained. Blank experiments carried out in the absence of the rhodium complex gave negative results.

Von Vugt *et al.*, have studied the oxygen uptake properties of $[RhCl(PPh_3)_3]$ in benzene solution as well as the catalytic oxidation of excess phosphine [121]. The rate of oxygenation of the metal complex, v, was expressed as shown in equation (75), in which P_{O_2} is the partial pressure of oxygen.

$$v = \frac{k[P_{O_2}]^{1.28}[RhCl(PPh_3)_3]}{[PPh_3]} \quad (75)$$

This is consistent with the reaction sequence suggested below, equation (76), in which the equilibrium between the *tris*phosphine and *bis*phosphine complexes has been established [124] and coordination of oxygen is postulated as the rate determining step.

$$Rh(PPh_3)_3Cl \overset{K}{\rightleftharpoons} Rh^{I}Cl(PPh_2)_2 + PPh_3 \quad (76)$$

$$Rh^{I}Cl(PPh_2)_2 \xrightarrow[k_2]{O_2} Rh^{II}Cl(PPh_3)_2(O_2^-)$$

$$Rh^{III}Cl(PPh_3)_2(O_2^{=}) \underset{k_{-3}}{\overset{k_3}{\rightleftharpoons}} Rh^{II}Cl(PPh_3)_2(O_2^-)$$

It has been found that 1.5 molecules of oxygen were taken up per molecule of $[RhCl(PPh_3)_3]$ [121–123]. Reaction (76), however, requires only 1 molecule of oxygen per rhodium. Von Vugt *et al.*, state, however, that the remaining 0.5 molecule of oxygen is used for the oxidation of one of the three phosphine ligands. Each additional molecule of triphenylphosphine gave rise to an additional uptake of 0.5 molecule of oxygen.

Oxygenated solutions of the rhodium complex showed the presence of an ESR signal the intensity of which remained constant during the period of oxygen uptake (15 min) and thereafter doubled within a few hours. Low temperature ESR studies gave two sets of g values one of which was ascribed to Rh(II) and the other to $(O_2)^-$. The occurrence of a steady state concentration of the $Rh(II)(O_2)^-$ species during oxygen uptake and the increase of this concentration by a factor of 2 after oxygen uptake is completed point to a slow establishment of equilibrium between the superoxo and the peroxo complex, equation (76). In the steady state $k_2[O_2][RhCl(PPh_3)_2]$ is approximately equal to $k_3[Rh^{II}Cl(PPh_3)_2(O_2^-)]$. From the final concentration of $Rh(II)(O_2)^-$, $K_3 = k_3/k_{-3}$ was estimated to be ~ 40.

A mechanism for the catalytic oxidation of triphenylphosphine was suggested, equation (77), in which the valence state of the catalytically active complex is not yet known.

$$
\begin{array}{ccc}
RhCl(PPh_3)_2(O_2) & \underset{}{\overset{2PPh_3}{\rightleftharpoons}} & RhCl(PPh_3)_4(O_2) \\
O_2 \uparrow k_2 & & \downarrow \\
RhCl(PPh_3)_2 & \underset{-2OPPh_3}{\rightleftharpoons} & RhCl(PPh_3)_2(OPPh_3)_2
\end{array} \qquad (77)
$$

Triphenylphosphine is oxidized easily in the presence of iridium complexes of the formula, $[IrX(CO)(PPh_3)_2]$, X = Cl, Br, I to give triphenylphosphine oxide [125]. The catalyst is recovered without decomposition after the reaction. The yield of triphenylphosphine oxide increases as the halogen, X, in the iridium complex is varied: Cl < I < Br. The fact that the iridium complexes coordinate easily with dioxygen in an oxygen atmosphere suggests that the reaction proceeds through a dioxygen complex. These authors propose the familiar scheme shown below, equation (78).

$$
trans\text{-}[IrX(CO)(PPh_3)_2] \xrightarrow{O_2} [IrX(CO)(PPh_3)_2(O_2)] \longrightarrow [IrX(CO)(PPh_3)_2(O_2)] \xrightarrow[+PPh_3]{-OPPh_3} trans\text{-}[IrX(CO)(PPh_3)_2] \qquad (78)
$$

8.4. Cobalt complexes

Cobalt complexes have been used to catalytically oxidize triphenylphosphine [126–128]. Schmidt and Yoke [126] have found that the sole product of the slow but quantitative oxidation of $[CoCl_2(PEt_3)_2]$ in several organic solvents is the phosphine oxide complex, equation (79).

$$CoCl_2(PEt_3)_2 + O_2 \longrightarrow CoCl_2(OPEt_3)_2 \qquad (79)$$

The reaction product was isolated and characterized. In its initial stages the reaction is first order in oxygen and first order in cobalt complex, equation (80).

$$\frac{d[O_2]}{dt} = k[CoCl_2(PEt_3)_2]\,P_{O_2} \qquad (80)$$

A free radical initiator azobisisobutyronitrile or an inhibitor, hydroquinone, have no effect on the rate of oxygen consumption. In any case a dissociative mechanism is precluded since autoxidation of uncoordinated phosphines is a radical process which gives mixed $R_nP(O)(OR)_{3-n}$ products. The mechanism suggested involves the formation and rearrangement of a cobalt dioxygen complex, equations (81) and (82), possibly *via* a dissociative oxygen insertion step of the type proposed by Halpern for platinum complexes.

$$CoCl_2(PEt_3)_2 + O_2 \rightleftharpoons CoCl_2(PEt_3)_2(O_2) \qquad (81)$$

$$CoCl_2(PEt_3)_2O_2 \longrightarrow CoCl_2(OPEt_3)_2 \qquad (82)$$

During the reaction in solution a redistribution, equation (83), to give a mixed ligand complex was observed with an equilibrium constant of about 10.

$$CoCl_2(PEt_3)_2 + CoCl_2(OPEt_3)_2 \rightleftharpoons 2CoCl_2(PEt_3)(OPEt_3) \qquad (83)$$

Although this system appears to possess all of the attributes necessary for the catalytic oxidation of phosphine to phosphine oxide, the catalytic activity of these cobalt complexes was not reported.

An unusual class of trinuclear cobalt complexes prepared by reaction of oxygen with solutions of bis(triphenylphosphine) cobalt(II) chloride in allylamine, AA, in the presence of benzotriazole, BT, have been shown to exhibit catalytic activity. Complex VIII, for example, weakly catalyzes the oxidation of triphenylphosphine at room temperature (80% after 3 days, ~ 20 turnovers) the process being accelerated by UV light.

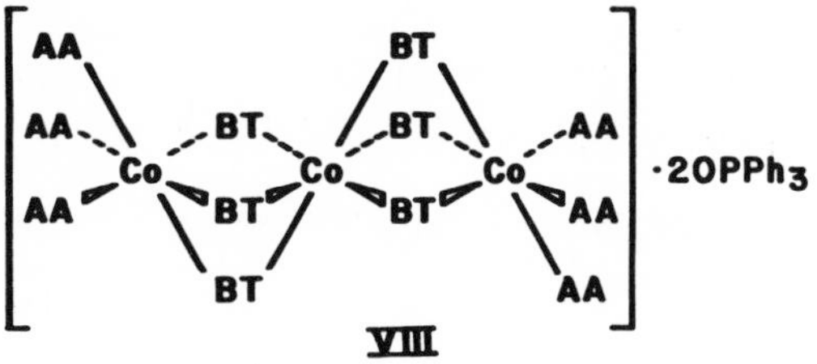

$$\left[(AA)_3Co(\mu\text{-}BT)_3Co(\mu\text{-}BT)_3Co(AA)_3\right]\cdot 2OPPh_3$$

VIII

A novel dioxygen complex has been found which is readily prepared, equation (84), by passing a stream of oxygen through a benzene solution of $[Co(CN)_2(PMe_2Ph)_3]$ for about 8 hr [128].

$$2Co(CN)_2(PMe_2Ph)_3 + O_2 \longrightarrow Co_2(CN)_4(PMe_2Ph)_5(O_2) + PMe_2Ph \qquad (84)$$

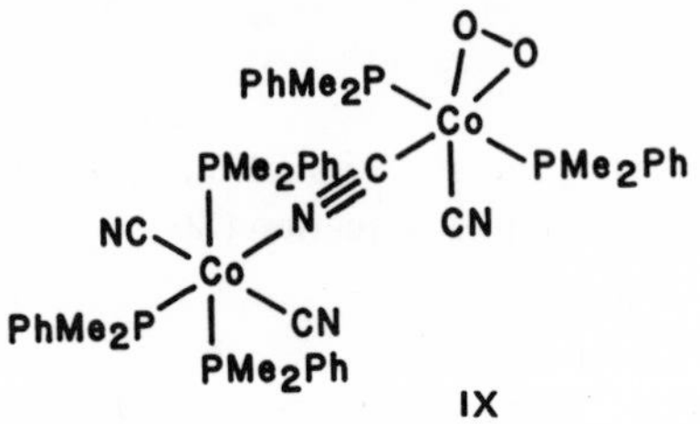

O O
PhMe2P Co PMe2Ph
PMe2Ph C
NC N CN
Co
PhMe2P CN
PMe2Ph

IX

X-ray crystallographic determination showed the dioxygen complex to have the structure, IX. Preliminary kinetic measurements on reaction (84) yielded a rate law of the form of equation (85) where $k \approx 1 \times 10^{-4}\ M^{-1}\,sec^{-1}$ in benzene at 25 °C.

$$\frac{-d[O_2]}{dt} = k_3[Co(CN)_2(PMe_2Ph)_3]^2[O_2][PMe_2Ph]^{-1} \qquad (85)$$

This result together with the unusual structure of IX lead the authors to suggest the following mechanism for reaction (84), according to which $k_3 = K_4 k_5$.

$$Co(CN)_2(PMe_2Ph_3)_3 + O_2 \overset{K_4}{\rightleftharpoons} Co^{II}(CN)_2(PMe_2Ph)_2(O_2) + PMe_2Ph$$
$$Co^{II}(CN)_2(PMe_2Ph)_2(O_2) + Co^{II}(CN)_2(PMe_2Ph)_3 \xrightarrow{k_5} IX \qquad (86)$$

According to this interpretation the binuclear complex IX, is formed through a CN-bridged inner sphere electron transfer reaction between $[Co^{II}(CN)_2(PMe_2Ph)_3]$ and the O_2-substituted complex, $[Co^{II}(CN)_2(PMe_2Ph)_2(O_2)]$. The driving force for this is reinforced by the ability of the O_2 ligand to remove electron density from the resulting "Co^{I}" center. The formal oxidation states of the final product, IX, accordingly, correspond to $[Co^{III}NCCo^{III}(O_2^{2-})]$. Assignment of the +3 oxidation states to the two Co atoms are consistent with the substitution inertness of IX.

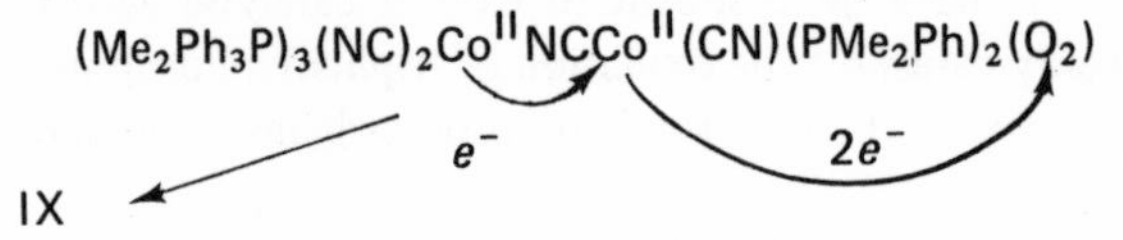

$$(Me_2Ph_3P)_3(NC)_2Co^{II}NCCo^{II}(CN)(PMe_2Ph)_2(O_2)$$

e^- $2e^-$

IX

In methanol solution the dioxygen complex, IX, reacted with added PMe_2Ph to form Me_2PhPO and regenerate $[Co(CN)_2(PMe_2Ph)_3]$ according to equation (87) with a rate law approximating equation (88), where $k_6 \approx 4 \times 10^{-5}\,M^{-1}\,sec^{-1}$.

$$Co_2(CN)_4(PMe_2Ph)_5(O_2) + 3PMe_2Ph \longrightarrow 2Co(CN)_2(PMe_2Ph)_3 + 2OPMe_2Ph \qquad (87)$$

$$\frac{-d[IX]}{dt} = k_6[IX][PMe_2Ph] \tag{88}$$

Taken together reactions (84) and (87) constitute a catalytic cycle for the oxidation of PMe_2Ph to $OPMe_2Ph$ by O_2 under the catalytic influence of $[Co(CN)_2(PMe_2Ph)_3]$. A significant feature of this system is that the dioxygen complex involved is coordinatively saturated whereas prior examples of oxidation of phosphines are facilitated by coordinative unsaturation through mechanisms involving attack of phosphine on the metal center.

8.5. Molybdenum oxo complexes

A novel type oxygen activation which has not been mentioned yet was reported by Barral, Bocard, Seree de Roch and Sajus [129–131]. These authors have shown that the oxo complexes: $[MoO_2(S_2CNR_2)_2]$, (R = ethyl, *n*-propyl or isobutyl) [132] catalyze the selective oxidation of tri-*n*-butylphosphine and triphenylphosphine to the corresponding phosphine oxides. Again, the fact that $OPBu_3$ is selectively formed from PBu_3 without products such as $OPBu_2(OBu)$, argues against a dissociative radical autoxidation mechanism and implies coordination catalysis.

In a typical experiment, oxygen reacts with 2 moles of triphenylphosphine in 1 litre of chlorobenzene containing 1.5×10^{-2} moles of $[MoO_2(S_2CN(n\text{-}Pr)_2)_2]$, X, at 40 °C quantitatively within 1 hr. In another case, triphenylphosphine was stirred in benzene solution containing $[MoO_2(S_2CNR_2)_2]$ under 1 Kg/cm^2 partial pressure of O_2 to give a 98% yield of Ph_3PO.

The probable steps of the catalytic reaction have been studied individually. It has been shown that the molybdenum dioxo complex, X, reacts with triphenylphosphine in an inert atmosphere to give triphenylphosphine oxide and the red molybdenum IV complex, $[MoO(S_2CN(n\text{-}Pr)_2)_2]$, equation (89), which can be separated by chromatography on Al_2O_3 in an oxygen free atmosphere.

$$MoO_2(S_2CN(n\text{-}Pr)_2)_2 + PPh_3 \longrightarrow MoO(S_2CN(n\text{-}Pr)_2)_2 + OPPh_3 \tag{89}$$

The initially red solution, however, rapidly develops a violet color. The isolated red molybdenum IV complex, $[MoO(S_2CNR_2)_2]$ reacts readily with molecular oxygen to regenerate the catalytically active yellow dioxo species, equation (90). Again, however, a violet color is noted during the course of this reaction.

$$MoO(S_2CN(n\text{-}Pr)_2)_2 + 1/2\ O_2 \longrightarrow MoO_2(S_2CN(n\text{-}Pr)_2)_2 \tag{90}$$

The violet color which is observed during reactions (89) and (90) was found to correspond to the formation of a bimolecular molybdenum complex: $[Mo_2O_3(S_2CN(n\text{-}Pr)_2)_4]$. The authors have shown that the binuclear complex can be formed by reaction of phosphine with X, equation (91), and from a disproportionation reaction between X and $[MoO(S_2CN(n\text{-}Pr)_2)_2]$, equation (92).

$$2MoO_2(S_2CN(n\text{-}Pr)_2)_2 + PPh_3 \longrightarrow 2Mo_2O_3(S_2CN(n\text{-}Pr)_2)_4 + OPPh_3 \tag{91}$$

$$MoO_2(S_2CN(n\text{-}Pr)_2)_2 + MoO(S_2CN(n\text{-}Pr)_2)_2 \rightleftharpoons Mo_2O_3(S_2CN(n\text{-}Pr)_2)_4 \quad (92)$$

The mechanism of formation of the binuclear complex in equation (91) probably occurs *via* a combination of equations (89) and (92). The equilibrium constant for reaction (92) was determined, equation (93).

$$K = \frac{[MoO(S_2CN(n\text{-}Pr)_2)_2][MoO_2(S_2CN(n\text{-}Pr)_2)_2]}{[Mo_2O_3(S_2CN(n\text{-}Pr)_2)_4]} = 4.1 \times 10^{-3}\ mol \cdot l^{-1} \text{ at } 41\,^{\circ}C \quad (93)$$

The scheme developed for the catalytic oxidation of triphenylphosphine by this novel system is shown in equation (94).

$$MoO_2(S_2CNR_2)_2 + MoO(S_2CNR_2)_2 \underset{}{\overset{K}{\rightleftharpoons}} Mo_2O_3S_2(CNR_2)_4 \quad (94)$$

($PR'_3 \xrightarrow{k_r} OPR'_3$ converting $MoO_2(S_2CNR_2)_2$ to $MoO(S_2CNR_2)_2$; $+ 1/2\ O_2\ k_0$ regenerating $MoO_2(S_2CNR_2)_2$)

Studies of the absorption of oxygen by solutions of the dinuclear complex containing increasing quantities of X permit the determination of the value of the constant $k_0 = 1.1 \times 10^{-2}\ s^{-1}$ ($P_{O_2} = 760$ mm Hg) at 41 °C for R = *n*-Pr. The value for the second order rate constant k_r is $2.3\ mol^{-1} \cdot l \cdot s^{-1}$ at 41 °C for the catalytic oxidation of triphenylphosphine in chlorobenzene.

8.6. Iron complexes

The oxidation of triphenylphosphine has also been investigated using iron complexes as catalysts [133]. Kinetics of this reaction were studied using $[Fe(mnt)_2]^-$ and $[Fe(mnt)_3]^{2-}$ where mnt^{2-} is *cis*-1,2-dicyanoethylene-1,2-dithiolate. Approximately 1 mol of the iron(IV) complex catalyzed the oxidation of 15 mol of triphenylphosphine compared to about 10 mol of triphenylphosphine for the iron(III) complex. The rate of oxygen uptake in the presence of either complex was found to be proportional to the concentrations of both triphenylphosphine and the iron complex but independent of oxygen pressure. No evidence was obtained for the formation of molecular oxygen complexes.

In addition to the oxidation of organic phosphines, phosphine itself PH_3, has been oxidized in the presence of metal complexes (cf. for example, [134] and [135]). However, since this review will deal primarily with the catalysis of organic reactions this topic will not be considered herein.

We have considered the oxidation of organic phosphines in some detail because as a group of reactions they constitute the most widely studied examples of coordination catalysis of oxidation. It seems clear that in the presence of many different metal complexes, oxygen is indeed catalytically activated and transferred to the organic substrate within the coordination sphere. For many complexes, phosphine oxidation reactions have much in common and concepts derived from this area of oxidation catalysis might well trigger ideas which will enable chemists to control the metal catalyzed addition of oxygen to other organic substrates as well.

9. OXIDATION OF ARSINES TO ARSINE OXIDES

More limited investigations of the oxidation of triorganoarsines in the presence of transition metal complexes have been carried out but it appears that many of the same complexes which catalyze phosphine oxidation are also effective for arsines. The ruthenium complex, $[Ru(NCS)(NO)(PPh_3)_2(O_2)]$, for example, has been found to be an efficient catalyst for the oxidation of triphenylarsine [79]; the initial rate of oxidation being three to four times greater than the corresponding rate for triphenylphosphine oxidation under the same conditions. However, the rate of this oxidation was rapidly diminished by decomposition of the catalyst so that in a short time (*ca.* 30 min) the catalyst was virtually ineffective.

Poddar and Agarwala report that Rh(I) complexes catalyze the oxidation of a diarsine but are less active for triphenylarsine oxidation [119]. When air was bubbled through a solution of $[RhCl(PPh_3)_3]$ (0.05 g) and EDA (ethylene-1,2-*bis*-diphenylarsine) (0.5 g) in refluxing toluene (40 ml) for 4 hr, a quantitative yield of EDAO (ethylene-1-diphenylarsine-2-diphenylarsine oxide) was obtained. When a similar reaction was carried out with triphenylarsine the corresponding arsine oxide was not obtained after 4 hr. After 70 hr, however, triphenylarsine oxide was obtained. Blank experiments, in which $[RhCl(PPh_3)_3]$ was not added, gave negative results.

When 30 ml of a solution of $[RhCl(PPh_3)_3]$ (0.3 g) and EDA (0.3 g) was refluxed for 2 hr (presumably in the presence of O_2) then concentrated to 10 ml and chilled at 0 °C overnight, a new complex crystallized. This complex contained one fully oxidized ligand, $EDAO_2$ (ethylene-1,2-*bis*-diphenylarsine oxide) and one unoxidized ligand, EDA. The structure of the complex $[RhCl(EDA)(EDAO_2)]$ was formulated as XI, below based on magnetic, spectroscopic (UV, visible and IR) and other physical studies.

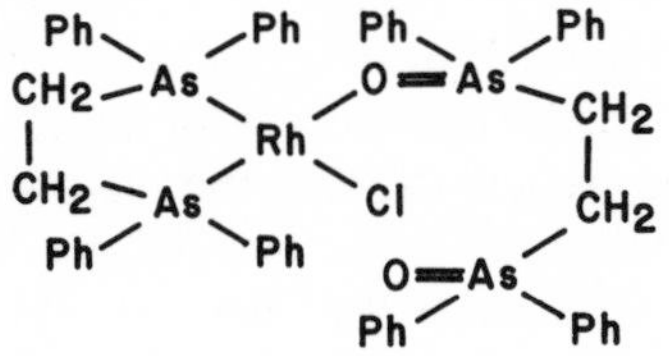

```
Ph     Ph  Ph      Ph
  \   /      \    /
 CH2—As       O=As
 |      \    /      \
 |       Rh          CH2
 |      /    \        |
 CH2—As       Cl     CH2
  /   \              /
Ph     Ph       O=As
               /    \
             Ph      Ph
```

XI

Solutions of triphenylarsine and $[Fe(mnt)_2]^-$ in acetonitrile take up oxygen slowly with formation of triphenylarsine oxide [133]. The rate of oxygen consumption by triphenylarsine was slow ($t_{1/2}$ = 10–50 hr), but the first-order rate plots were linear over more than two half-times. For $[Fe(mnt)_2]^- = 1.93 \times 10^{-2}$ M, $[Ph_3As] = 3.3 \times 10^{-2}$ M, $t_{1/2}$ = 36 hr in oxygen.

10. OXIDATION OF SULFIDES AND SULFOXIDES

10.1. Catalytic reactions with dioxygen

The examples of direct O_2 oxidation of sulfides catalyzed by metal complexes are even less abundant and less detailed than are oxidations of arsines. Examples from the patent literature include the removal of H_2S from waste gases by oxidation of H_2S in DMF solutions of metal complexes such as chloro (N,N′-ethylene*bis*(salicylideneiminato)cobalt [136] or the oxidation of mercaptans in the presence of alkali using Cu, Fe, Co, or Mn complexes with various azo compounds [137]. Avdeeva and Mashkina [138] have reported the oxidation of sulfides from a diesel fraction of Arlan petroleum at 120 °C and 50 atm in the presence of 1 weight % vanadylacetylacetonate to obtain 25–32 mole % sulfoxides. The selectivity was 40–60% and could be improved by the addition of butyraldehyde as a co-oxidant. Evidence for the catalytic transfer of dioxygen to dialkyl sulfides has not been reported. It has recently been reported [137*a*], however, that dibenzothiophene may be oxidized to dibenzothiophene-5,5-dioxide in high yield in the presence of the ruthenium complexes: $[Ru_3(CO)_{12}]$, $[Ru(acac)_3]$, and $RuCl_3$ at elevated temperatures.

Henbest and Trocha-Grimshaw have shown that sulfoxides may be oxidized to sulfones in the presence of iridium and rhodium complexes [139, 140]. Oxidations were studied by passing air through a solution of the sulfoxide in hot *iso*propanol containing 10% water in the presence of the catalyst. Sulfones were formed when chlorides of iridium or rhodium were used while chlorides of ruthenium, osmium and palladium were not effective. Under reaction conditions the chlorides should be converted wholly or partly to metal sulfoxide complexes.

Of the iridium complexes tried, the best catalyst was the hydride $[IrHCl_2(Me_2SO)_3]$. When this complex was used, dimethylsulfone was obtained in more than 90% yield after 15 hr. The solution remained clear yellow throughout the reaction. Oxidation proceeded slowly during the first 3 hr and then became more rapid suggesting that the hydride may be transformed into more catalytically active species. After reaction was complete the iridium was recoverable as a gummy substance, the infrared spectrum of which indicated the presence of metal sulfoxide groups but which gave no evidence for the presence of IrH or IrO_2 groups. The iridium dioxygen complex $[IrCl(CO)(PPh_3)_2(O_2)]$, was inactive as a catalyst for this reaction. The authors state, however, that this does not exclude the intervention of iridium peroxo complexes stabilized by sulfoxide ligands in the oxidation catalyzed by the hydride, $[IrHCl_2(Me_2SO)_3]$, which may be more reactive than the carbonyl phosphine complex.

The related rhodium complex $H[RhCl_4(Me_2SO)_2]$ was similarly effective as a catalyst for the oxidation of dimethylsulfoxide to the sulfone, equation (95).

$$R_2SO + 1/2\ O_2 \xrightarrow{H[RhCl_4(Me_2SO)_2]} R_2SO_2 \qquad (95)$$

$$R = CH_3, Ph, PhCH_2$$

This complex was used to examine the catalytic oxidation of two other sulfoxides. Diphenylsulfoxide gave a 65% yield of sulfone in 22 hr and dibenzyl sulfoxide gave a 30% yield of the corresponding sulfone after 48 hr. It appears that the ease of oxidation diminishes with the substituents in the order: Me > Ph > $PhCH_2$.

10.2. Catalytic reactions using hydroperoxides

Although little work has appeared concerning the reaction of molecular oxygen with dialkyl sulfides catalyzed by metal complexes, the homogeneous catalytic oxidation of dialkylsulfides with hydroperoxides is accomplished with ease. Zirconium salts have been shown to catalyze the conversion of sulfides to sulfones using hydrogen peroxide [141, 142]. Complexes of vanadium, molybdenum or titanium are capable of catalyzing the reaction of sulfides with alkylhydroperoxides [143–145]. Although in the absence of a catalyst the hydroperoxide oxidation of a sulfide gives a sulfoxide, quantitative yields of sulfones can be obtained in the presence of some group V and VI metal complexes. For example, oxidation of thioanisole with *t*-butylhydroperoxide in benzene in the presence of molybdenylacetylacetonate produces phenylmethylsulfone in 98% yield. The selectivity of this reaction is shown by reaction 96 in which allyl *n*-butyl sulfide is oxidized to the sulfone without appreciable oxidation of the olefinic group [144, 145].

$$\begin{matrix} CH_3{-}CH_2{-}CH_2{-}CH_2 \\ CH_2{=}CH{-}CH_2 \end{matrix} \Big\rangle S \xrightarrow[MoO_2(acac)_2]{t\text{-BuOOH}} \begin{matrix} CH_3{-}CH_2{-}CH_2{-}CH_2 \\ CH_2{=}CH{-}CH_2 \end{matrix} \Big\rangle SO_2 \qquad (96)$$

Diallylsulfide can also be oxidized selectively to the sulfone without oxidation of the unsaturated side chain [144]. Coordination of the sulfur is apparently strong enough to exclude the olefinic ligand from the reactive center since in the absence of the sulfur, olefins are readily epoxidized by *t*-BuOOH in the presence of molybdenum complexes. The product of hydroperoxide oxidation of sulfides depends on the reaction conditions [143]. Sulfoxides are obtained at temperatures below 50 °C using excess sulfide whereas sulfones are the predominant product at temperatures above 55 °C using excess hydroperoxide.

The reaction of di-*n*-butylsulfide with *t*-butyl hydroperoxide proceeds at a convenient rate in ethanol at 25 °C in the presence of catalytic quantities of bisacetylacetonato-oxovanadium(IV) [$VO(acac)_2$] affording di-*n*-butylsulfoxide in quantitative yield [146]. The kinetics of this reaction have been examined in detail. With $[t\text{-BuOOH}]_o/[VO(acac)_2] \geqslant 8$ and $[VO(acac)_2]_o = 1 \times 10^{-4}$ to $\sim 2 \times 10^{-3}$ M the reaction is first order in sulfide and first order in vanadium catalyst, equation (97).

$$v = k_3[\text{cat}]_o[t\text{-BuOOH}][n\text{-Bu}_2\text{S}] \qquad (97)$$

With an excess of sulfide, the rate dependence on *t*-BuOOH was found to be similar to the Michaelis–Menten equation for enzyme catalysis, equation (98).

$$v = k[\text{cat}]_o[t\text{-BuOOH}]_o/K' + [t\text{-BuOOH}]_o \tag{98}$$

Here K′ is a constant related to the apparent dissociation constant of the vanadium-hydroperoxide and vanadium-ethanol complexes and k is the limiting specific rate constant which could be observed when the intermediate complex concentration is maximal.

The formation of the active catalytic species from $VO(acac)_2$ and t-BuOOH in ethanol was studied in detail. From visible, UV spectral measurements and ESR data which was gathered the authors conclude that a red vanadium (V) complex is formed which is the catalytically active species.

Finally, the relative rates of the vanadium catalyzed oxidation of some nucleophilic substrates by t-BuOOH in ethanol at 25 °C were found to be $(n\text{-Bu})_2S(100) > PhS(n\text{-Bu})(58) > (n\text{-Bu})_2SO(17) >$ cyclohexene (0.2). These data suggest strongly that "electropositive oxygen" is being transferred from the metal-activated hydroperoxide to the substrate. Thus, a polar mechanism can be advanced. A radical mechanism is ruled out by the observation of reproducible kinetics, the simple rate law, the failure of radical traps to influence rate, and by the absence of products expected to arise from radical intermediates.

The sequence of reactions expressed by equations (99)–(104) is consistent with the observations made in this study. In fact, the kinetic first-order dependence on sulfide, hydroperoxide, and vanadium catalyst which was observed implies that one molecule of each reactant is involved in the transition state of the rate-determining step.

$$VO(acac)_2 \xrightarrow{Bu^tO_2H} V^{5+} \qquad k_{act}, \text{fast} \tag{99}$$

$$V^{5+} + Bu^tO_2H \underset{k_{-1}}{\overset{k_1}{\rightleftharpoons}} V^{5+}\text{---}O(Bu^t)\text{—}O\text{—}H \;(1) \qquad K_p \simeq 1/K_m \tag{100}$$

$$V^{5+} + EtOH \rightleftharpoons V^{5+}\text{---}O(Et)H \qquad K_A \simeq 1/K_j \tag{101}$$

$$(I) + R_2S \xrightarrow{k_{II}} R_2SO + \text{---}V^{5+}\text{---}O(Bu^t)(H) \quad \text{rate determining} \tag{102}$$

$$\text{---}V^{5+}\text{---}O(Bu^t)(H) \xrightarrow{EtOH} \text{---}V^{5+}\text{---}O(Et)(H) \qquad \text{---}V^{5+}\text{---}O(Bu^t)(H) \xrightarrow{Bu^tO_2H} \text{---}V^{5+}\text{---}O(Bu^t)\text{---}O\text{---}H$$

$$\text{---}V^{5+}\text{---}O(Et)(H) \underset{EtOH}{\overset{Bu^tO_2H}{\rightleftharpoons}} \text{---}V^{5+}\text{---}O(Bu^t)\text{---}O\text{---}H \tag{103}$$

$$K' = (K_m K_j + K_m[EtOH])\, K_j \tag{104}$$

11. OXIDATION OF NITROGEN COMPOUNDS

11.1. Isocyanides to isocyanates

The oxidation of alkylisocyanides to alkylisocyanates is catalyzed by a variety of metal complexes [82, 118, 147]. Included among the complexes which have been reported to catalyze this reaction are $[Ni(t\text{-}BuNC)_4]$, $[Ni(t\text{-}BuNC)_2(O_2)]$, and $[RhCl(Ph_3P)_3]$. When ether solutions of $[Ni(t\text{-}BuNC)_4]$ were contacted with oxygen at $-20\,^\circ C$ the dioxygen complex $[Ni(t\text{-}BuNC)_2(O_2)]$ crystallized from solution. Other catalysts for isocyanide oxidation are *bis*(cycloocta-1,5-diene) nickel $[Ni(1,5\text{-}C_8H_{12})_2]$, and *bis*(cycloocta-1,5-diene)cobalt.

The following example is typical of these reactions. Oxygen was introduced at ambient temperature into 1 M solution of *t*-BuNC in ether to which $[Ni(1,5\text{-}C_8H_{12})_2]$ (2 mole % based on isocyanide) had been added. Oxygen uptake ensued smoothly under atmospheric pressure and the reaction was apparently homogeneous. Distillation of the reaction mixture produced *t*-butylisocyanate in 60–70% yield. The yield of isocyanate corresponded to the amount of oxygen absorption. The oxidation pathway is formulated as shown below, equation (105).

$$Ni(1,5\text{-}C_8H_{12})_2 \xrightarrow{t\text{-}BuNC} Ni(t\text{-}BuNC)_4 \xrightarrow{O_2}$$

$$(t\text{-}BuNC)_2Ni(O_2) \xrightarrow{\text{excess } t\text{-}BuNC} t\text{-}BuNCO + Ni(t\text{-}BuNC)_4 \tag{105}$$

The authors note that oxidation of *t*-BuNC occurs much more rapidly with $[Co(1,5\text{-}C_8H_{12})_2]$ than with the analogous nickel complex.

11.2. Oxidative cleavage reactions

A variety of organic nitrogen compounds undergo metal catalyzed oxidative cleavage reactions with ease. Copper(I) complexes are particularly effective catalysts for many cleavage reactions. Aldehyde enamines for example are oxidatively cleaved in a selective manner using cuprous chloride as the catalyst under extremely mild conditions [148]. The products of the catalytic cleavage of an enamine by dioxygen are an amide and a ketone, equation (106).

$$R_2C{=}C(R)NR_2 + O_2 \xrightarrow{Cu(I)} R_2C{=}O + O{=}C(R)NR_2 \qquad (106)$$

Some representative reactions are listed in Table 7. Reactions are carried out at 0 °C in chloroform using oxygen and catalytic amounts of cuprous chloride. The quantities of catalysts reported vary from "a trace" to one tenth the quantity of enamine used. In the absence of the metal complex, aldehyde enamines are stable to molecular oxygen.

Table 7

COPPER(I) – CATALYZED OXIDATIVE CLEAVAGE OF ENAMINES

Reactant Enamine	Reaction Products		Yield %	Ref
	Ketone	Amide		
$(CH_3)_2C{=}CH{-}N(CH_2CH_2)_2O$ (morpholino)	$(CH_3)_2C{=}O$	$O{=}CH{-}N(CH_2CH_2)_2O$ (morpholino)	100	148
$C_6H_{10}{=}CH{-}N(CH_2CH_2)_2O$ (cyclohexylidene, morpholino)	$C_6H_{10}{=}O$ (cyclohexanone)	$O{=}CH{-}N(CH_2CH_2)_2O$ (morpholino)	100	148
$(CH_3)_2C{=}C(CH(CH_3)_2){-}N(CH_3)_2$	$(CH_3)_2C{=}O$	$O{=}C(CH(CH_3)_2){-}N(CH_3)_2$	80	148
$CH_3{-}C((CH_2)_2CO_2C_2H_5){=}CH{-}N(CH_2)_4$ (pyrrolidino)	$CH_3{-}C((CH_2)_2CO_2C_2H_5){=}O$	$O{=}CH{-}N(CH_2)_4$ (pyrrolidino)	> 62	149
$C_2H_5C((CH_2)_2CO_2C_2H_5){=}CH{-}N(CH_2)_4$ (pyrrolidino)	$C_2H_5C((CH_2)_2CO_2C_2H_5){=}O$	$O{=}CH{-}N(CH_2)_4$ (pyrrolidino)	> 58	149
$C_4H_9C((CH_2)_2CO_2C_2H_5){=}CH{-}N(CH_2)_4$ (pyrrolidino)	$C_4H_9C((CH_2)_2CO_2C_2H_5){=}O$	$O{=}CH{-}N(CH_2)_4$ (pyrrolidino)	> 57	149

The mild conditions under which the catalytic reaction occurs permits the rapid and selective cleavage of the enamine double bond in the presence of other sites of unsaturation in an organic molecule. A particularly elegant example of the selectivity of this reaction is shown in equation (107).

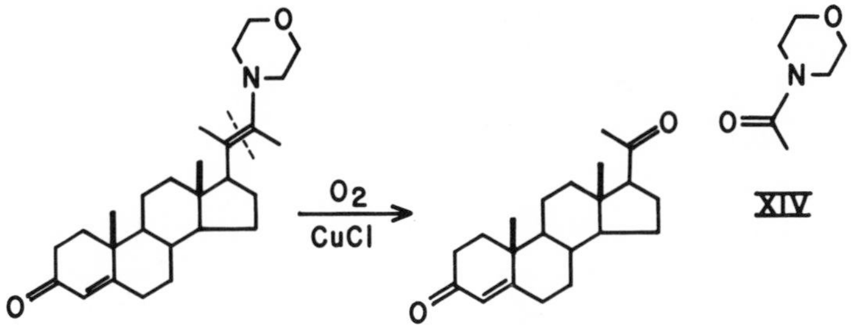

$$\mathrm{XII} \xrightarrow[\mathrm{CuCl}]{O_2} \mathrm{XIII} + \mathrm{XIV} \quad (107)$$

A chloroform solution of XII containing a catalytic amount of cuprous chloride was oxygenated to give progesterone, XIII and N-formylmorpholine, XIV, in quantitative yield. Reaction was complete in 4.5 hr at 0 °C. Cyclic enamines are readily cleaved to give a product which contains both a keto group and an amide function, equation (108).

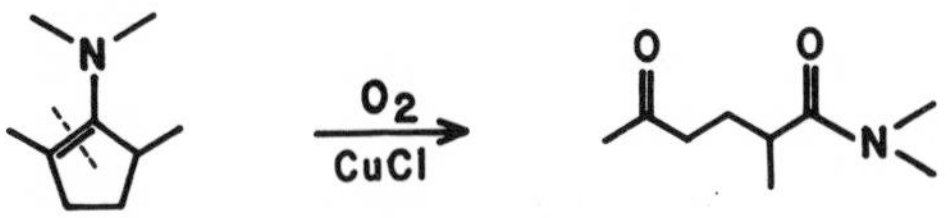

$$\xrightarrow[\mathrm{CuCl}]{O_2} \quad (108)$$

A kinetic study using enamine XII revealed the oxygenation rate to be first order in concentration of XII after an induction period, but also dependent on the amount of cuprous chloride present. The radical inhibitor, 2,6-di-*t*-butyl-*p*-cresol did not retard the reaction but pyridine and trimethylphosphite, both of which may compete with the enamine for coordination sites, inhibited oxygenation completely. The rate was also markedly dependent on the anion attached to copper. In dimethylformamide, chloroform or methylene chloride as solvents, cupric acetate, cuprous acetate, cupric nitrate and cupric sulfate were poor oxygenation catalysts. Cuprous cyanide and the chlorides and bromides of both oxidation states of copper, however, were active although cupric chloride and bromide gave traces of halogenated by-products. Recent studies [149*a*, *b*, *c*] have extended the synthetic utility of this reaction and have examined the nature of substituent effects in more detail.

Another remarkable oxidative cleavage reaction is the $CuCl_2$-catalyzed oxygenation of 1-pyrrolidino [*n*,*n*,0]-bicycloalkanes ($n = 3, 4, 5$), XV, XVI, XVII, to afford epoxyketones XVIII, XIX, and XX, equations (109), (110) and (111) [150].

XV a R = H, b R = CH$_3$ $\xrightarrow[CuCl_2]{O_2}$ XVIII a R = H, b R = CH$_3$ (109)

XVI $\xrightarrow[CuCl_2]{O_2}$ XIX (110)

XVII $\xrightarrow[CuCl_2]{O_2}$ XX (111)

In a typical example, 1.0 m mole $CuCl_2$ in 10 ml acetonitrile was added to a solution of 20 m mole 1-pyrrolidino [4,1,0]-bicycloheptane XV_a, while O_2 was bubbled through the solution at room temperature. After 24 hr, XVIII_a was produced in 28% yield. Under the same conditions the 2-methyl derivative, XV_b gave the 7-methyl epoxyketone, XVIII_b indicating that the epoxide ring is introduced at the opposite edge to the amino group in the cyclopropane ring. Other copper and cobalt halides were also effective catalysts for this reaction whereas $FeCl_3$ was not. In the absence of the metal salt, no oxygen absorption was observed.

In a preliminary ESR study of this reaction, the fact that the equimolar addition of XV_a to a DMF solution of $CuCl_2$ under argon quenched the Cu(II) paramagnetic signals, indicates occurrence of one electron transfer from XV_a to Cu(II) to give a cation radical and Cu(I). Copper(I) is then proposed to react with dioxygen to give the $O_2^{\overline{\cdot}}$ species which may react with the ring cleaved product as schematically depicted below, equation (112).

$(CH_2)_n$ + Cu(II) ⟶ $[\ (CH_2)_n\ ]^{\dot{+}}$ + Cu(I) $\xrightarrow{O_2}$ ⟶ N+ $(CH_2)_n$· + $O_2^{\overline{\cdot}}$ + Cu(II) ⟶ N–H + O $(CH_2)_n$ (112)

The catalyst formed when oxygen is added to copper(I) chloride in pyridine has found considerable utility in the oxidation of a large number of nitrogen compounds. The preparation of unsaturated nitriles from unsaturated amines can be accomplished in a catalytic manner using cuprous chloride in pyridine [15]. The oxidative cleavage of *o*-phenylenediamine catalyzed by cuprous chloride in pyridine takes place at room temperature to give *cis,cis*-mucononitrile, equation (113) [152].

$$C_6H_4(NH_2)_2 + O_2 \xrightarrow[\text{pyridine}]{\text{CuCl}} NC{-}CH{=}CH{-}CH{=}CH{-}CN \qquad (113)$$

The catalytic oxidation of dihydrazones to acetylenes [153] also proceeds smoothly in the presence of CuCl in pyridine, equation (114).

$$R(C{=}N{-}NH_2){-}(C{=}N{-}NH_2)R + O_2 \xrightarrow[\text{pyridine}]{\text{CuCl}} R{-}C{\equiv}C{-}R + 2N_2 + 2H_2O \qquad (114)$$

In a related reaction catechol is selectively oxidized to the monomethyl ester of *cis-cis*-muconic acid [152*b*]. This catalyst system has also been used for the oxidative dimerization of aromatic amines (Section 11.3) and for a number of other organic oxidations which will be subsequently covered.

A recent study [154] has examined the nature of the catalytically active brown solution which results from reaction of oxygen with pyridine solutions of copper(I) chloride. The stoichiometry of this reaction as determined by oxygen uptake experiments is $\Delta(Cu^I)/\Delta(O_2) = 4.0 \pm 0.2$. The reaction products were separated by gel permeation chromatography with pyridine as the eluent. Elution of the product mixture results in two bands; the first XXI, is brown and catalytically active, and the second, XXII, is green and inactive for the oxidative polymerization of phenols.

The retention volume for the green component, XXII, was identical to that of $[Cu(py)_2Cl_2]$. The formation of this product was confirmed by quantitative ESR measurements which were consistent with the overall reaction shown in equation (115).

$$4CuCl + O_2 + (\text{excess})py \longrightarrow \underset{\text{XXI}}{(py)_nCuOOCu(py)_n} + \underset{\text{XXII}}{2Cu(py)_2Cl_2} \qquad (115)$$

The identity of the brown catalytic component, XXI, was established by gravimetric determination of the sulfate formed from reaction with SO_2, equation (116), a reaction characteristic of transition metal peroxo complexes.

$$\text{XXI} + SO_2 \longrightarrow Cu_2SO_4 \qquad (116)$$

The existence of coordinated peroxide was confirmed by the detection of a polarographic wave (Pt electrodes) at -0.95 V for XXI in 0.1M tetramethylammonium perchlorate solution in pyridine and by the presence of a characteristic sharp band at $856\,cm^{-1}$ in the Raman spectrum in pyridine. Coordination of pyridine to the complex was confirmed by the ^1H nmr spectrum of this material.

Although pyridine solutions of the peroxo complex were stable over long periods of time, attempts to isolate a solid always resulted in the precipitation of black CuO. Thus, excess pyridine was necessary to stabilize the peroxo complex. Observations on the stoichiometric effect of varying copper/pyridine ratios, suggest that the minimum number of pyridine ligands required is 2. These results are consistent with the formulation of the catalytically active brown species, as $[(py)_nCuOOCu(py)_n]$ ($n \geqslant 2$) in the reaction of copper(I) chloride with oxygen in pyridine.

Although demonstration of oxidative dimerization activity for the complex, XXI, clearly does not necessarily imply activity in other oxidation reactions, it is of considerable importance that a copper dioxygen complex is formed in these systems.

Acid hydrazides are converted easily to carboxylic acids with molecular oxygen activated by copper salts [155], equation (117).

$$RCONHNH_2 + O_2 \xrightarrow[CH_3OH]{Cu(OAc)_2} RCOOH + N_2 + H_2O \tag{117}$$

The oxidation of hydrazine itself has been studied using tetrasulfophthalocyanine complexes of Co(II), Cu(II), Ni(II), Mn(II) and Fe(II) but only the Co(II) complex was found to possess catalytic activity [156]. Differences in catalytic activity were explained by the ability of the Co(II) complex to reversibly bind molecular oxygen. The kinetics of the autoxidation catalyzed by cobalt tetrasulfophthalocyanine can be described by the Michaelis–Menten law. The mechanism suggested for this reaction is shown in reaction sequence (118).

$$CoTSP \rightleftharpoons \begin{matrix} CoTSP(O_2) \\ CoTSP(N_2H_4) \end{matrix} \rightleftharpoons CoTSP(N_2H_4)(O_2) \longrightarrow CoTSP + N_2 + H_2O \tag{118}$$

Cobalt(III) acetate has been shown to possess activity in the oxidative cleavage of diamines such as benzidine and *o,o*-dianisidine to the corresponding quinones in acetic acid. A radical mechanism is probably involved [157].

Oxidative cleavage of cyclic ketonitrones [158] is catalyzed by ferric chloride while the oxidation of some complex oximes in the presence of $FeCl_3$ gave imidazoline-N-oxides [159].

An interesting stoichiometric oxidative cleavage reaction of an oxime with a palladium dioxygen complex has been observed [160]. The palladium dioxygen complex, $[Pd(PPh_3)_2(O_2)]$ has been shown to rapidly deoximate a variety of ketoximes in benzene at 25 °C. The yield of ketone formed was 98%. A 1,3-dipolar cycloaddition of the dioxygen complex to the ketoxime was proposed, equation (119).

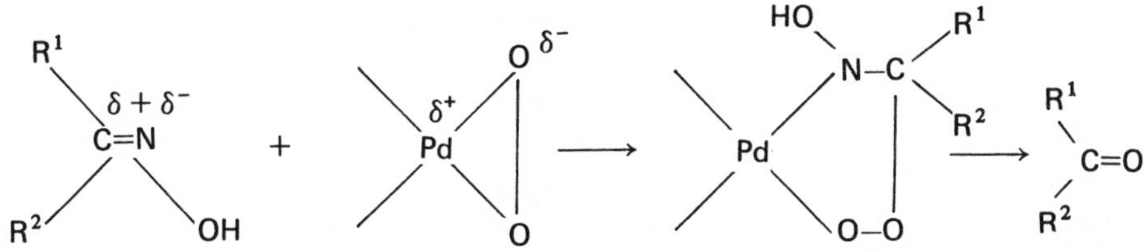

(119)

Other reactions have been observed in which oxygen interacts with organonitrogen ligand systems in a stoichiometric manner. The oxidative deamination of the alanine ligand occurs in cobalt complexes and this transformation has been related to the oxidative deamination of amino acids in metallo-enzyme systems [161]. Preliminary oxidation of an amino acid to the α-hydroxyaminoacid had been postulated as an intermediate. Recently it has been found that when the complex, XXIII is oxidized in absolute methanol the product is XXIV [161], equation (120). It is suggested that the α-hydroxy group arises from attack by the oxygen molecule rather than from coordinated water as had been previously postulated.

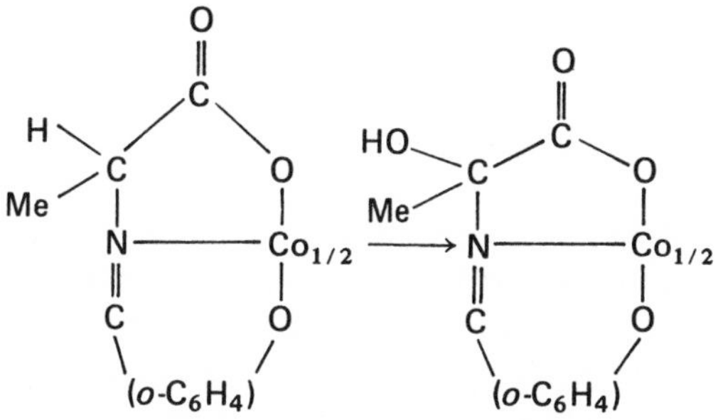

XXIII XXIV (120)

On the other hand, the oxidative deamination of amines in the coordination sphere of a molybdenum complex [162, 163] requires water. Treatment of the complex, $[(\pi\text{-}C_5H_5)_2Mo(SMe_2)Br]^+ PF_6^-$ with an excess of an amine, RR^1CHNH_2 having hydrogen on the α-carbon (β-H) in water at 60 °C for several hours results in formation of the complex, $[(\pi\text{-}C_5H_5)_2Mo(NH_2CHRR')]^+PF_6^-$ and the corresponding ketone or aldehyde RR′CO, equation (121).

$$[(\pi\text{-}C_5H_5)_2Mo(SMe_2)Br]^+ PF_6^- + 2RR'CHNH_2 + H_2O \longrightarrow [(\pi\text{-}C_5H_5)_2MoNH_2CHRR']^+ PF_6^- + RR'C=O + Me_2S + NH_4^+Br^- \quad (121)$$

Evidence was obtained for the mechanism suggested in reaction sequence (122). This reaction is stoichiometric in molybdenum and, although oxidation of the amine to a ketone occurs, dioxygen is not involved.

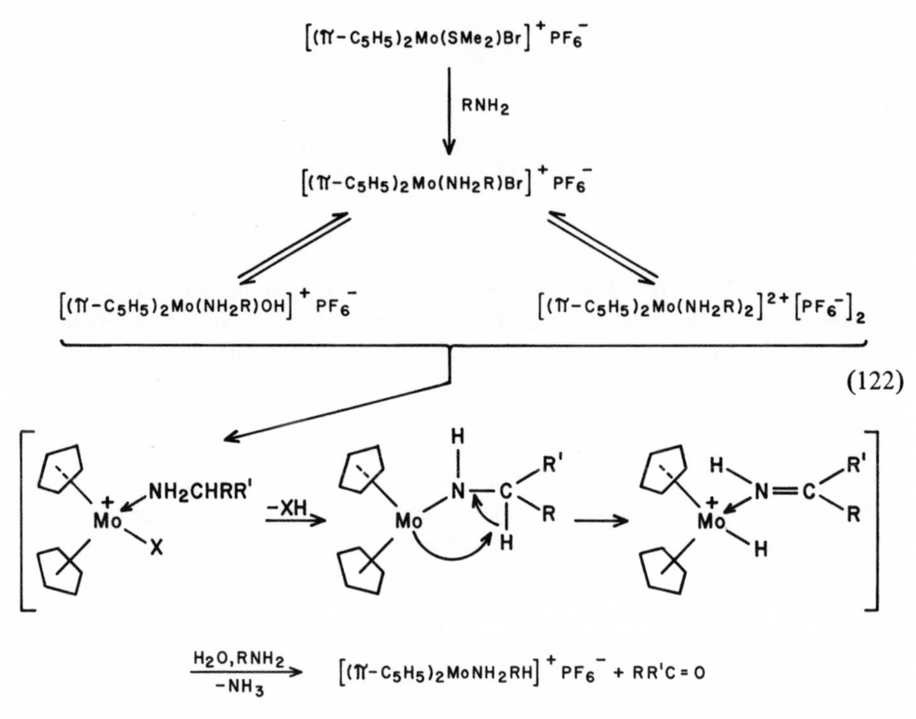

(122)

Oxidative dehydrogenation of ruthenium amine complexes occurs to give coordinated nitrile with some primary amines [164–166]. When aqueous solutions of $[Ru(NH_2CH_3)_6]^{2+}$ complexes are exposed to oxygen, $Ru(CN)_3 \cdot 3H_2O$ is formed. Both molecular oxygen and water are involved although an oxygenated organic product is not formed. The reaction stoichiometry, equation (123), is complex.

$$Ru(MeNH_2)_6Br_2 + 3.25O_2 \longrightarrow$$
$$Ru(CN)_3 \cdot 3H_2O + MeNH_3Br + MeNH_3OH + 2.5H_2O \quad (123)$$

Tetra- and penta-peptide complexes of Ni(II) have been shown to consume O_2 in neutral solutions as the metal ion catalyzes the oxidation of the peptide to give a number of products including amides of amino acids and of peptides, oxo acids and CO_2 [167]. A nickel cyclic amine complex has been shown to react with dioxygen [168] to give a 1:1 complex. When excess oxygen was added, attack on the ligand system occurred.

11.3. Oxidative dimerization of aromatic amines to azo compounds

Copper complexes in pyridine solution have been shown to be effective catalysts for the oxidative dimerization of aromatic amines to azo compounds [169]. A thorough treatment of oxidative dimerization reactions is beyond the scope of this review. However, recent reports give evidence for copper peroxo complexes [154] in these systems and for the intermediacy of a species of the type,

Table 8

OXIDATION OF AROMATIC AMINES CATALYZED BY Cu(I) IN PYRIDINE

$$2ArNH_2 + O_2 \longrightarrow ArN = NAr + 2H_2O$$

Reactant	Yield of Azo Compound, %	Reference
$C_6H_5NH_2$	89	174
$2\text{-}CH_3OC_6H_4NH_2$	16	172
$4\text{-}CH_3OC_6H_4NH_2$	93	174
$2\text{-}CH_3C_6H_4NH_2$	14	174
$3\text{-}CH_3C_6H_4NH_2$	70	172
$4\text{-}CH_3C_6H_4NH_2$	95	174
$3\text{-}ClC_6H_4NH_2$	26	174
$4\text{-}ClC_6H_4NH_2$	73	174
$3\text{-}BrC_6H_4NH_2$	29	174
$4\text{-}BrC_6H_4NH_2$	75	174
$3\text{-}IC_6H_4NH_2$	31	174
$3\text{-}IC_6H_4NH_2$	69	174
$3\text{-}NO_2C_6H_4NH_2$	25	171

$[Cu(py)_3(C_6H_5N^{\cdot}O)]^{2+}$, in one instance [170]. Since there is at least a possibility that catalytic oxygen activation may be occurring and since we have been concerned with copper–amine complexes as catalysts for oxygenations of amines and other organic substrates, consideration of this reaction seems warranted here.

Terentev and Mogilyansky [171] first reported that the air oxidation of aniline catalyzed by cuprous chloride in pyridine gave good yields of azobenzene. Kinoshita [172, 173] found that the actual oxidant was produced by air oxidation for a cuprous chloride pyridine complex. Table 8 lists some representative reactants and yields of azo compounds formed. Reaction occurs at room temperature when air or oxygen is bubbled into the reaction mixture.

It can be seen that yields of azo compounds produced from *p*-toluidine and *p*-anisidine are almost quantitative, whereas yields of azo compounds from *o*-toluidine and *o*-anisidine are poor. Kinoshita rationalized this difference on the basis of steric hindrance to copper amine complex formation exerted by the ortho substituents. It is of interest to note that the yield of azo compounds obtained in the free radical initiated autoxidation of aromatic amines are not significantly affected by the position of substituent groups on the aromatic ring [175, 176].

Another interesting observation was that hydrazobenzene could be almost quantitatively converted to azobenzene under reaction conditions. These facts led Kinoshita to propose a mechanism in which a copper aniline complex is formed. Oxidation then occurs in the coordination sphere to give the coordinated radical, $[C_6H_5NH]^{\cdot}$ which dimerizes to hydrazobenzene.

In a related system, EPR has been used to study the oxidative coupling reaction of aniline to give azobenzene catalyzed by $[Cu(py)_4(NO_3)_2]$ [170]. Addition of aniline to the catalytic complex results in progressive reduction of Cu(II) to Cu(I) which in turn can be reoxidized. An intermediate radical species proposed to be $[C_6H_5N^{\cdot}{=}O^{+}]$ has been observed. In this species – a radical cation produced by partial oxidation of aniline – the unpaired electron is almost exclusively localized on nitrogen and it has a predominant 2_{p_x} character. The mechanism, equation (123*a*) which was proposed for the oxidative coupling reaction differs considerably from the original mechanism put forward by Kinoshita [172].

$$Cu(py)_4^{2+} + C_6H_5NH_2 \rightleftharpoons Cu(py)_3^{+}(C_6H_5\dot{N}H_2)^{+} + py$$

$$\downarrow -H_2O \quad +O_2$$

$$Cu(py)_4^{2+} + 1/2C_6H_5N{=}NC_6H_5 \xleftarrow[-1/2\,O_2]{+py} Cu(py)_3^{+}(C_6H_5\dot{N}{=}O)^{+} \tag{123a}$$

Wöthrich and Fallab [177] have studied the kinetics of the oxidation of *o*-phenylenediamine by molecular oxygen in the presence of cupric ion in the pH range 5.5–7.0. Complex formation between cupric ion and *o*-phenylenediamine in aqueous solution is reversible when the solution is kept under nitrogen. On addition of O_2 the ligand molecules are oxidized to 3,5-dihydro-2-amino-3-iminophenazine.

The kinetics suggest that oxygen reacts with a 2:1 diamine:copper complex. The rate determining step is preceded by the formation of a dioxygen complex which decays, equation (124), with production of a free radical which undergoes further fast reactions to the final oxidation product.

(124)

These authors found [178] that the Cu(II)-catalyzed oxidation of *o*-phenylenediamine by O_2 is strongly influenced by organic and inorganic ligands. While chelating agents such as ethylenediamine and EDTA inhibit catalysis, reaction is accelerated under the influence of some monodentate ligands, e.g. Cl^-, Br^-, NH_3, pyridine and imidazole.

The oxidation of α-naphthylamine catalyzed by copper(II) stearate, $[CuSt_2]$, has also been reported [179]. Reactions were carried out by bubbling a continuous stream of oxygen into a solution of the amine and $[CuSt_2]$ in chlorobenzene at 115 °C. Oxygen concentration was varied by changing its partial pressure in mixtures with nitrogen. The rate of oxidation of the amine increased linearly with increased concentrations of amine, $[CuSt_2]$, and O_2, when each was present in low concentration. The rate did not depend on the concentration of each component when its concentration was high. Kinetic data agreed well with the reaction scheme shown in equation (125). Activation parameters were determined.

$$ArNH_2 + CuSt_2 \overset{K_1}{\rightleftharpoons} X_1$$

$$X_1 + O_2 \overset{K_2}{\rightleftharpoons} X_2 \longrightarrow \text{oxidation products} \tag{125}$$

The activation energy for oxidation of the amine to complex, X_2 was 14.7 kcal/mole and activation entropy was -35.3 entropy units. A mechanism similar to that postulated by Wörthrich and Fallab was proposed, equation (126).

$$\mathrm{ArNH_2\overset{|}{\underset{|}{Cu}}(O_2)} \longrightarrow \mathrm{ArNH\overset{|}{\underset{|}{Cu}}(OOH)} \qquad (126)$$

In recent years model compounds containing a copper–hydroperoxide linkage have been formed from reaction of copper(II) acetate and hydrogen peroxide [180]. In a typical experiment, 4 g of $[Cu(OAc)_2] \cdot H_2O$ in a minimum amount of water was added to 350 ml of an aqueous solution containing 20 ml of conc. H_2O_2. After stirring the solution for several minutes the brown precipitate formed was collected on a sintered glass funnel, washed with water, then acetone and air dried (yield, 1.4 g). The material was characterized by elemental analysis, $KMnO_4$ titration, iodometric analysis, infrared and optical spectrometry. Based on this and other data a polynuclear structure, XXV, was assigned to this compound.

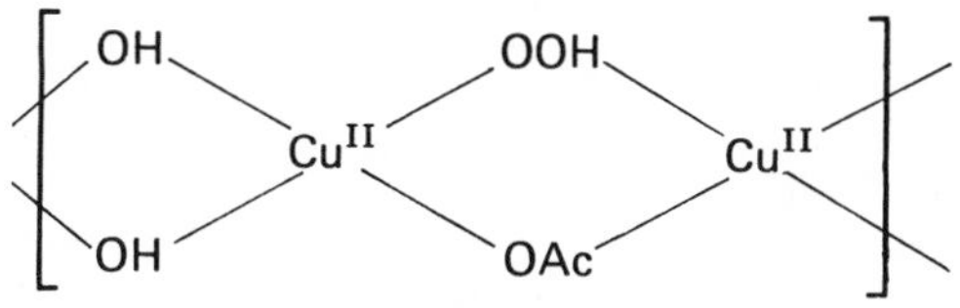

OH OOH
Cu^{II} Cu^{II}
OH OAc

XXV

Although much work still remains to be done before a complete and consistent mechanistic picture will emerge in these systems, it is clear that we are dealing with more than a classical free radical reaction and that coordination of oxygen and transfer to the substrate might both play a major role in the oxidation of aromatic amines.

11.4. Hydroperoxide oxidations

11.4.1. TERTIARY AMINES TO AMINE OXIDES

Tertiary amines react with organic hydroperoxides to give high yields of amine oxides in the presence of group VB and VIB transition metal complexes [181–183], Table 9. Four parameters affect this reaction: the metal complex used as catalyst, the reactivity and stability of the hydroperoxide, the solvent and steric effects in the amine.

Sheng and Zajacek [182] have shown that vanadium complexes are somewhat superior catalysts to molybdenum. Both give good selectivity to amine oxide but reaction rates are faster when vanadium complexes are used. Compounds of tungsten, niobium and tantalum were poorer catalysts whereas chromium, cobalt, manganese and iron complexes were ineffective.

Hydroperoxide reactivity was found to increase in the order: *t*-butyl hydroperoxide $<$ cumene hydroperoxide $\leqslant$ amylene hydroperoxide for reactions carried out

Table 9

CATALYTIC OXIDATION OF TERTIARY AMINES WITH HYDROPEROXIDES [182]

$$R_3N + R'OOH \xrightarrow{\text{catalyst}} R_3N \rightarrow O + ROH$$

Catalyst	Amine	Hydroperoxide	Solvent	T °C	T Hr	Conv. %	Yield %
$VO(acac)_2$	$(CH_3)_2N(n\text{-}C_{12}H_{26})$	*t*-Butyl	acetone	60	0.75	88	94
	$(CH_3)_2N(n\text{-}C_{12}H_{26})$	Cumene	acetone	60	0.75	89	91
	$(CH_3)_2N(n\text{-}C_{12}H_{26})$	Amylene	acetone	60	0.75	86	96
	$(C_4H_9)_3N$	Amylene	acetone	60	1.50	84	50
	$(C_2H_5)_3N$	Amylene	acetone	60	2.30	80	93
	$(CH_3)_2N((CH_2)_4OH)$	*t*-Butyl	acetone	60	0.50	84	80
	$(CH_3)_2N((CH_2)_4OH)$	Amylene	acetone	60	0.75	86	100
$Mo(CO)_6$	$(CH_3)_2N(n\text{-}C_{12}H_6)$	*t*-Butyl	*t*-BuOH	86	2.00	79	86
	$(CH_3)_2N(n\text{-}C_{12}H_6)$	Cumene	*t*-BuOH	86	1.00	100	73
	$(CH_3)_2N(n\text{-}C_{12}H_6)$	Amylene	*t*-BuOH	86	0.75	92	77
	$(C_4H_9)_3N$	*t*-Butyl	acetone	60	3.16	23	100
	$(C_2H_5)_3N$	Amylene	acetone	60	4.75	60	85
	$(CH_3)_2N((CH_2)_4OH)$	*t*-Butyl	*t*-BuOH	86	3.50	88	92
	$(CH_3)_2N((CH_2)_4OH)$	*t*-Butyl	acetone	60	3.75	58	100
	$(CH_3)_2N((CH_2)_4OH)$	Amylene	acetone	60	3.00	52	100
$W(CO)_6$	$(CH_3)_2N(n\text{-}C_{12}H_{26})$	*t*-Butyl	*t*-BuOH	86	5.00	69	51
$Cr(acac)_3$	$(CH_3)_2N(n\text{-}C_{12}H_{26})$	*t*-Butyl	*t*-BuOH	86	5.00	85	17
$Fe(acac)_2$	$(CH_3)_2N(n\text{-}C_{12}H_{26})$	*t*-Butyl	*t*-BuOH	86	3.00	90	10
$Mn(acac)_3$	$(CH_3)_2N(n\text{-}C_{12}H_{26})$	*t*-Butyl	*t*-BuOH	86	2.50	90	6

at moderate to low temperature. Stability of the hydroperoxide in the presence of the catalyst is also important. Two competing reactions occur: the decomposition of the hydroperoxide and the catalytic oxidation of the amine. The hydroperoxide stability decreases in the order *t*-butyl > cumene ⩾ amylene.

Rates are fastest in polar aprotic solvents such as THF, diethyl ether and acetone but are slower in alcohols such as *t*-butyl alcohol and methanol. Steric hindrance about the nitrogen also plays a role. The yields of amine oxide formed from tertiary amines increase in the order: $(CH_3)_2N(n\text{-}C_{12}H_{26}) > (C_2H_5)_3N > (C_4H_9)_3N$.

Titanium complexes are also effective in this reaction [145]. High yields of the N-oxide are formed from triethanolamine using $[Ti(i\text{-}OC_4H_9)_4]$ as the catalyst, equation (127), (82% yield).

$$(HO\text{-}CH_2CH_2)_3N \xrightarrow[Ti(i\text{-}OC_4H_9)_4]{t\text{-}BuOOH} (HOCH_2CH_2)_3N \rightarrow O \qquad (127)$$

The oxidation of nitrogen heterocycles by *t*-pentyl hydroperoxide in the presence

Table 10

MOLYBDENUM(V) – PROMOTED HYDROPEROXIDE OXIDATION OF NITROGEN HETEROCYCLES

N-Oxides	Yield %	N-Oxides	Yield %
Pyridine	100	Quinoline	100
2-Picoline	100	2-Methylquinoline	95
3-Picoline	100	Methylquinaldinate	90
2,4,6-Collidine	95	Benzo(*f*)quinoline	95
4,4′-Bipyridyl (dioxide)	80	Acridine	100
2-Phenylpyridine	90	Phenazine (dioxide)	90
Methylnicotinate	90	2,3,5,6-Tetramethylpyrazine (dioxide)	80
4-Acetamidopyridine	40	Papavarine	70

of molybdenum pentachloride has been studied in some detail [184–186]. Table 10 shows the relative ease with which these compounds may be converted to the N-oxide. The kinetics of a series of 3-substituted pyridines having the substituents X = H, CH_3, Br, CO_2CH_3, $NHCOCH_3$ and CN was studied in benzene [185]. The reactions were first order in both *t*-pentyl hydroperoxide and in amine. A linear relation was found between the reactivities of the 3-substituted pyridines and the values of the Taft σ constants. Reaction rate increases with decrease in the electron density on the reaction center; $\rho = +1.2$. The effective activation energy of the N-oxidation decreases with introduction of an electron accepting substituent from 9.1 kcal/mole for pyridine (X=H) to 7.7 kcal/mole for methylnicotinate (X=$NHCOCH_3$). Nicotinonitrile (X=CN) has an anomalously low rate of oxidation as compared with other derivatives of this series and deviates from the correlation. This is rationalized by formation of nitrile complexes thus excluding the pyridine nitrogen from coordination. The scheme shown below, equation (128), is proposed as the reaction mechanism.

Mo + RO_2H ⟶ Mo - - - - O–R (O–H) ⟶[3-X-pyridine] 3-X-pyridine N - - Mo - - O–R / O–H complex

↓

3-X-pyridine N ⟶ O + Mo - - - - OR(H) ⟵ 3-X-pyridine N^+–OH + Mo–$\bar{O}$R (128)

11.4.2. PRIMARY AROMATIC AMINES TO AZOXY COMPOUNDS OR NITRO COMPOUNDS

When primary aromatic amines are used instead of tertiary amines the products of reaction are either azoxy compounds [186] or nitro compounds [187]. In the presence of soluble complexes of titanium, aniline, *o*- and *p*-toluidine and *m*-nitro aniline are oxidized by *tert*-butyl hydroperoxide or cumyl hydroperoxide to the corresponding azoxy compounds in 92–99% yields [182], equation (129).

$$C_6H_5NH_2 \xrightarrow[\text{Ti(acac)}_3]{\text{Ph(CH}_3)_2\text{COOH}} C_6H_5\text{–N=N(}\rightarrow\text{O)–}C_6H_5 \tag{129}$$

In the presence of vanadium and molybdenum complexes, however, aniline and substituted anilines are reported to give rise to nitro compounds exclusively, equation (130) [187]. Corresponding complexes of tungsten and cobalt do not catalyze oxidation.

$$C_6H_5NH_2 \xrightarrow[\text{VO(acac)}_2]{(\text{CH}_3)_3\text{COOH}} C_6H_5NO_2 \tag{130}$$

Vanadium complexes and in particular [VO(acac)$_2$] are the most active catalysts for the oxidation of substituted anilines to nitro compounds. The effect of substituents upon reaction rate corresponds to a reaction involving an electron deficient transition state in that electron withdrawing groups decrease the rate and *vice versa*. The relative order of reactivity: *p*-Me > *m*-Me > aniline > *p*-Cl > *p*-Br > *m*-Cl > *m*-Br is the same as observed in electrophilic aromatic substitution. Straight line correlations between the log of the relative rates and Hammett σ or Brown σ^+ constants were obtained with ρ values of -1.42 and -1.97, respectively, indicating an electron deficient transition state in the rate determining step.

Kinetic studies were carried out at 66.1° in benzene-chlorobenzene using [VO(acac)$_2$] as catalyst. The stoichiometry of the reaction is given by equations (131), (132) and (133) where A is the aniline concentration, P the peroxide, B and C the probable oxidation intermediates, phenylhydroxylamine and nitrosobenzene, and D is nitrobenzene. A reaction first order in all three components and with zero concentration of reaction intermediates should follow the rate expression given in equation (134).

$$A + P \longrightarrow B \tag{131}$$

$$B + P \longrightarrow C \tag{132}$$

$$C + P \longrightarrow D \tag{133}$$

$$\frac{1}{P_0 - 3A_0} \ln \frac{P_0 A}{A_0 P} = kt \tag{134}$$

Kinetic plots were linear over 50% of the reaction which though indicative, may be fortuitous. Considering all of the data which was gathered, Howe and Hiatt [187] postulate the following reaction mechanism, equation (135).

(a) catalyst activation (rapid)

$$\text{(b) V + ROOH} \underset{k_{-1}}{\overset{k_1}{\rightleftharpoons}} \text{V} \cdots \text{O}(\textit{tert}\text{-Bu})(\text{OH}) \quad (k_1 \gg k_{-1})$$

$$\text{(c) V} \cdots \text{O}(\textit{tert}\text{-Bu})(\text{OH}) + \text{NH}_2\text{Ph} \xrightarrow{k_2} \text{OH}{-}\overset{+}{\text{N}}\text{H}_2\text{Ph} + \text{V} \cdots \text{O}^-(\textit{tert}\text{-Bu}) \quad (k_2 \ll k_1)$$

$$\text{(d) OH}{-}\overset{+}{\text{N}}\text{H}_2\text{Ph} + \text{V} \cdots \text{O}^-(\textit{tert}\text{-Bu}) \longrightarrow \text{OH}{-}\text{N(Ph)(H)} + \text{V} \cdots \text{OH}(\textit{tert}\text{-Bu}) \quad \text{(rapid)} \qquad (135)$$

$$\text{(e) V} \cdots \text{OH}(\textit{tert}\text{-Bu}) + \textit{tert}\text{-BuO}_2\text{H} \rightleftharpoons \text{V} \cdots \text{O}(\textit{tert}\text{-Bu})(\text{OH}) + \textit{tert}\text{-BuOH} \quad \text{(rapid reversible)}$$

$$\text{(f) PhNHOH} + 2\,\textit{tert}\text{-BuO}_2\text{H} \longrightarrow \text{PhNO}_2 + 2\,\textit{tert}\text{-BuOH} + \text{H}_2\text{O} \quad \text{(rapid)}$$

11.4.3. ALIPHATIC PRIMARY AMINES TO OXIMES

In the absence of a metal catalyst, alkyl hydroperoxides react with primary amines to give either ketimines or aldimines [188]. In the presence of molybdenum-, vanadium-, or titanium naphthenate, however, the oxidation of cyclohexylamine with cumene hydroperoxide [189] is reported to give cyclohexanone oxime in good yield, equation (136).

$$\text{C}_6\text{H}_{11}{-}\text{NH}_2 \xrightarrow[\text{Catalyst}]{\text{ROOH}} \text{C}_6\text{H}_{10}{=}\text{N}{-}\text{OH} \qquad (136)$$

11.4.4. IMINES TO OXAZIRIDINES

The oxidation of N-alkyl or N-aryl aldimines or ketimines with *t*-amyl hydroperoxide in the presence of either $MoCl_5$ or $Mo(CO)_6$ leads to oxaziridine formation, equation (137) [184]. In benzene solution the reaction proceeds rapidly and yields are high (80–95%). The catalytic procedure is considerably more convenient than conventional methods of oxaziridine preparation.

$$\begin{matrix} R_1 \\ R_2 \end{matrix}\!\!>\!C{=}N{-}R_3 \xrightarrow[\text{Mo-catalyst}]{t\text{-amyl hydroperoxide}} \begin{matrix} R_1 \\ R_2 \end{matrix}\!\!>\!C\overset{O}{\frown}N{-}R_3 \qquad (137)$$

11.4.5. NITROSOBENZENE TO NITROBENZENE

The reaction of nitrosobenzene with *t*-butyl hydroperoxide in benzene or cyclohexane is initiated by hydrocarbon-soluble β-diketonates of a number of transition metals [190]. In all cases the reaction product is nitrobenzene which is formed rapidly and quantitatively, equation (138).

$$t\text{-BuOOH} + C_6H_5NO \xrightarrow{\text{catalyst}} C_6H_5NO_2 + t\text{-BuOH} \qquad (138)$$

The induction periods observed, the inhibition by traces of 2,6-di-*t*-butyl-4-methylphenol, and the fractional reaction orders appearing in the rate laws suggested a series of radical chain processes.

Catalytic action by the pivaloylmethane chelate of Co(II), $[Co(dpm)_2]$ appears to occur *via* a two step initiation sequence in which the 2+ state of the metal reacts with hydroperoxide yielding $t\text{-BuO}^{\cdot}$, equation (139), after which the 3+ state reacts with a second hydroperoxide, yielding $t\text{-BuOO}^{\cdot}$, equation (140).

$$\text{Co(II)} + t\text{-BuO}_2\text{H} \xrightarrow{k_1 \sim 10^2\,M^{-1}\,\text{sec}^{-1}} \text{Co(III)OH} + t\text{-BuO}\cdot \qquad (139)$$

$$\text{Co(III)} + t\text{-BuO}_2\text{H} \xrightarrow[\text{rapid}]{k_2} \text{Co(II)} + H^+ + t\text{-BuOO}\cdot \qquad (140)$$

The oxidation of Co(II) in the absence of nitrosobenzene has been found to be first order in each reaction component, equation (141).

$$\frac{d\text{Co(II)}}{dt} = \frac{-2d[t\text{-BuOOH}]}{dt} = 2.0 \times 10^2[\text{Co(II)}][t\text{-BuOOH}] \qquad (141)$$

The oxidation of nitrosobenzene catalyzed by $[Co(dpm)_2]$ closely follows the rate expression (142), where $[Co]_T$ = total concentration of added cobalt.

$$\text{rate}_{\text{obsd}} = k[\text{Co}]_T^{1/2}[t\text{-BuO}_2\text{H}][\text{PhNO}] \qquad (142)$$

The rate of initiation, R_i, resulting from a combination of both reactions (139) and (140) is given in equation (143).

$$R_i = 2k_1k_2/(k_1 + k_2)[\text{Co}]_T[t\text{-BuOOH}] \qquad (143)$$

Since the overall rate is proportional to $R_i^{1/2}$ and includes an additional factor of $[t\text{-BuOOH}]^{1/2}$ arising from the propagation sequence, the combined initiation mechanism is in accord with the observed rate law. A similar situation involving Mn(II) and Mn(III) species – both active in bimolecular initiation with *t*-BuOOH – applies to the reaction catalyzed by $[Mn(acac)_3]$ which obeys the same type of rate law as $[Co(dpm)_2]$.

Kinetic data indicate that catalysis by $[VO(dpm)_2]$ is complicated by the formation of a complex, $[VO(dpm)(t\text{-BuO}_2H)]$ ($K_{assn} = 1.9 \times 10^3$ M^{-1} at 25 °C) which may undergo unimolecular homolysis forming *t*-BuO˙, or, alternatively may react with a second hydroperoxide again forming *t*-BuO˙, equation (144). Cyclization of vanadium occurs between the 4+ and the 5+ states.

$$V^{IV} \xrightarrow[K = 1900]{BuO_2H} V^{IV}P \begin{cases} \xrightarrow{-BuO\cdot} V^{V}OH \\ \quad\Updownarrow P \\ \xrightarrow[-BuO\cdot]{P} PV^{V}OH \longrightarrow V^{IV} + BuO_2^{\cdot} + H_2O \end{cases} \tag{144}$$

11.4.6. NITROSAMINES TO NITRAMINES

The oxidation of nitrosamines to nitramines using alkyl hydroperoxides and Mo complexes, equation (145), can be accomplished in up to 80% yields over long reaction times [184].

$$R_1R_2N\text{—}NO \xrightarrow[MoCl_5]{ROOH} R_1R_2N\text{—}NO_2 \tag{145}$$

Thus from nitrosomorpholine, nitromorpholine was formed. Nitrosodimethylamine and nitrosopiperidine gave N-nitrodimethylamine and N-nitropiperidine, respectively.

11.4.7. AZOBENZENES TO AZOXYBENZENES

Molybdenum complexes catalyze the hydroperoxide oxidation of azo compounds to azoxy compounds, equation (146), in high yields [191]. Both $[MoO_2(dpm)_2]$ (dpm = dipivaloylmethane) and $[Mo(CO)_6]$ have been used as catalysts and a range of substituents (R_1, R_2) were employed.

$$R_1\text{—}C_6H_4\text{—}N{=}N\text{—}C_6H_4\text{—}R_2 \xrightarrow[\text{Mo-catalyst}]{ROOH} R_1\text{—}C_6H_4\text{—}N{=}N(\rightarrow O)\text{—}C_6H_4\text{—}R_2 \tag{146}$$

In both the oxidation of *p*-methoxyazobenzene, XXVI, and of *p*-nitroazobenzene, XXVII, using $[MoO_2(dpm)_2]$ as the catalyst, attack occurs preferentially at N_α, where α is the most activated nitrogen. Thus conjugative activation operates on XXVI and conjugative deactivation operates on XXVII, with the net result that either a strongly electron donating or a strongly electron attracting substituent at one para position directs electrophilic attack to that nitrogen adjacent to the unsubstituted ring.

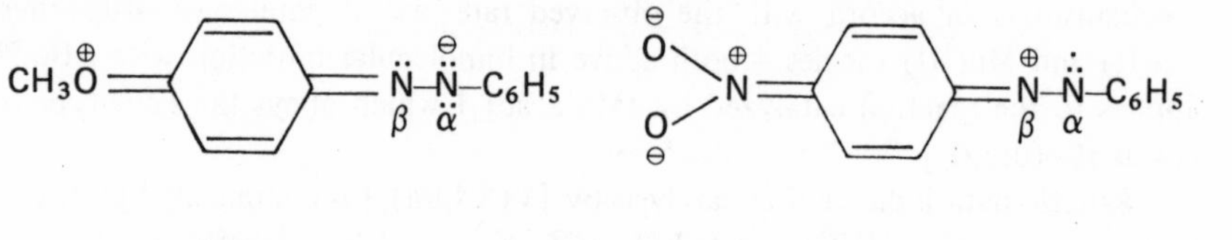

XXVI

XXVII

With $[Mo(CO)_6]$ as the catalyst the hydroperoxide oxidation of the monomethoxy compound, XXVI, yields a mixture of nearly equal quantities of α- and β-azoxy products. This change in selectivity suggests that variation in the oxidation state and ligand environment has resulted in different mechanisms for the Mo(O) and Mo(VI) complexes. Johnson and Gould suggest transition states XXVIII and XXIX for reactions catalyzed by $[MoO_2(dpm)_2]$ and $[Mo(CO)_6]$ respectively.

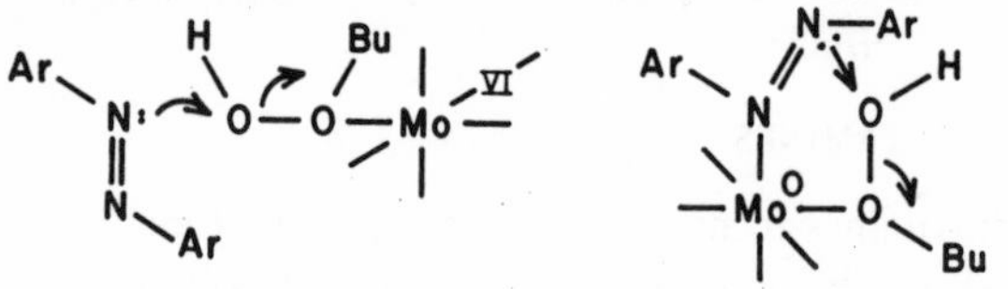

XXVIII

XXIX

12. CARBON MONOXIDE

12.1. Reactions in non-aqueous media

Carbon monoxide is readily oxidized in the coordination sphere of a number of transition metal complexes. In many cases the product of reaction is a carbonate complex which is formed irreversibly, thus precluding the possibility of a catalytic transformation. In Section 5 the reaction between CO and platinum dioxygen complexes was shown to give carbonate complexes. The reaction between iridium, ruthenium and osmium carbonyl complexes and dioxygen to give coordinated carbonate was discussed in Section 6.

Some rhodium carbonyl complexes, on the other hand, undergo a different reaction with molecular oxygen. The reaction of the complex, $[Rh_2(PPh_3)_4(CO)_4]$ with molecular oxygen in aromatic hydrocarbon solvents produced a rhodium complex containing coordinated CO_2, equation (48) [110]. Information was obtained concerning the mechanism of this reaction by examining the effect of introducing heavy oxygen into the reaction. The carbon dioxide complex was shown to contain $^{16}OC^{16}O$, $^{16}OC^{18}O$ and $^{18}OC^{18}O$. In addition, ^{18}O – containing triphenylphosphine oxide was also formed. Therefore, this reaction was considered to occur as shown in equation (147).

$$Rh_2(PPh_3)_4(CO)_4 + O_2 \longrightarrow Rh_2(PPh_3)_3(CO)_2(CO_2) + OPPh_3 + CO \quad (147)$$

The relative amounts of $^{16}OC^{16}O$, $^{18}OC^{16}O$ and $^{18}OC^{18}O$ in the carbon dioxide complex were found to be about 1, 2, and 1, respectively while the ratio of $^{16}OPPh_3$ to $^{18}OPPh_3$ was 1:1. Such a distribution led Iwashita and Hayata [110] to propose a reaction mechanism in which coordinated O_2 and two coordinated CO ligands form an intermediate having four equivalent oxygens and two equivalent carbon atoms.

Since rhodium carbon dioxide complexes are not as stable as the corresponding carbonate complexes with respect to CO, catalysis is possible. For example thermal decomposition of the complex $[Rh_2(PPh_3)_3(CO)_2(CO_2)] \cdot C_6H_6$, begins at 90 °C and the complex gives up one mole of CO_2 per rhodium atom [192]. Furthermore, treatment of a benzene solution of $[Rh_2(PPh_3)_3(CO)_2(CO_2)_2] \cdot C_6H_6$ with carbon monoxide gave a rhodium carbonyl complex and 2 moles of carbon dioxide, equation (148).

$$Rh_2(PPh_3)_3(CO)_2(CO_2)_2 + CO \longrightarrow Rh_2(PPh_3)_2(CO)_4 + PPh_3 + 2CO_2 \quad (148)$$

It is interesting to note that passing air through a benzene solution of the di-carbon monoxide complex leads to the mono-carbon monoxide complex, $[Rh_2(PPh_3)_3(CO)_2(CO_2)]$ reported by Iwashita and Hayata [110].

Shortly after the report of the reaction of a rhodium carbonyl complex with O_2 to give coordinated CO_2, the catalytic oxidation of CO to CO_2 was reported in a similar system [193]. Kiji and Furukawa reported that both $[RhCl(CO)(PPh_3)_2]$ and $[RhCl(CO)(Me_2SO)_2]$ catalyzed the oxidation of CO in benzene, ethanol, or dimethylsulfoxide. However, the rate was slow and the catalytic efficiency was poor. Reactions catalyzed by $[RhCl(CO)(Me_2SO)_2]$ in benzene gave approximately 11 m moles of CO_2 per m mole of metal complex. The authors describe the reaction as involving: (1) coordination of molecular oxygen, (2) reaction with CO in the coordination sphere, and (3) displacement of the product by unreacted CO.

The rhodium cluster compound, $[Rh_6(CO)_{16}]$, has been found to be a far more efficient catalyst for CO oxidation than were the complexes discussed above [194]. It was found that this catalyst retains its initial activity even after 12,500 m moles of CO_2 have been produced per m mole of $[Rh_6(CO)_{16}]$.

In a typical reaction, $[Rh_6(CO)_{16}]$ ($\sim 10^{-2}$ m mol) was suspended in 10 ml of solvent in a glass vial and the mixture stirred at 100 °C under a pressure of 34 atm. (35 ml volume). The ratio of $CO:O_2$ was 2:1 and the reaction was complete within 24 hr. The catalyst could be recovered unchanged from the reaction. The catalytic reaction was shown to take place in solution since $[Rh_6(CO)_{16}]$ was catalytically inactive in the absence of solvent or as suspension in hexane. Acetone, cyclohexanone dimethoxyethane, or dimethylsulfoxide were all found to be suitable solvents. However, yields were highest in acetone. Verification of the homogeneity of the active species was obtained by cooling the reaction vessel, filtering the solution, returning the filtrate to the reaction vessel and proceeding with catalysis. Oxidation continued in the homogeneous solution.

When $[Rh_2Cl_2(CO)_4]$ was used as catalyst a small amount of CO was converted to CO_2 corresponding to formation of small amounts of $[Rh_6(CO)_{16}]$. With $[Rh_4(CO)_{12}]$ catalysis was reported to be facile due to its ready conversion to $[Rh_6(CO)_{16}]$.

In contrast to $[Pt(PPh_3)_2(O_2)]$ which forms a carbonato complex with CO, the nickel complex, $[Ni(t\text{-}BuNC)_2(O_2)]$, reacts with CO at 20 °C to give $[Ni(CO)_2(t\text{-}BuNC)_2]$ and CO_2 [89]. Catalytic activity, however, was not reported.

12.2. Reactions in aqueous media

A variety of metal complexes are known to catalyze oxidation of CO to CO_2 in aqueous media [195–217]. The mechanism of CO_2 formation in these cases, however, is quite different and may not involve oxygen atom transfer to coordinated CO. One mechanism which appears to be operative in a number of cases involves attack by water on CO within the coordination sphere to give CO_2 and a reduced metal complex. The function of oxygen may therefore be merely to oxidize the metal complex back to the catalytically active state. In other instances oxygen activation and transfer to coordinated CO have been postulated to occur in aqueous media.

The oxidation of CO by dioxygen is catalyzed by cupric acetate in aqueous solution [195]. The stoichiometry of this reaction was originally described as shown in equations (149) and (150), addition of which gives the observed overall reaction: $CO + 1/2\ O_2 \rightarrow CO_2$.

$$2Cu(II) + 3CO + H_2O \longrightarrow 2Cu(I)CO + CO_2 + 2H^+ \tag{149}$$

$$2Cu(I)CO + 1/2\ O_2 + 2H^+ \longrightarrow 2Cu(II) + 2CO + H_2O \tag{150}$$

Kinetic studies carried out by Byerley and Lee [195] lead these authors to postulate a different reaction pathway.

The rate of this reaction expressed as the rate of disappearance of oxygen, was found to be proportional to the concentrations of O_2, CO, and Cu(II), and inversely proportional to the hydrogen ion concentration [195], equation (151).

$$\frac{-d[O_2]}{dt} = k_{expt} \frac{[O_2][CO][Cu(II)]}{[H^+]} \tag{151}$$

$$k_{expt} = 7.88 \times 10^{-6}\ M^{-1} s^{-1} \text{ at } 120\ °C,\ [NaOAc] = 0.25\ M$$

A reaction mechanism consistent with this rate law was proposed by Byerley and Lee [195], equations (152)–(154).

$$Cu(II) + CO + H_2O \underset{\text{(rapid equilibrium)}}{\overset{K_a}{\rightleftharpoons}} [Cu\text{—}COOH]^+ + H^+ \tag{152}$$

$$[Cu\text{—}COOH]^+ + O_2 \xrightarrow{k_a} Cu(II) + CO_2 + HO_2^- \tag{153}$$

$$[Cu\text{—}COOH]^+ + HO_2^- + 2H^+ \xrightarrow[\text{fast}]{} Cu(II) + CO_2 + 2H_2O \tag{154}$$

The rate law derived from this reaction sequence corresponds to equation (155).

$$\frac{-d[O_2]}{dt} = k_a[Cu\text{-}COOH][O_2] = k_a K_a \frac{[Cu(II)][CO][O_2]}{[H^+]} \qquad (155)$$

The authors note that the intermediate CO insertion product, $[CuCOOH]^+$, has analogs in a number of similar reactions of silver(I) [196, 197], cobalt(II) [198], nickel(II) [199] and mercury(II) [200], among others.

James, Rempel and Rosenberg [201–203] as well as Stanko, Petrov and Thomas [204] have studied the carbonylation of rhodium halides in aqueous acid solutions under mild conditions. Reactions appear to proceed *via* a rhodium(III) carbonyl complex which is believed to react with water to form an insertion product prior to CO_2 formation, equation (156).

$$Rh^{III} \xrightarrow[k_1]{CO} Rh^{III}(CO) \xrightarrow{H_2O} [RhCOOH] \longrightarrow Rh^I + CO_2 + 2H^+ \qquad (156)$$

$$Rh^I \xrightarrow{CO} Rh^I(CO)_2$$

In the carbonylation of aqueous hydrochloric acid solutions of $RhCl_3 \cdot 3H_2O$ the intermediate rhodium(III) carbonyl complex, $[RhCl_5(CO)]^{2-}$, was detected which gave rise to the rhodium(I) anion $[Rh(CO)_2Cl_2]^-$. The reaction was found to be autocatalytic and kinetic data fit the rate law shown in equation (157).

$$\frac{-d[CO]}{dt} = k_1[Rh^{III}][CO] + k_2[Rh^{III}][Rh^I] \qquad (157)$$

The first term of the rate law is concerned with the production of the autocatalytic species $[Rh(CO)_2Cl_2]^-$, for which equation (156) represents the proposed mechanism. Further production of Rh(I) is autocatalytic due to an efficient reduction pathway which is thought to proceed through a $[Rh^I\text{- - - -}Cl\text{- - - -}Rh^{III}]$ bridged species, equation (158).

$$\left.\begin{matrix}[Rh(CO)_2Cl_2]^- \\ [RhCl_5(H_2O)]^{2-}\end{matrix}\right\} \longrightarrow [(H_2O)Cl_2(CO)_2Rh^I\text{–}Cl\text{–}Rh^{III}Cl_4(H_2O)]^{3-}$$

$$\downarrow$$

$$Rh(I) + [Rh(CO)_2Cl_3(H_2O)] \longleftarrow [(H_2O)Cl_2(CO)_2Rh^{III}\text{–}Cl\text{–}Rh^ICl_4(H_2O)]^{3-}$$

$$[Rh(CO)_2Cl_3(H_2O)] \longrightarrow CO_2 + 2H^+ + [Rh(CO)Cl_3]^{2-} \qquad (158)$$

The reaction of $[Rh(CO)_2Cl_2]^-$ with dioxygen in 3M HCl initially forms $[Rh(III)COCl_5]^{2-}$ by a path thought to involve formation of a dioxygen complex, equation (159).

$$Rh^{I}(CO)_2 + O_2 \xrightarrow{k_1} [Rh(CO)_2(O_2)] \longrightarrow [Rh^{III}(CO)(CO_3)] \xrightarrow{2H^+} Rh^{III}(CO) + CO_2 + H_2O \qquad (159)$$

The observed kinetics indicate that further $Rh^{III}(CO)$ is autocatalytically produced according to the sequence shown in equations (160)–(162).

$$Rh^{III}(CO) + Rh^{I}(CO)_2 \xrightarrow{k_2} [(CO)Rh^{III}\text{- - -}Cl\text{- - -}Rh^{I}(CO)_2] \longrightarrow Rh^{I}(CO) + Rh^{III}(CO)_2 \qquad (160)$$

$$Rh^{III}(CO)_2 + H_2O \longrightarrow Rh^{I}(CO) + CO_2 + 2H^+ \qquad (161)$$

$$2Rh^{I}(CO) + O_2 + 2H^+ \longrightarrow 2Rh^{III}(CO) + 2H_2O \qquad (162)$$

Aqueous solutions of a variety of cobalt complexes have been reduced by CO with formation of CO_2 [198, 205–207]. In most cases intermediate insertion products containing the Co–COOH linkage are postulated. Costa and co-workers [205] have shown that hydroxo complexes of the type [HO–Co(III)(chel)H_2O] (chel = *bis*(salicylaldehyde)ethylenediiminato(salen); *bis*(*ortho*-hydroxyacetophenone)ethylenediiminato(oiafen)) react with CO. Interestingly, when reaction is carried out in neutral or alkaline water THF solutions at room temperature in the dark, hydrogen is evolved, equation (163), together with CO_2 and the corresponding Co(II) derivative.

$$HO\text{–}Co(III)(chel)H_2O \xrightarrow[H_2O]{CO} Co(II)(chel) + CO_2 + \tfrac{1}{2}H_2 \qquad (163)$$

A metal hydride intermediate was suggested. The authors note that $[Mn(CO)_6]^+$ reacts with water to yield the hydride $[HMn(CO)_5]$ and carbon dioxide [208]. In a very recent communication [208*a*], it has been reported that both rhenium and manganese carbonyls having the formula $[LM(CO)_5]^+$ are attacked by nucleophiles to give insertion products, and that facile isotopic exchange occurs between $H_2{}^{18}O$ and the ^{16}O of coordinated CO groups.

Low spin Co(II) derivatives of salen or oiafen do not react with CO in neutral or alkaline solution in the absence of oxygen. In the presence of O_2, however, reaction occurs. The intermediate [HOOC–Co(chel)H_2O] complexes are postulated as reaction intermediates but were not stable enough to be isolated. However, when reaction was run in alcohol rather than H_2O, the more stable [ROOC–Co(chel)H_2O] complexes could be isolated.

Bayston and Winfield [205] report that two moles of aquocobalamin oxidize one mole of CO to CO_2 in the absence of O_2. These authors found further that aquocobalamin catalyzed the reaction of dioxygen with CO to give CO_2 in aqueous solutions at pH values between 6 and 9. An intermediate insertion product having the Co–COOH grouping was proposed in this system also.

Nicholson, Powell and Shaw [210, 211] have postulated carbon monoxide insertion into a Pd–OH bond to explain the oxidation of CO to CO_2 by sodium chloro-palladite [212] containing allyl chloride, equation (164). This method in fact, was found to constitute a good preparative method for the formation of π-allyl palladium complexes from a variety of allylic chlorides.

$$\mathrm{HO{-}Pd(Cl)(CO)(CH_2{=}CH{-}CH_2Cl)} \longrightarrow \left[\mathrm{H{-}O{-}C({=}O){-}Pd(Cl)(CH_2{=}CH{-}CH_2{-}Cl)}\right] \longrightarrow$$

$$\mathrm{Cl{-}Pd(\pi\text{-}C_3H_5)} + H^+ + CO_2 + Cl^- \tag{164}$$

More recently Likholobov, *et al.*, have shown [213] that palladium phosphine complexes catalyze the oxidation of CO at 40 °C in aqueous dioxane, provided that an acid, HX ($X = CH_3COO^-$, CF_3OO^-, NO_3^-, BF_4^-, ClO_4^-, etc.) whose anion is a weakly coordinating ligand, is present. These authors suggest a mechanism involving (*a*) coordination of dioxygen, (*b*) oxygen atom transfer to CO, (*c*) decomposition of the carbonate complex by HX and (*d*) reduction of Pd(II) to Pd(O) by CO, equation (165).

$$(PPh_3)_2Pd \xrightarrow[(a)]{+O_2} Pd(PPh_3)_2(O_2) \xrightarrow[(b)]{CO} (PPh_3)_2Pd(O_2C{=}O) \xrightarrow[(c)]{+2HX} (PPh_3)_2PdX_2 + CO_2 + H_2O$$

$$(PPh_3)_2PdX_2 \xrightarrow[(d)]{+CO,\ H_2O} (PPh_3)_2Pd + 2HX + CO_2 \tag{165}$$

Mixtures of palladium and copper salts have proven to be effective catalysts for the oxidation of CO to CO_2 [214–216]. For example, if CO is bubbled through aqueous solutions of Li_2PdCl_4 (0.001 M) and a mixture of $CuCl_2$ and a Cu(II) non-halide salt with a total Cu(II) concentration from 0.1–2 M at 10–50 °C, CO_2 is produced [214]. More complex systems containing $FeCl_3$, $CuCl_2$ and $PdCl_2$ have also been found to be effective catalysts for air oxidation of CO [215, 216]. A number of other mixtures of metal complexes have also been used for oxidizing CO to CO_2 using both metallic and non-metallic oxidants other than O_2 [217–220].

12.3. Oxidative dimerisation

In the presence of dehydrating agents, such as trialkyl orthoformates the $PdCl_2$/$CuCl_2$ redox system catalyzes the reaction of CO, O_2 and ROH to form dialkyl oxalates [221, 222], equations (167) and (168).

$$2CO + 2ROH + 1/2\ O_2 \xrightarrow[CuCl_2]{PdCl_2} RO_2C\text{—}CO_2R + H_2O \qquad (167)$$

$$H_2O + (RO)_3CH \longrightarrow 2ROH + HCO_2R \qquad (168)$$

If the dehydrating agent is not present, then CO_2 is formed and no oxalates are found. Ferric chloride can be used in place of copper(II) chloride as a cocatalyst, but the alcohol is oxidized to an acid in a competing reaction to some extent.

In typical reactions, 6 m moles of $PdCl_2$, 200 ml of absolute ethanol, 200 ml of triethylorthoformate, and 30–120 m moles of cocatalyst were stirred under 1000 psig of CO at 125 °C in a 0.5 gal. autoclave. Oxygen was admitted in 10–20 psig increments until a total of 150–500 psig had been added. When 37 m moles of $CuCl_2$ was the cocatalyst, diethyl oxalate was formed in 47% yield. In this case a 34% yield of diethylcarbonate was also obtained. At lower CO pressures, increased amounts of dialkylcarbonates are produced. The mechanism proposed for dialkyl oxalate formation is given in equation (169).

$$(ROH)(OC)Pd< \xrightarrow{CO} (OC)Pd(C(=O)OR)< + H^+ \rightleftharpoons (OC)Pd(H)(C(=O)OR)< \xrightarrow{ROH} RO\text{—}C(=O)\text{—}C(=O)\text{—}OR + H\text{—}Pd(H)< \qquad (169)$$

Fenton and Steinwand suggest a catalyst regeneration scheme as shown in equation (170), which differs from the frequently described, equations (171) and (172) regeneration reaction for this system [221].

$$\mathrm{H_2Pd(\mu\text{-}Cl)_2Cu(\mu\text{-}Cl)_2Cu(\mu\text{-}Cl)_2PdH_2 \xrightarrow{O_2} (HO)(H)Pd(\mu\text{-}Cl)_2Cu(\mu\text{-}Cl)_2Cu(\mu\text{-}Cl)_2Pd(OH)(H)}$$

$$\Big\downarrow\!\!\uparrow\; 2(RO)_3CH \qquad (170)$$

$$\mathrm{(ROH)(OC)Pd(\mu\text{-}Cl)_2Cu(\mu\text{-}Cl)_2Cu(\mu\text{-}Cl)_2Pd(ROH)(CO) \xleftarrow[2ROH]{2CO} (H_2O)((RO)_3CH)Pd(\mu\text{-}Cl)_2Cu(\mu\text{-}Cl)_2Cu(\mu\text{-}Cl)_2Pd(OH_2)(HC(OR)_3)}$$

$$\mathrm{Pd^{\circ} + 2CuCl_2 \longrightarrow Cu_2Cl_2 + PdCl_2} \qquad (171)$$

$$\mathrm{CuCl_2 + 2HCl + 1/2\ O_2 \longrightarrow 2CuCl_2 + H_2O} \qquad (172)$$

Gaenzler, Klaus and Schroeder report [222] that esters of oxalic acid can be produced in 90% selectivity at 60–70 °C at 120 atmospheres pressure. In a typical reaction for dimethyl oxalate synthesis, an autoclave is charged with $PdCl_2$ (1.5 g), LiCl (0.75 g), anhydrous $CuCl_2$ (10 g) and methanol (400 g). The reactor (one-fourth filled) is pressurized with CO (100 atm.) and O_2 (20 atm.) and warmed to 60°C. An exothermic reaction ensues and 37 g of dimethyl oxalate are distilled from the reaction mixture. The selectivity to the ester is 90% and the yield based on CO charged to the reactor is 9.65%.

13. ALDEHYDES

Aldehydes readily undergo oxidation to the corresponding acid on standing in air [223]. This reaction generally proceeds by a chain mechanism as shown in equations (173)–(175) for benzaldehyde. The peracid which is initially formed often oxidizes benzaldehyde to benzoic acid.

$$\mathrm{PhCHO + R^{\cdot} \longrightarrow Ph\dot{C}O + RH} \qquad (173)$$

$$\mathrm{Ph\dot{C}O + O_2 \longrightarrow PhCO\dot{O}_2} \qquad (174)$$

$$\mathrm{PhCO\dot{O}_2 + PhCHO \longrightarrow PhCOO_2H + Ph\dot{C}O} \qquad (175)$$

The effect of metal ions on the oxidation of aldehydes has long been recognized [223–226]. In a number of instances, metal ions such as Co^{+3} or Fe^{+3} can convert the aldehyde to a radical *via* the electron transfer reaction, equation (176).

$$\mathrm{M^{n+} + RCHO \longrightarrow R\dot{C}O + M^{(n-1)+} + H^+} \qquad (176)$$

In these instances the role of the metal is chiefly to assist in the formation of the radical species which initiate autoxidation.

The oxidation of benzaldehyde in the presence of cobaltous acetate has been studied in detail by Marta, Boga and co-workers [226–230]. A radical chain mechanism was involved and inhibition of this reaction both by β-naphthol and by cobaltous ion at high concentration, have been observed. The initiation step was found to involve the decomposition of a $Co^{2+}(PhCO_3H)$ complex to give Co^{3+} species which were the reactive intermediates. The rate constant and heat of formation of the $Co^{2+}(PhCO_3H)$ complex were determined. Bawn and Jolley [225] have shown that at low Co^{2+} concentration, the oxidation of benzaldehyde follows rate law, equation (177).

$$\frac{-d[O_2]}{dt} = k[C_6H_5CHO]^{3/2}[Co(OAc)_2]^{1/2} \tag{177}$$

The oxidation of benzaldehyde in the presence of metal compounds has been studied in recent years by several groups of workers. Cobalt salts of di- and tribasic inorganic and organic acids were poor catalysts for the oxidation of benzaldehyde to perbenzoic acid but cobalt salts of mono-basic acids were effective [231]. The catalytic activity of halides was found to increase in the order: $CoF_2 \ll CoCl_2 \ll CoBr_2 < CoI_2$ [231]. The catalytic activity of cobalt complexes varies widely with the nature of the anionic ligand and depends upon the ability of the complex not only to catalyze oxidation but also on its ability to participate in the radical chain propagation [231*b*]. The addition of KBr had no effect on the oxidation of benzaldehyde at 25 °C in acetic acid in the presence of either $CoBr_2$ or $Co(OAc)_2$ [232]. Coordination of peroxidic species in cobalt catalysts was observed during oxidation [233]. When carried out in acetic anhydride, the oxidation of benzaldehyde catalyzed by acetates of Co, Cu, Mn, and Ni gave benzoyl acetate, as the principle product [234].

Recent work concerning the oxidation of a variety of aldehydes in the presence of metal compounds such as cobalt, manganese, iron and copper acetates, nitrates, halides and the like [235–239*b*] have, for the most part corroborated earlier concepts [223–226] regarding free radical reaction pathways. High yields of peracetic acid have been formed from oxygen and acetaldehyde using iron 2,4,6-trinitroresorcinol monophenolate as the catalyst [236]. Although iron compounds give large amounts of peracetic acid [236, 237] manganese complexes catalyze the oxidation of acetaldehyde to acetic acid in good yield [238]. The catalytic effects of metal naphthenates on the oxidation of acetaldehyde in acetone at 20 °C showed that rates of oxygen uptake were in the order: $Fe > Co \gg Mn > Ni > Cr, V >$ none but that some metal naphthenates retarded oxidation [239]. The selectivity to peracetic acid as a function of the metal naphthenate used decreased in the order Fe, Co, none $\gg$ Ni, Cr, V, Cu $\gg$ Mn. This order was found to be due to the relative rates of oxidation and peracetic acid decomposition catalyzed by the metal complexes [239]. Evidence was presented which suggested that the catalytic activity of cobalt naphthenate was due to its ability to abstract an H atom from acetaldehyde and to form a peroxy radical from peracetic acid [239]. The authors [239] suggest, on the other hand, that the activity of iron naphthenate might be based on oxygen activation by the metal complex. It has been shown that during acetaldehyde oxidation, the transition of cobalt from the di- to the trivalent state occurs by reaction with peracetic acid [239*a*] and with an intermediate, α-hydroxyperacetate [239*b*].

Propionic acid was prepared selectively from propionaldehyde using manganese diacetate [240]. The catalytic acitivity of nitrates of Cu(II), Fe(II), Zn(II) and Mn(II) toward the oxidation of propionaldehyde was improved by the addition of 2,2′-bipyridyl (bipy) in quantities from 1–4 moles per mole of metal ion [241, 242]. The catalytic activity of copper complexes was in the order $[Cu(bipy)_3]^{2+} \simeq [Cu(bipy)_2]^{2+} \gg Cu^{2+}$. Furthermore, the presence of bipyridyl increased the selectivity of the copper catalyzed oxidation to peracetic acid. For example, oxidation of propionaldehyde in acetic acid at 30 °C in the presence of $[Cu(bipy)_2(NO_3)_2]$ gave the peracid in 58% yield and the acid in yields of 32–40%. When reaction was run using $Cu(NO_3)_2$ in the absence of bipyridyl, the peracid was formed in 9% yield while the yield of carboxylic acid was 80%. The nature of the amine was varied and the catalytic activity of the copper complex toward oxidation of propionaldehyde varied with the amine as follows: bipyridyl, phenanthroline $\gg$ none, pyridine $>$ 2,9-dimethyl-1,10-phenanthroline $>$ quinoline $>$ ethylenediamine. The rate equations for oxidation in the presence of $Cu(NO_3)_2$ and $[Cu(bipy)_2(NO_3)_2]$ differed substantially and the apparent activation energies were 8.2 and 10.1 kcal/mole, respectively [241].

Mixtures of perbutyric and butyric acid were formed during oxidation of butyraldehyde in the presence of $Co(OAc)_2$ [243]. The oxidation of *n*-butyraldehyde [243–245] and 2-ethyl hexaldehyde [246] are second order in aldehyde and follow expected radical pathways. Isobutyraldehyde was oxidized using ferrocene derivatives as catalysts [247, 248]. In a typical experiment isobutyric acid was produced in 97% selectivity at 60% conversion of the aldehyde. Cobalt(II) and manganese(II) acetylacetonates have been used in a similar manner to prepare phenyl acetic acid from phenylacetaldehyde in good yield [249, 250].

Copper-iron-polyphthalocyanine [251, 252] showed a specific catalysis for the oxidations of saturated aldehydes and substituted benzaldehydes with oxygen. The catalytic reaction was solvent dependent so that tetrahydrofuran, ethanol, acetonitrile, ethyl acetate and anisole inhibited benzaldehyde oxidation while oxidation occurred readily in benzene or acetone. Benzaldehyde was catalytically oxidized with copper-iron-polyphthalocyanine and oxygen to give a quantitative yield of a mixture of perbenzoic (61%) and benzoic (39%) acids. Reaction was carried out at 30 °C and atmospheric pressure of oxygen and exhibited no induction period. By contrast *p*-methyl and *p*-chlorobenzaldehyde had induction periods of 8 and 15 min respectively while no oxidation of *p*-substituted benzaldehydes was observed when the *para*-substituent was: NO_2, OH, OCH_3, or $N(CH_3)_2$.

Inoue, Kida and Imoto [252] found that the oxidation of unsaturated aldehydes such as cinnamaldehyde and acrolein proceeded much more slowly than did oxidation of the saturated substrates in the presence of copper-iron-polyphthalocyanine. As in the case of the saturated acids the products were a mixture of the peracid and the corresponding carboxylic acid. Other groups have recently investigated the oxidation of unsaturated aldehydes in the presence of metal complexes [253–260]. Methacrylic acid and acetic acid were formed in the copper naphthenate catalyzed oxidation of methacrolein [255]. The oxidation of acrolein to acrylic acid was catalyzed by Co, Ni, Mn and Cu acetates [256]. It was found that at concentrations of acrolein in excess of 50%, substantial co-polymerization of acrolein and acrylic acid occurs [256]. In the polymerization reaction during the cobalt laurate catalyzed oxidation of acrolein, a large portion (~ 86%) of the polymer was formed by radical polymerization of the acyl radical, not *via* the peracid radical [257]. Metal naphthenates or acetylacetonates were used to catalyze the oxidation of acrolein in solvents such as benzene, cyclohexane or tetrahydrofuran at 20–45 °C under atmospheric or increased oxygen pressure [258]. The products identified were acrylic acid, peracrylic acid, acetic acid, CO_2, CO and a copolymer of acrylic acid and acrolein. The highest yield of acrylic acid was found with manganese acetylacetonate as the catalyst. The kinetics of the cobalt naphthenate-catalyzed oxidation as well as inhibition studies using hydroquinone were consistent with a radical mechanism in which chain initiation was assumed to occur as shown in equation (178).

$$CH_2 = CHCHO + Co^{3+} \longrightarrow CH_2 = CH\dot{C}O + Co^{2+} + H^+ \qquad (178)$$

It was found that water coordinated to the cobalt catalyst increased the rate of the oxidation of acrolein but free water decreased the rate [259]. A six coordinate intermediate having water, acrolein and acrylic acid (or an oligomer of acrolein and acrylic acid) in the coordination sphere was detected [258].

It has been shown [260–263] that α, β- and β, γ-unsaturated aldehydes and ketones can be oxidized specifically at the γ-carbon atom by oxygen (1 atm) at room temperature in an alkaline methanol solution in the presence of catalytic amounts of cupric pyridine complexes. For example, crotonaldehyde was oxidized to butenal which reacts with one molecule of methanol to give methoxysuccinaldehyde, equation (179).

$$CH_3CH{=}CHCHO \xrightarrow{O_2} [OCHCH{=}CHCHO] \xrightarrow{CH_3OH} OCHCH(OCH_3)CH_2CHO \qquad (179)$$

Kinetic studies suggested that the reaction comprises the following consecutive steps (1) rate determining proton abstraction from the substrate by the base yielding a dienolate anion; (2) interaction between the dienolate anion and the Cu(II) complex yielding a radical which is either free or coordinated to the Cu(I) complex; (3) oxygenation of the coordinated radical leading to a hydroperoxide anion and the original Cu(II) complex.

A cupric ion-initiated chain oxidation of substrate does not occur under reaction conditions [260]. In the absence of oxygen the cupric complex yields a stoichiometric amount of the dienoxy radicals which combine to give dehydrodimers. The active cupric complex was shown to have a methoxide ligand and only three pyridine ligands.

When the base is ammonia and the catalyst is $[RhCl(CO)_2]_2$, oxidation of acrolein or crotonaldehyde gives acrylonitrile or crotonitrile as the reaction product [264]. For example, when a methanolic solution of dirhodium-μ-tetracarbonyl chloride containing crotonaldehyde is treated with ammonia and oxygen, crotonitrile can be detected [265].

In a recent report it has been shown that the dioxygen complexes formed from several square planar d^8 or d^{10} metal complexes were catalytically active for the reaction of dioxygen with benzaldehyde to give perbenzoic and benzoic acids [265–266]. Reactions were carried out at 30 °C in benzene solution under an oxygen atmosphere. The catalytic activity of the metal complexes which were investigated was found to be in the order: $[RhCl(CO)(PPh_3)_2] > [Pd(PPh_3)_2(O_2)] \simeq [Pd(PPh_3)_4] > [RhCl(PPh_3)_3] > [RhCl(PPh_3)_2]_2 > [IrCl(CO)(PPh_3)_2] \simeq [Pt(PPh_3)_2(O_2)]$.

The kinetics of the oxidation of benzaldehyde in the presence of $[Pd(PPh_3)_2(O_2)]$ was studied [265]. The initial rate of oxygen consumption was described by equation (180), where C_1 and C_2 are constants. The reaction was inhibited by free phosphine and markedly retarded by perbenzoic acid while added benzoic acid had no effect.

$$-\left(\frac{dO_2}{dt}\right)_0 = \frac{C_1[Pd(PPh_3)_2(O_2)]_0[PhCHO]_0 P_{O_2}}{1 + C_2[PhCHO]_0} \tag{180}$$

The authors [265] note that it is unlikely that peracids should have a retardation effect if this reaction were a radical chain process because peracids are known to act as initiators or re-oxidizing agents. They note that no retardation by perbenzoic acid was observed in the oxidation of benzaldehyde when cobalt acetate was used as a catalyst. Thus, it is suggested [265] that the course of this reaction proceeds *via* an intermediate complex containing both oxygen and benzaldehyde in the coordination sphere. The explanation offered for retardation by perbenzoic acid was that competitive coordination of the perbenzoic acid with the metal center interferes with coordination of benzaldehyde to form the active intermediate.

The catalytic activity of $[Pd(PPh_3)_4]$ after preoxidation treatment was similar to that of $[Pd(PPh_3)_2(O_2)]$ suggesting the participation of the oxygen complex in the oxidation reaction. The reaction product of the platinum dioxygen complex $[Pt(PPh_3)_2(O_2)]$ with benzaldehyde: $[Pt(PPh_3)_2(PhCHO_3^=)]$ was found to have catalytic activity comparable to the platinum dioxygen complex itself. This encouraged the authors [265] to propose a mechanism, equations (181)–(184), in which such an intermediate participates.

$$Pt(PPh_3)_2(O_2) + PhCHO \underset{k_{-1}}{\overset{k_1}{\rightleftharpoons}} Pt(PPh_3)_2(O_2)(PhCHO) \tag{181}$$

$$Pt(PPh_3)_2(O_2)(PhCHO) \underset{k_{-2}}{\overset{k_2}{\rightleftharpoons}} (Ph_3P)_2Pt\langle O{-}O{-}CHPh{-}O\rangle \tag{182}$$

$$(Ph_3P)_2Pt\langle O{-}O{-}CHPh{-}O\rangle + O_2 \xrightarrow{k_3} Pt(PPh_3)_2(O_2) + PhCO_3H \tag{183}$$

$$PhCO_3H + PhCHO \underset{k_{-4}}{\overset{k_4}{\rightleftharpoons}} PhC(=O){-}O{-}O{-}CH(OH){-}Ph \xrightarrow{k_5} 2PhCOOH \tag{184}$$

Assuming stationary states in this reaction mechanism from step 1 to step 3, the rate equation (185), is derived for the initial reaction, where k_1, k_{-1}, k_2, k_{-2}, and k_3 are the rate constants, $K_1 = k_1/k_{-1}$ and $K_2 = k_2/k_{-2}$. If step 3 is assumed to be rate determining (i.e., $k_{-2} \gg k_3$), equation (185) is consistent with the experimental equation (180).

$$V_0 = \frac{K_1K_2k_3[ML_2(O_2)]_0[PhCHO]_0P_{O_2}}{1 + K_1(1 + K_2)[PhCHO]_0 + (1 + (k_2/k_{-1}) + K_1[PhCHO]_0)(k_3/k_{-2})P_0} \qquad (185)$$

The oxidation of formaldehyde has been studied in aqueous solution [267–270]. Copper(II) was found to oxidize formaldehyde according to reaction (186). In these cases, it is not necessary to involve oxygen activation by the metal center.

$$2Cu(II) + HCHO + H_2O \longrightarrow 2Cu(I) + HCOOH + 2H^+ \qquad (186)$$

14. KETONES

The oxidation of ketones in the presence of metal compounds is a well known reaction [223]. The usual products of oxidation are a carboxylic acid and an aldehyde, equation (187).

$$R\text{—}\underset{\substack{\|\\O}}{C}\text{—}R' + O_2 \xrightarrow{\text{catalyst}} RCHO + R'COOH \qquad (187)$$

Methyl ethyl ketone, for example is oxidized under mild conditions in the presence of a number of metal complexes in aqueous solution [271–274]. The products of reaction are acetaldehyde and acetic acid. Komissarov and Denisov [272–274] have shown that an iron(III)-*o*-phenanthroline complex [272] and a copper(II) pyridine complex [274] catalyze this reaction. In the proposed reaction mechanisms [272, 274] it is suggested that the enolate ion from the ketone is incorporated into the coordination sphere of the metal complex where electron transfer occurs to yield a radical which is attacked by dioxygen, equation (188). In the absence of molecular oxygen, aqueous iron(III) is capable of further oxidizing the radical to form butane 2,3-dione, equation (189) [271].

$$CH_3\underset{\substack{\|\\O}}{C}CH_2CH_3 \rightleftharpoons CH_3C{=}\underset{}{CHCH_3}\ (\text{with OH on C}) \xrightarrow[-(M^{(n-1)+},\ H)^+]{+M^{n+}} CH_3\underset{\substack{\|\\O}}{C}\dot{C}HCH_3$$

$$\xrightarrow{O_2} CH_3\underset{\substack{\|\\O}}{C}\text{—}\underset{\substack{|\\OO^\cdot}}{CH}CH_3 \xrightarrow[-M^{n+}]{+M^{(n-1)+}} CH_3\underset{\substack{\|\\O}}{C}\text{—}\underset{\substack{|\\OO^-}}{CH}CH_3 \xrightarrow{H^+} CH_3\underset{\substack{\|\\O}}{C}\text{—}\underset{\substack{|\\OOH}}{CH}CH_3 \longrightarrow CH_3CHO + CH_3COOH \qquad (188)$$

$$CH_3C(=O)\dot{C}H{-}CH_3 + 3Fe(III) + H_2O \longrightarrow CH_3C(=O){-}C(=O)CH_3 + 3Fe(II) + 3H^+ \qquad (189)$$

When the copper(II)-catalyzed oxidation of methyl ethyl ketone was carried out in acetonitrile, the major products were again acetaldehyde and acetic acid [275]. Butane 2,3-dione was formed as a result of anaerobic oxidation in this case as well. A kinetic study was consistent with a mechanism similar to reaction sequence (188).

Manganese(III) palmitate was found to be a selective catalyst for the conversion of dialkyl ketones to carboxylic acids at temperatures of 110 °C or greater [276]. A single carboxylic acid could be formed from symmetrical ketones [276]. The ease with which manganese complexes convert aldehydes to carboxylic acids (Section 13) is utilized in this conversion.

The Mn(III)-catalyzed autoxidation of acetophenone to benzoic acid and formaldehyde was studied in acetic acid at temperatures from 50 to 110 °C [277] under 1 atm. of oxygen. Oxidation of acetophenone in CH_3COOD as well as acetophenone-ω-d_3 in CH_3COOH proceeded at the same rate as the H/D exchange under the same conditions but in the absence of oxygen. In both oxidation reactions, the unoxidized ketones retained their isotopic compositions. Moreover, substituent effects on oxidations of acetophenones were found to be the same as observed for H/D exchange ($\rho \approx -0.7$). Den Hertog and Kooyman [277] conclude that these data show that enolization is the rate determining step in these oxidation reactions.

A variety of transition metal compounds have been shown to exhibit catalytic activity in the oxidation of acetophenone [278]. A strong synergistic effect in the catalytic activities of mixtures of cobalt and manganese stearates, propionates and acetates, has been observed [279]. It has been proposed [279] that the synergism is due to formation of multinuclear complexes having sites capable of coordinating and activating dioxygen as well as the enolate ion. It is suggested that in these cases, the oxygen transfer reaction may occur within the coordination sphere of the multinuclear complex [279].

The oxidation of cyclohexanone is a reaction which has been the subject of considerable study over the years. Continued research in this area has given rise to many recent patents and papers. The product of the oxidation reaction is rather dependent on the metal complex which is used as a catalyst. When manganese(III) complexes are used the major reaction product is adipic acid [280–288]. Selectivity to adipic acid is about 70% in most cases. When copper(II) complexes are used, 5-formylvaleric acid predominates [289, 290] whereas iron complexes catalyze the formation of ϵ-caprolactone [291, 292] in up to 56% yield. In fact, liquid phase air oxidation of 2-methyl-cyclohexanone at 100 °C in the presence of copper stearate gave ϵ-methyl-ϵ-caprolactone [292*a*]. Reaction scheme (190) shows the predominant reaction pathways.

$$O_2 + \text{cyclohexanone} \longrightarrow \begin{cases} \xrightarrow{Mn(III)} HOOC(CH_2)_4COOH \\ \xrightarrow{Cu(II)} HOOC(CH_2)_4CHO \\ \xrightarrow{Fe(II)} \epsilon\text{-caprolactone} \end{cases} \qquad (190)$$

The oxidation of cyclohexanone in the presence of cobalt complexes has been extensively studied [282, 293–297]. Both adipic acid and ϵ-caprolactone were formed. Oxidation was found to take place *via* the decomposition of the primary oxidation product, 2-hydroperoxy cyclohexanone [282, 293]. Cobalt(III) stearate, [$CoSt_2OH$] was found to form a 1:1 complex with cyclohexanone which decomposed to Co(II) and free radicals [295]. Thus, the suggested involvement of cobalt in this oxidation [293, 295] can be summarized in equations (191) and (192), while oxidation can take place *via* reactions such as (193), (194) [293] and (195) [282]. It is of interest that co-oxidation of acetaldehyde and cyclohexanone in the presence of cobalt naphthenate yields acetic acid and ϵ-caprolactone [299–301].

$$\text{cyclohexanone} + Co(III)St_2OH \rightleftharpoons \text{complex} \longrightarrow \text{2-oxocyclohexyl radical} + CoSt_2 \qquad (191)$$

$$\text{2-oxocyclohexyl radical} + O_2 \longrightarrow \text{2-OO}^{\cdot}\text{-cyclohexanone} \xrightarrow[-R^{\cdot}]{RH} \text{2-OOH-cyclohexanone} \qquad (192)$$

$$\text{2-OOH-cyclohexanone} + CoSt_2 \longrightarrow \text{2-O}^{\cdot}\text{-cyclohexanone} + CoSt_2OH \qquad (193)$$

$$\text{2-OOH-cyclohexanone} \longrightarrow [\text{HO, O–O bridged cyclohexane, H}] \longrightarrow COOH(CH_2)_4CHO \qquad (194)$$

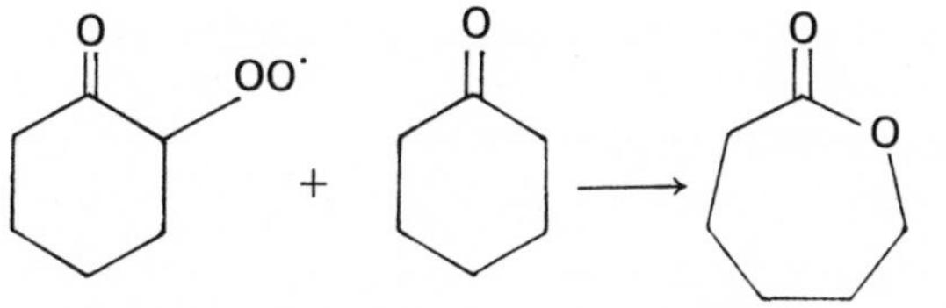

(195)

It was postulated that acetaldehyde is oxidized to peracetic acid which is known [302] to oxidize cyclohexanone to ϵ-caprolactone, equation (196).

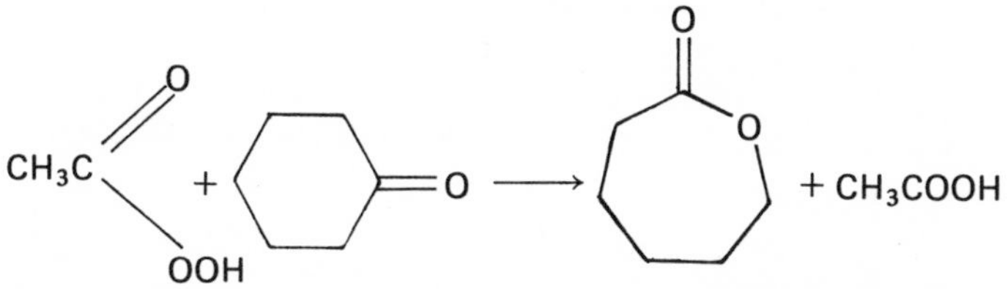

$$CH_3C(=O)OOH + C_6H_{10}O \longrightarrow C_6H_{10}O_2 + CH_3COOH \quad (196)$$

Finally, although cobalt complexes alone give several products, mixtures of cobalt and vanadium complexes give high yields (~ 76%) of adipic acid [296–298].

The rhodium cluster compound $[Rh_6(CO)_{16}]$, also catalyzes the oxidation of cyclohexanone to adipic acid at 100 °C in homogeneous solution [194]. The reaction pathway was not investigated and in view of the reaction of some group VIII metal dioxygen complexes with ketones, a study of such complexes as catalysts for ketone oxidation could prove interesting.

Cyclohexanone may be oxidized by hydrogen peroxide in the presence of iron(II) chelates to give dodecanoic acid [303]. Although Fe(II) is oxidized to Fe(III) in the process, it may be reduced catalytically for reuse, equation (197).

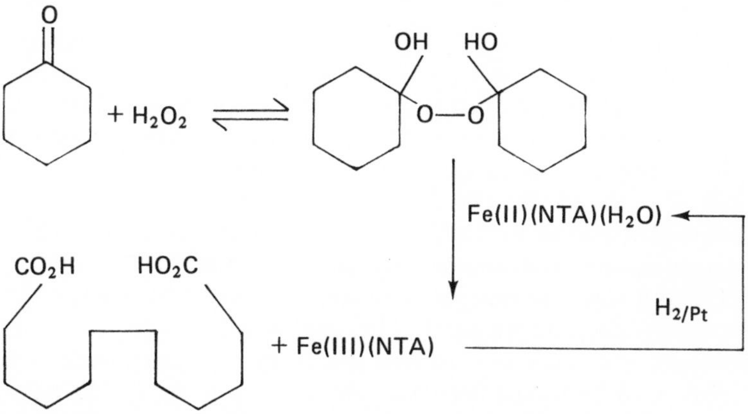

$$C_6H_{10}O + H_2O_2 \rightleftharpoons \text{(HO)}C_6H_{10}\text{–O–O–}C_6H_{10}\text{(OH)} \xrightarrow{Fe(II)(NTA)(H_2O)} HO_2C(CH_2)_{10}CO_2H + Fe(III)(NTA); \quad Fe(III)(NTA) \xrightarrow{H_2/Pt} Fe(II)(NTA)(H_2O) \quad (197)$$

α-Diketones and *o*-quinones are readily oxidized [304] by the oxygen adduct of cobaltocene [305, 306], equations (198) and (199). These reactions are in fact oxidations by coordinated peroxide rather than reactions of ligated dioxygen since the cobaltocene oxygen adduct has an oxygen bridge between the cyclopentadienyl ligands. Oxidation results in the cleavage of the carbon–carbon bond of the α-diketone to give the corresponding cobalticinium carboxylate from which the carboxylic acid can be liberated in good yield with HCl, equation (200).

$$2(\pi\text{-}C_5H_5)_2Co + O_2 \longrightarrow CpCo\text{-}(C_5H_5)(H)\text{-}O\text{—}O\text{-}(H)(C_5H_5)\text{-}CoCp \quad (198)$$

$$CpCo\text{-}(C_5H_5)(H)\text{-}O\text{—}O\text{-}(H)(C_5H_5)\text{-}CoCp + RCOCOR \longrightarrow 2[(\pi\text{-}C_5H_5)Co]^+[RCOO]^- \quad (199)$$

$$R = Ph, CH_3$$

$$[(\pi\text{-}C_5H_5)_2Co]^+[RCOO]^- \xrightarrow{HCl} RCOOH + [(\pi\text{-}C_5H_5)_2]^+Cl^- \quad (200)$$

In this manner benzil was converted to benzoic acid quantitatively and acenaphthene quinone was oxidized to naphthalene-1,8-dicarboxylic acid in 80% yield. Although these reactions are very facile and occur at a very low temperature (− 50 to 15 °C) cobaltocene was irreversibly oxidized to a cobalticinium salt and a catalytic system was apparently not developed.

15. ALCOHOLS

The reaction of alcohols and dioxygen in the presence of metal complexes often gives rise to aldehydes or ketones [307]. An early patent reports that methanol may be converted to formaldehyde in 80% yield by reaction with oxygen at 25 °C in the presence of copper–amine complexes [308]. Gas phase oxidation of 2-propanol to acetone and water was catalyzed by β-Cu-phthalocyanine [309].

Savitskii and co-workers found that aldehydes or ketones could be prepared by liquid phase oxidation of alcohols in the presence of vanadium chelates [310]. CIDNP was used to study the oxidation of 2-propanol using titanous ion with hydrogen peroxide [311]. A variety of radical species were identified in this reaction [311].

Cobalt complexes have frequently been used as catalysts for the oxidation of alcohols. Secondary alcohols were oxidized to ketones in the liquid phase in the presence of mixtures of Co(II) and Co(III) acetates at 60 °C in nearly 80% yield [312]. Primary alcohols, on the other hand, gave high yields of the corresponding carboxylic acids in the presence of $Co(OAc)_3$ [313]. Reaction proceeded stepwise through the aldehyde to the acid [314]. Chromium acetate could also be used to catalyze the oxidation of primary alcohols to carboxylic acids at temperatures below 100 °C but manganese and ferrous naphthenates were inactive under these conditions [314].

Evidence has been presented [315] that the oxidation of phenylmethylcarbinol catalyzed by $CoBr_2$ takes place without the formation of free radicals.

Solutions of [Co(salen)] in 2-propanol absorb oxygen slowly at 60–70 °C, and acetone is formed [316]. When triphenylphosphine is added, the oxidation of the alcohol is greatly accelerated and goes at an appreciable rate even at room temperature. In the course of alcohol oxidation, triphenylphosphine is oxidized to triphenylphosphine oxide to the extent of 10–20% of the yield of acetone. Catalyst systems comprising mixtures of triphenylphosphine and [Co(salen)] also catalyze the oxidation of primary aliphatic alcohols to the corresponding aldehydes but reactions are slower than oxidations of secondary alcohols. Radical inhibitors such as tri-*tert*-butylphenol and N-phenyl-2-naphthylamine did not affect reaction rates.

It was found that while 2-propanol solutions of [Co(salen)] reacted with oxygen irreversibly, in the presence of triphenylphosphine oxygen uptake took place in a reversible manner [315]. Kinetic investigations of both the alcohol oxidation reaction and the pre-reaction equilibrium suggest the formation of a labile pre-reaction complex of [Co(salen)] with triphenylphosphine, oxygen and the alcohol. The rate determining step of the reaction is postulated to involve an outer sphere hydrogen migration from the alcohol to the dioxygen ligand, equation (201). Further steps in the reaction were not delineated.

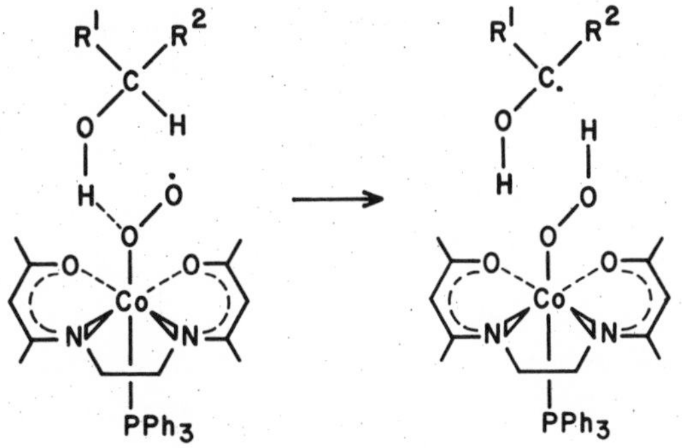

(201)

In an interesting series of reports [317–319] ferrous ion-hydrogen peroxide oxidation of cyclohexanol in acetonitrile has been shown to occur with remarkable regioselectivity to give predominantly *cis*-1,3-cyclohexanediol. While these results appear inconsistent with a mechanism involving free hydroxyl radical, they are in accord with a scheme mediated instead by an iron species. Hydrogen abstraction through a cyclic transition state, equation (202), explains the high reactivity of cyclohexanol and the selectivity of the reaction. In a typical experiment, 30% H_2O_2 (3.68 equiv.) was added to a solution of cyclohexanol (0.19 m) and perchloric acid (0.1 m) in 90% $CH_3CN–H_2O$ containing $Fe(ClO_4)_2$ (0.19 m) at −18°C. At 6% conversion the diol yield is 66% and the selectivity for *cis*-1,3-diol is over 70%.

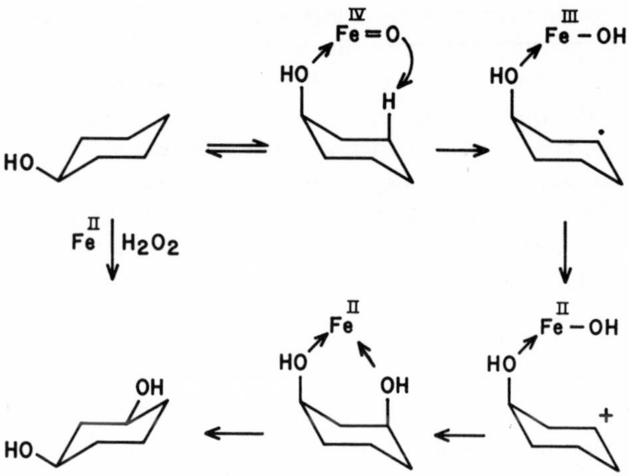

(202)

Carboxylic acids can be obtained from the oxidation of vicinal diols in the presence of cobalt complexes [320–322]. α-Ketols have been oxidized to α-diketones by Cu(II) in buffered aqueous pyridine and the reaction exhibits a rate-controlling enolization step [323]. The catalytic effect of Fe(III) and Fe(II) on the autoxidation of acetoin in aqueous $HClO_4$ showed that the formation of an iron(III) enolate was the slow step [324]. The oxidation of monosaccharides in the presence of catalysts such as copper phthalocyanine or cobalt glutamate in the presence of alkaline compounds gave oxidation products which were useful sequestering agents [325]. Thus cleavage to the acid salts occurred.

16. OXIDATION OF OLEFINS

Of the many substrates which have been oxidized in the presence of transition metal complexes, one of the most extensively studied group of compounds has been olefinic hydrocarbons. The obvious incentive for the pursuit of this research is the identification of catalysts which could convert the abundant olefinic hydrocarbons into more valuable oxygen-containing derivatives under mild conditions in high selectivity. Since hydrocarbon-soluble complexes of the transition metals have been successfully applied to the catalytic addition of other small molecules to olefinic substrates, attempts have been made to activate and catalytically transfer dioxygen to olefins in a similar manner. It has been difficult, however, to exclude other pathways by which oxygen can interact with olefins and to achieve selective reactions. In cases where selective oxidations are achieved it is often hard to decide whether products arise *via* coordination catalysis or autoxidation pathways.

Hydroperoxide intermediates are formed wherever possible and free radical reactions usually result. Because of their prevalence during homogeneous catalytic oxidation of olefins the first part of this section will deal with the reactions which hydroperoxides undergo in the presence of metal complexes. Next we will consider metal catalyzed interactions between hydroperoxides and olefins. With this background the nature of reactions of oxygen with olefins in the presence of transition metal complexes will be discussed.

16.1. Reactions of hydroperoxides in the presence of metal complexes

Since comprehensive reviews of this subject [326, 327, 327*b*] have appeared recently, our treatment of this subject will be restricted to a number of systems having relevance to the homogeneous catalytic oxidation of olefins. Metal catalyzed reactions of hydroperoxides which occur in the absence of olefins will be considered first. It has been shown [326, 327, 327*b*] that often the decomposition of a hydroperoxide follows quite a different pathway when the metal complex is present as a stoichiometric reagent than when only catalytic quantities of the metal compound are present [326, 328]. Our primary emphasis here for the sake of simplicity will be on catalytic reactions rather than stoichiometric ones. When hydroperoxide is in excess over the metal catalyst, radical induced decomposition of hydroperoxide may become a preferred reaction especially with tertiary hydroperoxides. This simplified treatment admittedly has its drawbacks inasmuch as steady state concentrations of hydroperoxide in some autoxidations are less than catalyst concentrations. For more detailed accounts of metal-ion-hydroperoxide reactions references [326], [327], and [327*b*] should be consulted.

16.1.1. TERTIARY HYDROPEROXIDES

The majority of investigations on metal compound catalyzed reactions of alkyl hydroperoxides have been carried out with *tert*-butyl hydroperoxide ([326] and references cited therein). Indictor and Jochsberger [329] reported that whereas the acetylacetonates of Cr(III), V(III), Co(II), Co(III), Mn(III) and Mn(II) catalyzed the decomposition of *tert*-butyl hydroperoxide under mild conditions, Fe(III), Ni(II), Zn(II) or Zr(IV) acetylacetonates were ineffective.

Hiatt, Irwin and Gould [328] studied the decomposition of *tert*-butyl hydroperoxide in the presence of cobaltous and cobaltic stearates (St), octanoates (Oct) and acetylacetonates as well as iron phthalocyanine. They found that the acetylacetonates of Ni(II), Co(III) and Fe(III) were inert toward *tert*-butyl hydroperoxide at room temperature. In chlorobenzene or alkanes at 25–45 °C, half lives for decomposition of 0.1 M *tert*-butyl hydroperoxide by 10^{-4} M catalyst ranged from 1–10 min with the active catalysts. Products included approximately 88% *tert*-butyl alcohol, 11% di-*tert*-butyl peroxide, 1% acetone and ~ 93% O_2. These authors reported that in general, the choice of metal ion, as long as it can undergo a facile one-electron redox reaction, had little effect on products or reaction rates [328].

In chlorobenzene [328] over the concentration range: $[t\text{-BuOOH}]_0 = 0.05$–0.50 M; [catalyst] = 0.01–0.45 mM, the rate expression was first order in each reactant, equation (203), where M = $CoSt_2$, $CoSt_3$, $CoOct_2$ or FePCN.

$$-d[t\text{-BuOOH}]/dt = k[t\text{-BuOOH}][M] \qquad (203)$$

Steady state equilibrium mixtures of Co(II) and Co(III) were rapidly established. Richardson [331–335] obtained similar results using cobalt 2-ethylhexanoate in chlorobenzene. The uniformity of products regardless of the catalyst used and the bimolecular rate expressions suggested a radical induced decomposition of the usual type initiated by metal–hydroperoxide interactions, equations (204) and (205) [328].

$$M^n + RO_2H \longrightarrow M^{n+1} + RO^{\cdot} + OH^- \tag{204}$$

$$M^{n+1} + RO_2H \longrightarrow M^n + RO_2^{\cdot} + H^+ \tag{205}$$

Complexing between catalyst and hydroperoxide [330] probably occurs prior to equations (204) and (205) and this step may be retarded by introduction of compounds which are superior ligands. Thus reaction in alcohol–chlorobenzene mixtures are 100 × slower than in chlorobenzene alone [328].

$[Co(Oct)_2]$-catalyzed decompositions of *tert*-butyl hydroperoxide in refluxing pentane (38 °C) differed little in inital rates or products from comparable decompositions in chlorobenzene, however, rate expressions were first order in *t*-BuOOH and second order in $[Co(Oct)_2]$. In refluxing cyclohexane, radical attack on solvent becomes important and new competing reactions were introduced.

Richardson [331–333] has studied the catalytic decomposition of *tert*-butyl hydroperoxide by cobalt acetate in acetic acid. The principle products were again *tert*-butyl alcohol and oxygen. Results of kinetic studies, ESR measurements and deuterium isotope effects are in agreement with a mechanism involving attack by *tert*-butylperoxy radical on a dimeric cobalt–hydroperoxide complex, equation (206) [333].

$$ROO^{\cdot} + Co(II)_2(ROOH) \longrightarrow RO^- + ROH + Co(III)Co(II) + O_2 \tag{206}$$

Richardson [334] also examined the decomposition of *tert*-butyl hydroperoxide in the presence of copper(II) 2-ethylhexanoate in chlorobenzene at 50 °C. The product composition was approximately 87% *tert*-butyl alcohol, 11% di-*t*-butyl-peroxide, 1–2% acetone and a large amount of oxygen. Reaction is 0.55 order in copper salt and less than first order dependence on hydroperoxide was observed. Trapping experiments with 2,6-di-*t*-butyl-*p*-cresol indicate a radical mechanism. The kinetic data indicate a mechanism involving a hydroperoxide complex, $[Cu(II)_2(ROOH)_2]$, and NMR spectral evidence was obtained for axial hydroperoxide ligands [334].

Rhodium and iridium complexes are effective for the decomposition of *tert*-butyl hydroperoxide. In the presence of $[RhCl(PPh_3)_3]$, *t*-BuOOH decomposes to give *t*-BuOH (93%) and O_2 (88%) with small amounts of di-*t*-butylperoxide and acetone [335]. A variety of iridium complexes, $[IrX(CO)L_2]$ (X = Cl, Br, I; L = PPh_3, $AsPh_3$, PPh_2Me), have also been found to catalyze decomposition of *t*-BuOOH predominantly to the alcohol and oxygen [336]. Stable alkylperoxyiridium complexes of the formula $[IrX(OOBu^t)_2(CO)L_2]$ (X = Cl, L = PPh_3, $AsPh_3$, or PPh_2Me and X = Br, L = PPh_3 or $AsPh_3$) were isolated from reaction mixtures. The alkylperoxyiridium complexes were found to be secondary products and not catalytic intermediates since they did not catalyze the rapid decomposition of *t*-BuOOH.

The palladium complex, $[Pd(PPh_3)_4]$, was found to catalyze the decomposition of *t*-BuOOH in chlorobenzene at 35 °C [337]. The rate of oxygen evolution was found to be directly proportional to the catalyst concentration. The rate of radical production (R_r) was measured by the inhibition method [338] using 2,6-di-*t*-butyl-4-methylphenol. The chain length of the reaction $[dO_2/dt/R_r]$ was found to be 10 in agreement with other studies [339, 340]. Thus, decomposition is a normal radical induced reaction as in the case of catalysis by cobalt compounds.

The decomposition reactions which have been discussed have been catalytic reactions wherein each metal center decomposes many hundreds of hydroperoxide molecules as opposed to stoichiometric reactions in which the ratio is nearer 1:1. The relative unimportance of metal ion reactions which occur with stoichiometric quantities of metal compounds [341–344] must be due largely to the low concentrations of the catalyst [328].

When α,α-dimethylbenzyl hydroperoxide was added to a catalytic amount of *trans*-$[IrCl(CO)(PPh_3)_2]$ in the absence of solvent, a rapid exothermic reaction occurred to give 2-phenylpropan-2-ol and oxygen as the major products, equation (207) [345]. Only traces of acetophenone and α-methylstyrene were detected.

$$2PhMe_2COOH \xrightarrow{\text{cat.}} 2PhMe_2COH + O_2 \tag{207}$$

The complex $[RhCl(PPh_3)_3]$ also catalyzes this reaction but at a much reduced rate [345, 348]. Cobalt octanoate was also effective [327] giving the alcohol in 95% yield and forming acetophenone in 5% yield. Both *bis*-(1,10-phenanthroline) copper(II) and $[Co(acac)_2]$ are also known to cause decomposition of α,α-dimethylbenzyl and *tert*-butyl hydroperoxides to oxygen and the corresponding alcohols [346, 347]. It was suggested [345] that in these cases alkylperoxyl radicals are formed *via* the general Haber–Weiss equations (204) and (205), and that these then undergo a bimolecular self reaction to alkoxyl radicals (and hence alcohols) and oxygen [349–353], equation (208).

$$\begin{array}{lcl} 2ROO^{\cdot} \longrightarrow [ROOOOR] \longrightarrow [RO^{\cdot} + O_2 + {}^{\cdot}OR]\ \text{cage} \\ 2ROO^{\cdot} \xleftarrow{2ROOH} \qquad 2RO^{\cdot} + O_2 \qquad ROOR + O_2 \\ + \\ 2ROH \end{array} \tag{208}$$

It is interesting to note that potential ligands such as phenol or pyridine which are rather poor radical scavengers completely inhibit peroxide decomposition by $[IrCl(CO)(PPh_3)_2]$ or $[RhCl(PPh_3)_3]$ [345]. This suggests that an important prerequisite for hydroperoxide decomposition by Ir(I) or Rh(I) may be the availability of a vacant coordination site on the metal and that the inhibitory action of potential ligands may be to block this site by coordination through lone pair electrons on the N or O atoms. Thus, as in the case of cobalt complexes, a hydroperoxide complex may be formed prior to decomposition.

16.1.2. SECONDARY HYDROPEROXIDES

The presence of hydrogen on the carbon bearing the hydroperoxide group results in increased yields of carbonyl compounds formed during metal catalyzed decomposition reactions [326]. Thus, most secondary hydroperoxides give relatively high yields of ketones in reaction mixtures. For example, the decomposition of *sec*-butyl hydroperoxide in the presence of $[Co(Oct)_2]$ at 38 °C in pentane gives *sec*-butyl alcohol (61%), oxygen (70%) and methylethylketone (36%) [328]. Cobalt, copper, nickel, iron or manganese laurates accelerate the decomposition of 3-hydroperoxyhexane at 80 °C to give 23% alcohol and 15% ketone in 20 hr [353]. The decomposition of cyclohexyl hydroperoxide in the presence of ferrous sulfate, cupric sulfate or manganese laurate yields a mixture of cyclohexanol and cyclohexanone [353, 354].

An investigation of the kinetics of the decomposition of cyclohexyl hydroperoxide at 60–70 °C in the presence of vanadyl acetylacetonate was recently carried out [355]. Cyclohexanol and cyclohexanone were formed in roughly a 1:1 ratio. The initial rate of decomposition was first order in initial concentrations of hydroperoxide and vanadium complex at $[ROOH] \leqslant 5 \times 10^{-2}$ M and $[VO(acac)_2] \leqslant 1 \times 10^{-3}$ M. The initial rate of decomposition changed from first order in [ROOH] to zero order giving evidence of complex formation prior to hydroperoxide decomposition. Using a chemiluminescence method the authors [355] concluded that only about 20% of the cyclohexyl hydroperoxide which decomposed gave free radicals.

When solutions of vanadyl acetylacetonate and cyclohexyl hydroperoxide were mixed, the signal of the paramagnetic V^{4+} rapidly disappeared indicating a one electron transfer reaction, $V^{4+} \rightleftharpoons V^{5+}$ [355]. The limiting step of the reaction was the reaction of V^{5+} with cyclohexyl hydroperoxide, ROOH. The authors [355] noted that the rate of decomposition depended little on the ligand system of the initial catalyst. The color change which occurred suggested the formation of vanadium peroxo complexes. Free acetylacetone was librated from $[VO(acac)_2]$ and the active catalyst was supposed to be $[ROVO(O_2)]$. A decomposition scheme, equations (209)–(214) was postulated to account for these and other observations made on this system [355].

The decomposition of cyclohexylhydroperoxide was also studied in the presence of molybdenum and chromium complexes [356]. The decomposition of cyclohexylhydroperoxide in benzene catalyzed by $[MoO_2(acac)_2]$, has many characteristics of the $[VO(acac)_2]$-catalyzed reaction [355]. The ketone/alcohol ratio in the product was ~ 1 and the kinetic pattern of reaction is similar. When chromium(III) acetylacetonate is used, however, there is a substantial difference. The chromium complex selectively converts cyclohexyl hydroperoxide to cyclohexanone. It is suggested that in this case the extent of release of free radicals to the solution is small [356]. The ketone/alcohol ratio in this case is ~ 13.7. The predominant formation of cyclohexanone on decomposition of cyclohexyl hydroperoxide in the presence of $[Cr(acac)_3]$ is no doubt related to the much higher yield of ketone obtained in cyclohexane oxidation in the presence of chromium complexes than observed when Mo or V compounds are used as catalysts [356].

$VO(acac)_2 \xrightarrow[-\,HO^{\cdot}]{+\,ROOH} VO(OR)(acac)_2 \xrightarrow[-\,\text{acetylacetone}]{+\,ROOH}$ (209)

$\left[ROO{-}V(O)(OR)(acac) \right] \longrightarrow ROV{=}O + (acac)^{\cdot} + RO^{\cdot}$

$RO{-}V(O){=}O + ROOH \rightleftharpoons RO{-}V(O)(O_2) + ROH$ (210)

$RO{-}V(O)(O_2) + ROOH \rightleftharpoons (RO)(HO)V(O)(OR)(O_2)$ (211)

$(RO)(HO)V(O)(OR)(O_2) \longrightarrow ROH + O_2 + RO{-}V(O){=}O$ (212)

$\longrightarrow RO^{\cdot} + HO^{\cdot} + RO{-}V(O)(O_2)$ (213)

$\longrightarrow H_2O + \text{ketone} + RO{-}V(O)(O_2)$ (214)

The kinetics of the decomposition of tetralin hydroperoxide was examined in the presence of $[CoSt_2]$ [357]. Evidence was presented for the formation of an intermediate complex, $[CoSt_2(ROOH)]$ which decomposes largely without forming free radicals. Thus, the introduction of a radical inhibitor did not influence the kinetics of hydroperoxide decomposition and also the rate of consumption of inhibitor was two orders of magnitude lower than the rate of consumption of the hydroperoxide. The first observation ruled out the existence of a chain mechanism and the second observation showed that decomposition of tetralin hydroperoxide to free radicals was in the order of 1%. The authors conclude that decomposition of the $[CoSt_2(ROOH)]$ complex occurs to give equal amounts of 1-tetralone and 1-tetralol without furnishing free radicals to the solution [357]. In the decomposition of tetralin hydroperoxide catalyzed by $[Ni(acac)_2]$ it was also found that the rate of formation of free radicals was only 2–3% of the overall hydroperoxide decomposition rate [358]. Similarly, the decomposition of α-phenylethylhydroperoxide catalyzed by $[NiSt_2]$ or $[Ni(acac)_2]$ generates few free radicals [359, 360], although this is dependent on reaction temperature [361]. Evidence was obtained by EPR for the formation of a copper–hydroperoxide complex during the initial stages of the decomposition of α-phenylethylhydroperoxide in the presence of copper(II)*bis*-(α-thiopicolinanilide) [362].

16.1.3. PRIMARY HYDROPEROXIDES

Primary alkyl hydroperoxides are decomposed in the presence of many metal salts to give mainly alcohols, aldehydes and in some cases carboxylic acids [326]. Cobalt(II) octanoate catalyzes the decomposition of *n*-butyl hydroperoxide in pentane at 38 °C to give *n*-butyl alcohol (67%), oxygen (70%) and *n*-butyraldehyde (32%) [328], equation (215).

$$CH_3CH_2CH_2CH_2OOH \xrightarrow{Co(Oct)_2} CH_3CH_2CH_2CH_2OH + O_2 + CH_3CH_2CH_2CHO \tag{215}$$

The decomposition of *n*-decyl hydroperoxide in the presence of cobalt complexes has been studied in detail by Bulgakova, Skibida and Maizus [363, 364] and found to proceed *via* a radical pathway at 50 °C. The reaction mechanism depended on the nature of the anionic ligand in this case. Thus, when cobalt(II) acetylacetonate was used as the catalyst a linear dependence on $[Co(acac)_2]$ concentration was observed. This and other evidence suggested to these authors [364] that decomposition of *n*-decyl hydroperoxide in the presence of $[Co(acac)_2]$ proceeds *via* monomolecular decomposition of a mononuclear complex: $[Co(acac)_2(ROOH)]$, equations (216) and (217). Inhibition studies indicated that while decomposition of *n*-decyl hydroperoxides gave radicals, a chain mechanism was not involved.

$$n\text{-}C_{10}H_{21}OOH + Co(acac)_2 \rightleftharpoons n\text{-}C_{10}H_{21}OOH{\cdot}Co(acac)_2 \longrightarrow Co(acac)_2OH + n\text{-}C_{10}H_{21}O^{\cdot} \tag{216}$$

$$n\text{-}C_{10}H_{21}OOH + Co(acac)_2OH \rightleftharpoons n\text{-}C_{10}H_{21}OOH{\cdot}Co(acac)_2OH$$

$$\longrightarrow Co(acac)_2 + n\text{-}C_{10}H_{21}O_2^{\cdot} + H_2O \qquad (217)$$

When cobalt(II) stearate was used as the catalyst, however, the rate of hydroperoxide decomposition was a linear function of the square of the [$CoSt_2$] concentration. [364]. Thus, the pathway shown in equations (218) and (219) was proposed [364].

$$n\text{-}C_{10}H_{21}OOH + CoSt_2 \rightleftharpoons n\text{-}C_{10}H_{21}OOH\ CoSt_2$$

$$\xrightarrow{CoSt_2} CoSt_2OH + CoSt_2 + n\text{-}C_{10}H_{21}O^{\cdot} \qquad (218)$$

$$n\text{-}C_{10}H_{21}OOH{\cdot}CoSt_2 + CoSt_2OH \longrightarrow 2CoSt_2 + H_2O + n\text{-}C_{10}H_{21}O_2^{\cdot} \qquad (219)$$

In the decomposition of *n*-decyl hydroperoxide catalyzed by copper(II) stearate, the formation of radicals was preceded by the formation of the complex: [$CuSt_2(ROOH)_2$] [365].

In summary, alkyl hydroperoxides are readily decomposed in the presence of catalytic quantities of transition metal complexes. In most cases the predominant reaction products are the corresponding alcohol and oxygen. Carbonyl compounds are formed in varying yields depending on the nature of the hydroperoxide. Tertiary alkyl hydroperoxides often decompose by a radical chain process, but non-chain radical processes as well as molecular processes which do not liberate large numbers of radicals occur frequently when secondary or primary hydroperoxides are catalytically decomposed. It appears that in many cases, a metal hydroperoxide complex is formed prior to decomposition.

Allylic hydroperoxides are likely intermediates during olefin oxidation in the presence of many metal complexes. A recent study by Arzoumanian, Blanc, Metzger and Vincent [366] has elucidated the mechanism of the decomposition of cyclohexenyl hydroperoxide in benzene catalyzed by [$RhCl(PPh_3)_3$]. The decomposition products are shown in equation (220).

OOH-cyclohexene $\xrightarrow[\text{Benzene}]{RhCl(PPh_3)_3}$ OH-cyclohexene + cyclohexenone (O) $+ O_2 + H_2O +$ polymers (220)

Inhibition studies showed that the consumption of inhibitor was half that of hydroperoxide indicating that no molecular decomposition occurred. Furthermore, the superposition of decomposition curves with or without inhibitor gave evidence that the process did not occur *via* a free radical mechanism but rather each hydroperoxide molecule was catalytically decomposed to give one free radical.

The existence of the cyclohexenyloxy radical was shown by ESR spectroscopy and by trapping with 2-methyl-2-nitrosopropane to give the corresponding nitroxide. Oxygen was believed to be formed *via* the cyclohexenylperoxy radical according to equation (221).

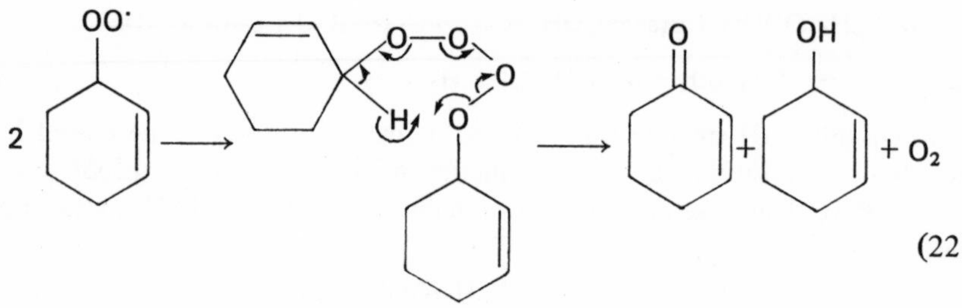

(221)

Kinetic studies are in agreement with a monoelectronic Haber–Weiss mechanism involving rhodium(I), rhodium(II) and rhodium(III) species, equations (222)–(225).

$$\mathrm{Rh(I) + ROOH \longrightarrow Rh(II)OH + RO\cdot} \tag{222}$$

$$\mathrm{Rh(II)OH + ROOH \longrightarrow Rh(I) + ROO\cdot + H_2O} \tag{223}$$

$$\mathrm{Rh(II)OH + ROOH \longrightarrow Rh(III)(OH)_2 + RO\cdot} \tag{224}$$

$$\mathrm{Rh(III)(OH)_2 + ROOH \longrightarrow Rh(II)OH + H_2O + ROO\cdot} \tag{225}$$

16.2. Reactions of hydroperoxides with olefins catalyzed by metal complexes

The nature of the metal catalyzed reactions of hydroperoxides with olefins can be divided into two general types. The reaction pathway which is preferred depends on the metal complex which is used as the catalyst. One group of reactions are characterized largely by homolytic bond cleavage resulting in products of free radical processes. Complexes which tend to promote this type of reaction are those of groups VIII and IB, in particular complexes of iron, cobalt and copper. The second group of reactions involve heterolysis of the O–O bond and result largely in the formation of an epoxide and an alcohol from reaction of the olefin with the hydroperoxide, equation (226). Transition metal complexes which catalyze this reaction preferentially are those of groups IVB, VB and VIB, especially complexes of Mo, W, V, Cr and Ti.

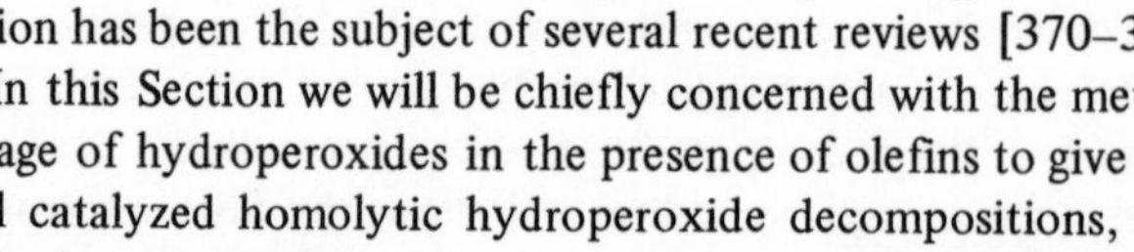

$$\mathrm{ROOH + {>}C{=}C{<} \xrightarrow{Catalyst} ROH + {>}C{-}C{<}\ (epoxide,\ \text{-}O\text{-}\ bridge)} \tag{226}$$

Reaction (226) has found considerable utility in the commerical production of propylene oxide from propylene [367] and as a result has stimulated intense industrial interest and patent activity [367–369]. The general scope and utility of this reaction has been the subject of several recent reviews [370–373].

In this Section we will be chiefly concerned with the metal catalyzed heterolytic cleavage of hydroperoxides in the presence of olefins to give epoxides. In contrast to metal catalyzed homolytic hydroperoxide decompositions, this reaction appears to clearly be an example of coordination catalysis. If the hydroperoxides initially formed during olefin autoxidation could be selectively reacted in this manner, selective catalytic reactions between dioxygen and olefins can be envisioned.

Table 11

METAL CATALYZED EPOXIDATION OF OLEFINS BY HYDROPEROXIDES

Olefin	Hydro-peroxide	Temp °C	Catalyst	Time (Hrs)	Conver-sion of Hydro-peroxide	Selec-tivity to Epoxide, %	Ref.
2,4,4-Trimethyl-1-Pentene	*tert*-Butyl	25	$MoO_2(acac)_2$	96	17	100	[373]
			$VO(acac)_2$	168	46	100	
			$V(acac)_3$	168	40	100	
			$Cr(acac)_3$	96	53	100	
			$Co(acac)_3$	96	25	30	
			$Co(acac)_2$	96	48	20	
			$Cu(acac)_2$	96	33	25	
			$Mn(acac)_3$	120	47	15	
			$Mn(acac)_2$	120	44	15	
2-Methyl-2-Pentene	Cumene	110	$MoO_2(acac)_2$	3/4	90	86	[374]
			$Mo(CO)_6$	3/4	90	83	
			$VO(acac)_2$	1	90	75	
Cyclohexene	*tert*-Butyl	90	$MoO_2(acac)_2$	1	98	94	[375]
			$Mo(CO)_6$	2	98	94	
			$W(CO)_6$	18	95	89	
			$Ti(OBu)_4$	20	80	66	
			$VO(acac)_2$	2.5	96	13	
			$TiO(acac)_2$	18	40	38	
			$Cr(acac)_3$	2	98	2	
Octene-1	*tert*-Butyl	90	$Mo(CO)_6$	3	88	64	[375]
			$W(CO)_6$	70	71	34	
			$Ti(OBu)_4$	24	34	10	
			$VO(acac)_2$	4	95	2	

The most important factor affecting the selectivity of the epoxidation reaction (226) is the choice of metal complex used as the catalyst [374–377]. Table 11 summarizes the results of several studies which indicate that in general, molybdenum complexes are superior catalysts for this reaction. The lower selectivity for several of the catalysts listed in Table 11 is due to competing metal catalyzed hydroperoxide decomposition *via* homolytic bond cleavage under reaction conditions. Sheldon and Van Doorn have shown that half times for decomposition of *tert*-butyl hydroperoxide in benzene at 90 °C were in the order: $[Co(Oct)_2] \gg [Cr(acac)_3] > [VO(acac)_2] > [Mo(CO)_6] \gg [W(CO)_6] > [Ti(OBu)_4]$. On the other hand, the relative rates of epoxide formation in reactions of *tert*-butyl hydroperoxides with cyclohexene in benzene at 90°C were in the order: $[Mo(CO)_6] \gg [VO(acac)_2] \gg [Ti(OBu)_4] > [W(CO)_6]$. Thus, the relative rates of homolytic decomposition pathways and heterolytic epoxidation for any given complex determine the epoxide selectivity.

The structure of the olefin is also a very important factor in determining the selectivity of the epoxidation reaction. The extent of alkyl substitution on the double bond markedly affects the rate of the epoxidation reaction. The greater the number of electron donating alkyl groups on the double bond, the more reactive is the olefinic substrate [374–377]. The reactivities of olefins with hydroperoxides in the presence of complexes of groups IVB, VB, and VIB are in the order: 2-methylpentene-2 > ocetene-2 > octene-1 [374–377].

It is interesting to note that little or no epoxide was obtained from either styrene or α-methylstyrene under conditions in which simple olefins gave good yields (*tert*-butyl hydroperoxide, MoO_3, 100 °C) [372]. Instead, the products were a ~ 4/1 mixture of α-phenylpropionaldehyde and acetophenone. Both products apparently resulted from the decomposition of the epoxide during reaction. At 60 °C, however, epoxides are formed in good yield [378].

It has also been reported [372] that *cis*- and *trans*-2-octenes react with similar rates in MoO_3 catalyzed epoxidations and that *cis-trans*-isomerization does not occur. A *cis*-olefin gives rise to a *cis*-epoxide and a *trans*-olefin yields a *trans*-epoxide [372, 375].

The structure of the hydroperoxide does not seem to have as great an effect on the ease of epoxidation as does the olefin structure. It has been reported, however, [375] that the epoxidation reaction is enhanced by electron withdrawal from the hydroperoxide. Thus, it was found that in the epoxidation of octene-2 in the presence of $[Mo(CO)_6]$, reaction rate varied with the hydroperoxide in the order: *p*-nitrocumene hydroperoxide > cumene hydroperoxide > *tert*-butyl hydroperoxide [375].

In order to observe rapid rates and high epoxide selectivity, the conditions under which reaction (226) is run must be within fairly restricted limits. In most instances, an excess of olefin over hydroperoxide will result in more efficient use of hydroperoxide and thus in greater selectivity [370]. In general, the lower the temperature, the less radical decomposition of hydroperoxide and the higher the selectivity. The maximum temperature at which each metal complex may be run without a large amount of radical decomposition varies with the metal center. For molybdenum catalysts epoxide selectivities of 98% can be achieved at 100 °C but fall to 75–80% at 130 °C. For vanadium complexes the maximum temperature for selective operation is ~ 80 °C and for chromium it is below 60 °C [370].

The kinetics, stereochemistry, mechanisms and synthetic applications of the epoxidation reaction, equation (226), have been studied extensively for a number of homogeneous transition metal catalysts in recent years. In some instances changes in the metal complexes have been carefully examined during epoxidation. We will, therefore, consider several of these systems in more detail at this point.

16.2.1. EPOXIDATION WITH MOLYBDENUM COMPLEXES

A number of kinetic studies have been made on the epoxidation of olefins in the presence of molybdenum complexes. It has been reported that the $Mo(CO)_6$-catalyzed reactions of 1-octene [375], 2-octene [375], styrene [378], and cyclohexene [379] with *tert*-butyl hydroperoxide in benzene appear to be first order in each component, equation (227).

$$\text{Epoxidation Rate} = k[t\text{-BuOOH}][\text{Olefin}][\text{Mo(CO)}_6] \qquad (227)$$

Free energy relationships for substituted styrenes showed considerable scatter but a $\rho\sigma$ plot gave $\rho = -1.4 \pm 0.6$ (95% confidence levels). Thus, it appears that the rate determining step involves electrophilic attack upon the alkene [378].

With styrene as solvent the reaction is first order in hydroperoxide but displays a complex dependence on molybdenum concentration. The apparent order in molybdenum complex is 3/2 at low concentrations and 1/2 order at high Mo concentrations.

More detailed studies of the oxidation of propylene [380, 381] and 1-octene [382] have shown that the actual rate equation is of the Michaelis–Menten type with the co-product alcohol being the competitive inhibitor. The fact that alcohols inhibit the molybdenum catalyzed epoxidation of olefins is well known [377, 378, 383]. In addition, it has been noted that induction periods are observed [378, 382] during which time the active catalyst, a molybdenum(VI) species [384, 379] is formed.

The kinetics of the $[MoO_2(acac)_2]$-catalyzed epoxidation of cyclohexene by *tert*-butyl hydroperoxide in cyclohexane have also been examined [385]. Reactions are nearly first order in catalyst but of nonintegral order in olefin with rates more nearly proportional to [olefin] at low concentrations than at high. A similar behavior was noted for hydroperoxide. This kinetic behavior was ascribed to the formation of molybdenum hydroperoxide and molybdenum olefin complexes. Kinetic evidence for the formation of a wide variety of metal–olefin and metal–hydroperoxide complexes has been reported during the molybdenum versatate-catalyzed epoxidation of cyclohexene with ethyl benzene hydroperoxide, and equilibrium constants for their formation were calculated [386].

Sheldon has considered the competing process of homolytic decomposition of hydroperoxides during the epoxidation of olefins with *tert*-butyl hydroperoxide in the presence of molybdenum complexes. It was found that homolytic decomposition of the hydroperoxide is initiated by electron transfer reactions of Mo(V) and Mo(VI) complexes with the hydroperoxide giving rise to free radical species. Reaction rates and products of hydroperoxide decomposition were dependent on the solvent and on the presence or absence of an olefin. The rates and selectivities of epoxidation were highest in polychlorinated hydrocarbons and very poor in coordinating solvents such as alcohols or ethers [387].

The relative rates of $[Mo(CO)_6]$-catalyzed epoxidation of olefins by *tert*-butyl hydroperoxide in benzene were found to be in the order: allyl chloride (0.1) $\ll$ 1-octene (1.0) $<$ styrene (1.3) $\ll$ 2-methyl-1-heptene (9) $<$ cyclohexene (13) $<$ *cis*-2-octene (14), norbornene (14) $\ll$ 2-methyl-2-heptene (75). Thus, electron donation in the form of alkyl groups about the double bond accelerates the reaction rate whereas electron withdrawal tends to retard it [379].

On the basis of kinetic evidence, structure-reactivity-relationships, induction periods and inhibition by alcohols, a schematic mechanism for the molybdenum catalyzed epoxidation of olefins has been put forward by several groups of workers [374–387] which accomodates most of the existing data and until recently has received general acceptance, equations (228)–(232).

$$Mo^{n} \longrightarrow Mo^{VI} \tag{228}$$

$$Mo^{VI} + ROOH \rightleftharpoons Mo^{VI}(ROOH) \tag{229}$$

$$Mo^{VI} + ROH \rightleftharpoons Mo^{VI}(ROH) \tag{230}$$

$$Mo^{VI}(ROOH) + C{=}C \xrightarrow{k_3} Mo^{VI} + ROH + \underset{\diagdown O \diagup}{C{-}C} \tag{231}$$

This mechanism involves initial formation of the catalytically active species, equation (228), coordination of hydroperoxide by the Mo(VI) center, equation (229), or alcohol (inhibition), equation (230) and rate determining oxygen transfer, equation (231). Equation (232) is the rate equation which was derived for this reaction sequence [382] in which Kp is the dissociation constant for the molybdenum–hydroperoxide complex and Keq. is the equilibrium constant for the alcohol–molybdenum complex. In the case where $1 - \text{KeqKp}$ is negligible, equation (232) becomes equation (233) and this explains the apparent first order dependence on hydroperoxide [384].

$$\frac{-d(\text{ROOH})}{dt} = \frac{k_3(\text{C=C})}{[\text{Kp}/(\text{ROOH})] + \text{KeqKp}[(\text{ROOH})_0/(\text{ROOH})] + (1 - \text{KeqKp})} \tag{232}$$

$$\frac{-d(\text{ROOH})}{dt} = \frac{k_3(\text{C=C})(\text{ROOH})}{\text{Kp} + \text{KpKeq}(\text{ROOH})_0} \tag{233}$$

The stereochemistry of molybdenum catalyzed epoxidation of olefins has been studied in some detail. Early work [375] indicated that *cis*-olefins give *cis*-epoxides and *trans*-olefins give *trans*-epoxides stereoselectively, equations (234) and (235).

$$PhC(CH_3)_2OOH + (CH_3)(H)C{=}C(CH_3)(H) \xrightarrow{Mo(CO)_6} (CH_3)(H)\underset{\diagdown O \diagup}{C{-}C}(CH_3)(H) \tag{234}$$

$$PhC(CH_3)_2OOH + (CH_3)(H)C{=}C(H)(CH_3) \xrightarrow{Mo(CO)_6} (CH_3)(H)\underset{\diagdown O \diagup}{C{-}C}(H)(CH_3) \tag{235}$$

The stereochemistry of the epoxidation of cyclic olefins containing no complexing groups is determined solely by steric factors. In a large number of instances [388,

389], equations (236)–(239), reactions were reported to be completely stereoselective [388, 389].

t-amyl hydroperoxide
$MoCl_5$ or $Mo(CO)_6$ (236)

t-amyl hydroperoxide
$MoCl_5$ or $Mo(CO)_6$ (237)

t-amyl hydroperoxide
$MoCl_5$ or $Mo(CO)_6$ (238)

t-amyl hydroperoxide
$MoCl_5$ or $Mo(CO)_6$ (239)

For example, the *tert*-amyl hydroperoxide-molybdenum complex is presumed to attack α-pinene selectively from the least hindered side to give only one isomeric epoxide [388], equation (240).

$-Mo^{VI}$
$-ROH$ (240)

When the olefin contains complexing groups, these groups usually direct the steric course of the reaction. 2-Cyclohexene-1-ol, for example, is epoxidized by hydroperoxides in the presence of molybdenum complexes to give predominantly *cis*-1,2-epoxycyclohexan-3-ol [390–392], equation (241), presumably *via* complexing of the hydroxyl group with the metal center.

OH + ROOH ⟶ [R–O–Mo, HO, OH, C=C] ⟶ OH, O, cis- (241)

Reaction (241) was selective with very little hydroperoxide decomposition and appeared to obey a rate law which was first order in hydroperoxide, olefin and molybdenum complex similar to equation (227) [393]. A number of additional examples of molybdenum catalyzed epoxidation reactions in which a complexing group has directed the stereochemical course of the reaction have appeared in the literature [389, 391, 394, 395], equations (242)–(247).

OH + *t*-BuOOH $\xrightarrow{Mo(CO)_6}$ OH, O, *cis*- (242)

OAc + *t*-BuOOH $\xrightarrow{Mo(CO)_6}$ OAc, O, *cis*- (243)

OH + *t*-BuOOH $\xrightarrow{Mo(CO)_6}$ O, OH (erythro) (244)

CH_3, CH_2OAc + *t*-PentOOH $\xrightarrow{Mo(CO)_6}$ CH_3, O, CH_2OAc (245)

(246) [Scheme: CH_3-substituted cyclohexene with $CH(CH_3)CH_2OAc$ substituent + t-PentOOH →($Mo(CO)_6$) epoxide]

(247) [Scheme: CH_3-substituted cyclohexene with $C(CH_3)=CH_2$ substituent + t-PentOOH →($Mo(CO)_6$) epoxide]

The nature of the oxygen transfer step, equation (231), in the epoxidation mechanism has been considered by several authors [375, 376, 378, 383]. Two pathways have been suggested. The first involves an activated complex having a three membered ring, equation (248), and the second involves 1,3-dipolar addition of the molybdenum hydroperoxide complex to the olefin through an activated complex having a five membered ring, equation (249). Although definitive information is not available, the lack of structure sensitivity (*i.e.*, to ring size) in epoxidation of cyclic olefins prompts Su, Reed and Gould to favor equation (248).

(248) [Scheme: Mo(VI)–O(R)–O–H + C=C → [three-membered ring activated complex] → Mo(VI)–O(R)H + epoxide]

(249) [Scheme: Mo(VI)–O(R)–O–H with coordinated C=C → [five-membered ring activated complex] → Mo(VI)–O(R)H + epoxide]

Sheldon has determined the structure of the catalytic species in a number of molybdenum catalyzed epoxidation reactions [396]. It is interesting that the active catalytic species isolated from the reaction mixtures of the molybdenum-catalyzed epoxidations of a variety of different olefins with *tert*-butyl hydroperoxide were all found to be Mo(VI)-1,2-diol complexes. The structure of the diol complex was found to be directly related to the structure of the olefin being epoxidized and was independent of the type of molybdenum complex originally added as the catalyst. The Mo(VI)-diol complexes formed during the epoxidation of propene or octene-1 catalyzed by either $[Mo(CO)_6]$ or $[MoO_2(acac)_2]$ had structures XXX and XXXI, those formed during cyclohexene epoxidation had structure XXXII and those complexes isolated from epoxidation of 2,3-dimethyl-2-butene and 2,4,4-trimethylpentene-1 had structures XXXIII and XXXIV respectively, formed from dimerization of the monomeric 1,2-diol complexes.

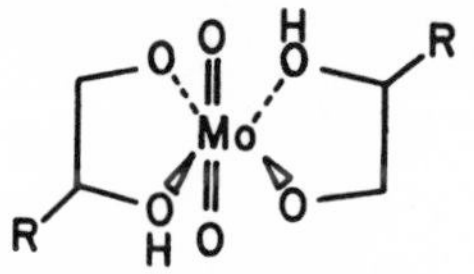

XXX R = CH_3

XXXI R = n-C_6H_{13}

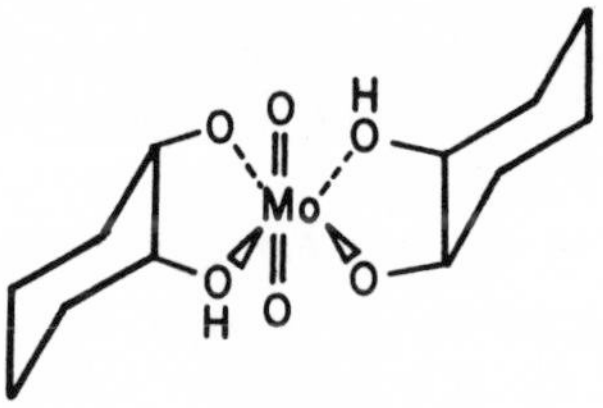

XXXII

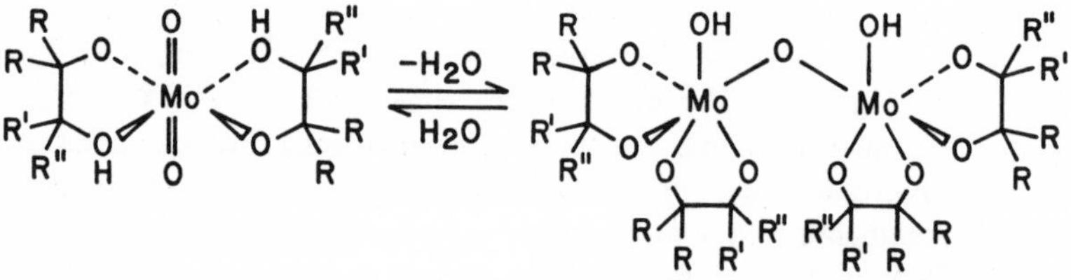

$-H_2O$ / H_2O

XXXIII R = R' = R'' = CH_3

XXXIV R = H, R' = CH_3, R'' = $(CH_3)_3CCH_2$

Thus, the rate of epoxidation depends on the structure of the olefin and, after an initial period during which the rate changes, it becomes independent of the structure of the molybdenum compound originally added [379]. When $[Mo(CO)_6]$ was used as a catalyst a short induction period was observed. Although initial rates varied depending on the molybdenum compound originally added, final constant rates were the same regardless of whether $[Mo(CO)_6]$, $[MoO_2(acac)_2]$ or the Mo(IV) diol complex was used. This implies that the 1,2-diol complex was the active catalytic species during most of the reaction.

Sheldon [396] suggested that the 1,2-diol complexes form *in situ* by reaction of an oxo molybdenum complex with the epoxide formed from the olefin and hydroperoxide. Kaloustian, Lena and Metzger [397] have recently shown that at low temperatures *tert*-butyl hydroperoxide reacts with $[MoO_2(acac)_2]$ exothermically to form permolybdic acid and at higher temperatures to give molybdic acid. In fact, molybdic acid is formed by reaction of either $[MoO_2(acac)_2]$, $MoCl_5$, molybdenum naphthenate or molybdenum blue with several hydroperoxides [397]. Furthermore, molybdic acid was found to react with cyclohexene oxide to give XXXI or its dimer.

Thus, the sequence of events leading to the formation of the final catalytic species are thought to be (*a*) oxidation of the original molybdenum complex with loss of its original ligands to give molybdic acid, equation (250), then (*b*) reaction of molybdic acid with the epoxide formed during early stages of epoxidation to give the 1,2-diol complex, equation (251).

$$MoL_x + ROOH \longrightarrow H_2MoO_4 \cdot H_2O \qquad (250)$$

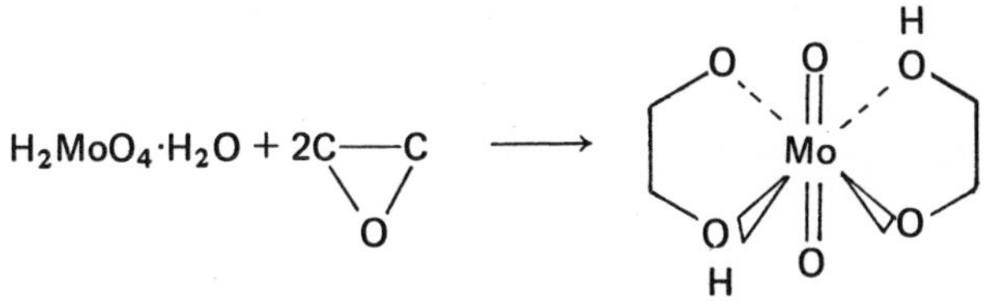

$$H_2MoO_4 \cdot H_2O + 2C\text{—}C \text{ (epoxide)} \longrightarrow \qquad (251)$$

Sheldon suggests that oxygen transfer occurs from a hydroperoxide complex of the Mo(VI)-1,2-diol system with protonation of the oxo group, equation (252).

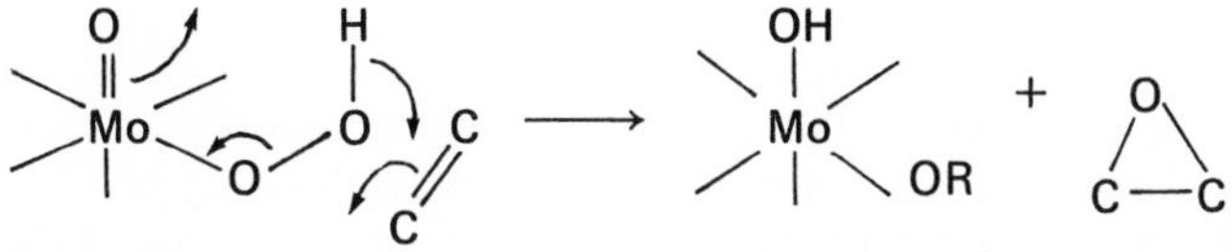

(252)

He stresses that Mo(VI)-1,2-diol complexes are not the only catalytically active Mo(VI) compounds; nor are they necessarily more active than other Mo(VI) complexes. In fact $[MoO_2(acac)_3]$ was initially more active than the 1,2-diol complex, however, the ligands around the molybdenum are destroyed *via* oxidation by the hydroperoxide and are replaced by 1,2-diol ligands. The 1,2-diol ligands are similarly destroyed but may be continually regenerated by reaction with epoxide formed *in situ*.

A completely different mechanism of epoxidation has obtained some support in the recent literature. This mechanism originated with the observation by Mimoun, Seree de Roch, and Sajus [95] that the covalent molybdenum VI peroxo complexes, $[MoO(O_2)_2L_x]$, reacted with olefins to form epoxides in high yield, equation (253). These Mo(VI) peroxo complexes are readily prepared from MoO_3, the neutral ligand and aqueous hydrogen peroxide [398–400].

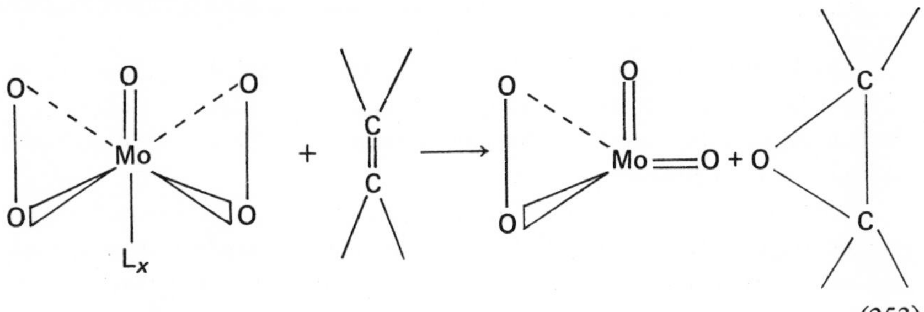

(253)

The relative reactivities of olefins with the peroxo Mo(VI) complexes are in the order: 1-octene < styrene < 2-octene < cyclohexene ≃ 2-methylbutene-1 < 2-methylbutene-2 < tetramethylethylene. Thus, the relative reactivities of olefins in reaction (253) are similar to the catalytic epoxidation of these substrates using hydroperoxides. It has been proposed [95] that the epoxidation of an olefin by $[MoO(O_2)_2(HMPA)]$ occurs in two steps, reversible coordination of olefin, equation (254), and irreversible decomposition of the olefin complex to give epoxide, equation (255).

$$MoO(O_2)_2L + \text{olefin} \underset{k_{-1}}{\overset{k_1}{\rightleftharpoons}} MoO(O_2)_2L\ (\text{olefin}) \qquad (254)$$

$$MoO(O_2)_2L\ (\text{olefin}) \xrightarrow{k_2} O{=}M(L)(O_2) + \text{epoxide} \qquad (255)$$

Arakawa, Moro-oka and Ozaki [401] have determined coordination constants ($K = k_1/k_{-1}$) of olefins with the peroxo molybdenum(VI) complex, $[MoO(O_2)_2HMPA]$, and rate constants k_2 for the decomposition of the olefin complexes for 22 mono olefins according to the mechanism given by equations (254) and (255). The results of this study are shown in Table 12. Surprisingly, the coordination constant increases in this treatment with the extent of alkyl substitution on the double bond – a tendency opposite to that for group VIII metal complexes. Kinetic data are consistent with the original suggestion of Mimoun that oxygen transfer occurs through formation of a five membered metalocycle in a 1,3-dipolar addition mechanism, equation (256).

$$MoO(O_2)_2(C{=}C) \longrightarrow [\text{Mo–O–O–C–C five-membered ring}] \longrightarrow O(O_2)Mo{=}O + \text{C–C epoxide} \qquad (256)$$

The kinetics of the epoxidation of allyl alcohol by peroxo complexes, $[MoO(O_2)_2L_x]$, has been studied in 1,2-dichloroethane between 30 and 60 °C [402]. The reaction rate dependence on the allyl alcohol concentration was described by the Michaelis–Menten equation.

Using labeling experiments, Sharpless and co-workers [403] have found that the epoxide oxygen arises exclusively from the peroxo oxygens as expected from equation (256). These authors, however, suggest alternative mechanisms for the oxygen transfer step which proceed *via* transition states such as XXXV or XXXVI.

Table 12

KINETICS OF $[MoO(O_2)_2HMPA]$ – CATALYZED EPOXIDATION OF OLEFINS[a,b]

Type of substitution	Olefin	Rate constant $k_2 \times 10^3 (s^{-1})$[c]	Coordination constant K(1/mol)[c]	Specific rate at 1.0 mol/l-olefin $k \times 10^5 (s^{-1})$[c]
	C_4			
1-Butene		0.25	0.16	3.4
cis-2-Butene		0.87	0.27	21.5
trans-2-Butene		0.44	0.18	5.3
Methyl propene		0.72	0.34	21.0
	C_5			
1-Pentene		0.20	0.14	2.5
3-Methyl-1-Butene		0.17	0.15	2.2
cis-2-Pentene		1.11	0.23	16.0
trans-2-Pentene		0.43	0.17	4.9
2-Methyl-1-Butene		0.53	0.24	11.5
2-Methyl-2-Butene		1.65	0.39	42.8
Cyclopentene		0.81	0.23	16.4
	C_6			
1-Hexene		0.19	0.13	2.0
3,3-Dimethyl-1-Butene		0.16	0.15	1.8
2-Methyl-1-Pentene		0.50	0.32	12.1
2-Methyl-2-Pentene		1.69	0.51	57.8
2,3-Dimethyl-2-Butene		1.96	0.80	86.2
Cyclohexene		0.89	0.25	17.8
	Others			
4-Methyl Cyclohexene		0.86	0.25	17.7
1-Methyl Cyclohexene		1.22	0.33	30.4
2,4-Dimethyl-2-Pentene		1.00	0.16	10.2
2,4,4-Trimethyl-2-Pentene		0.50	0.10	4.8
2,3-Dimethyl-2-Pentene		2.08	0.77	94.7

[a] Data from Table 1, reference [401].
[b] Reactions at 20 °C in ethylene dichloride.
[c] See equations (254) and (255).

XXXV

XXXVI

A recent report [404] has shown that addition of *tert*-butyl hydroperoxide to $MoCl_5$ in ethanol caused a blue to yellow color change and eventually precipitated a stable compound identified by IR, UV and elemental analysis as a diperoxo (MoO_5)

Table 13

ISOMERIC COMPOSITION OF EPOXIDATIONS OF Δ^4 AND Δ^5 STEROIDS[a]

Steroid epoxidized	Isomer compositions			
	t-AmOOH, $MoCl_5$		MoO_5(HMPA)	
	5α, 6α	5β, 6β	5α, 6α	5β, 6β
3-β-Acetoxyandrost-5-en-17-one	20	80	100	0
3-β-Acetoxycholest-5-ene	45	55	80	20
3-β-Acetoxypirost-5-ene	35	65	95	5
3-β-Acetoxypregna-5,16-dien-20-one	35	65	90	10
3-β-Acetoxy-N-acetyl-Δ^5-solasodine	35	65	95	5
3-(Ethylenedioxy)spirost-5-ene	0	100	0	100

[a] Data from reference [407].

complex. This complex epoxidized cyclohexene prompting the authors to support a previously suggested [405] mechanism in which a peroxo complex is formed *in situ* which is the actual epoxidizing agent in the molybdenum catalyzed epoxidation of olefins.

This hypothesis received recent support in a report by Arakawa and Ozaki [406] who found that when the MoO_3-catalyzed epoxidation of cyclohexene by *tert*-butyl hydroperoxide was carried out in the presence of hexamethylphosphoramide, HMPA, an intermediate peroxo complex could be isolated. Addition of 2,2′-bipyridyl to the reaction mixture caused the peroxo complex, XXXVII, to precipitate. A similar peroxo complex could not be isolated from MoO_3-catalyzed epoxidations carried out in the absence of HMPA suggesting that a different mechanism may be operative for epoxidations in the absence of HMPA.

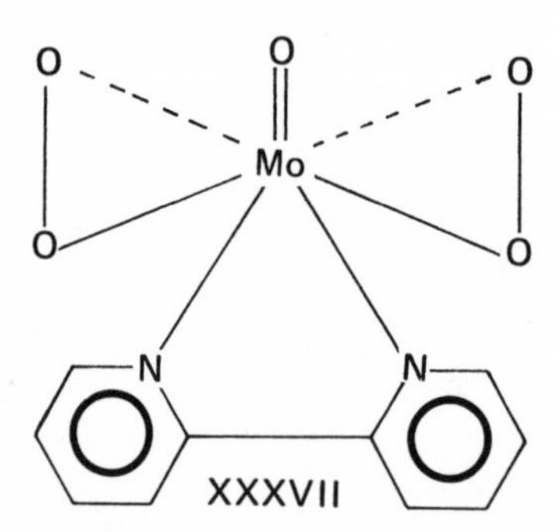

Additional evidence for the existence of different mechanisms for epoxidations by peroxo molybdenum complexes and catalytic epoxidations by hydroperoxides was furnished by recent stereochemical studies [407]. It was found that the epoxidation of several steroids with [MoO_5(HMPA)] gave a different isomer from the $MoCl_5$-catalyzed epoxidation with *tert*-amyl hydroperoxide, Table 13. Furthermore, reactions of $[MoO_5(HMPA)]$ appear to be determined more by steric factors than by the presence of polar groups within the molecule, equation (257).

(257)

From the discussions of this Section it is obvious that, although a large amount of work has been carried out, the mechanisms of molybdenum catalyzed olefin epoxidation are still not completely clear. It is expected, however, that with the intense effort occurring in many laboratories at the present time, current inconsistencies will be rapidly resolved.

16.2.2. EPOXIDATION WITH VANADIUM COMPLEXES

The epoxidation of olefinic hydrocarbons without other coordinating groups is $\sim 10^2$ times slower in the presence of vanadium complexes than with molybdenum catalysts. Nonetheless, the reaction of *tert*-butyl hydroperoxide with an olefin such as cyclohexene in the presence of $[VO(acac)_2]$, $[V(acac)_3]$, $[V(oct)_3]$ or $[VO(n\text{-}BuO)_3]$, is nearly quantitative at 84 °C [408, 386]. Rate laws are consistent with reaction *via* rate determining attack of olefin on a vanadium (V)-hydroperoxide complex. Epoxidations were first order each in olefin and in catalyst but exhibited a Michaelis-like dependency on hydroperoxide, equation (258), where k is a limiting specific rate (at very high ratios of hydroperoxide to catalyst), $[V_0]$ is the total concentration of added vanadium, and Kp is the association constant for the vanadium(V) complex presumed to be the active intermediate.

$$\text{rate} = \frac{k[\text{olefin}][V_0]}{(1/[\text{ROOH}]\,\text{Kp}) + 1} \tag{258}$$

Association constants for the complexes and activation parameters associated with the various vanadium catalysts were found to be different [386] indicating that these catalysts retain, at least in part, their ligand environments when converted to the catalytically active species. Rates of cyclohexene epoxidation in the presence of *tert*-butyl alcohol follow a more complex rate law indicating competitive inhibition by alcohol through two alcohol vanadium complexes [408]. The mechanism postulated [408, 386] for the epoxidation of cyclohexene catalyzed by vanadium complexes is given by equations (259)–(264).

$$VO(acac)_2 \xrightarrow{t\text{-}BuO_2H} -V^V- \quad \text{Catalyst activation; rapid, irreversible} \tag{259}$$

$$-V^V- + t\text{-}BuO_2H \rightleftharpoons -V^V-O(Bu)-O-H \quad \text{Complex formation; rapid, reversible} \tag{260}$$

$$-V^V-O(Bu)-O-H + \text{cyclohexene} \xrightarrow{k} H-\overset{+}{O}(\text{cyclohexane}) + -V^V-O(Bu)^- \quad \text{Heterolysis; rate-determining} \tag{261}$$

$$H-\overset{+}{O}(\text{cyclohexane}) + -V-\overset{-}{O}Bu \longrightarrow O(\text{cyclohexane}) + -V-O(Bu)-H \quad \text{proton transfer; rapid} \tag{262}$$

$$-V-O(Bu)-H + BuO_2H \rightleftharpoons -V-O(Bu)-O-H + BuOH$$

Ligand exchange; rapid, reversible (263)

$$-V^V- \underset{}{\overset{t\text{-}BuOH}{\rightleftharpoons}} -V-O(Bu)-H \underset{}{\overset{t\text{-}BuOH}{\rightleftharpoons}} -V(BuOH)_2 \quad \text{Inhibition; reversible} \tag{264}$$

In contrast to the results obtained with simple olefins, olefins containing alcohol functionality were epoxidized much more rapidly in the presence of vanadium complexes than with molybdenum [409, 410]. The efficiency of the vanadium catalyzed epoxidation of allyl alcohol has been rationalized on the basis of an intermediate complex having a geometry which places the electron-deficient oxygen of the hydroperoxide in the vicinity of the double bond, equation (265).

Table 14

EPOXIDATION OF 2-CYCLOHEXENE-1-ol BY *tert*-BUTYL HYDROPEROXIDE[a,d]

Catalyst	Reaction time, hrs	Conversion of t-BuOOH, %	Yield of t-BuOH, %[b]	Conversion of ene-ol	Yield of epoxyol, %[c]		Ratio *cis-/trans-*
					cis	*trans*	
$VO(acac)_2$	2.5	83	98	69	99	1	99
$V(acac)_3$	2.0	100	89	70	96	1	99
$MoO_2(acac)_2$	2.0	63	76	45	40	19	2.1
$Mo(acac)_3$	3.0	75	61	49	39	21	1.9
$Mo(CO)_6$	1.5	52	87	36	64	29	2.2
$W(CO)_6$	9.5	45	82	40	54	23	2.3

[a] *tert*-Butyl hydroperoxide, 5.0 g, was added to a solution of 0.10 g of the catalyst in 6.0 g of 2-cyclohexene-1-ol. The reaction mixture was stirred at 75 °C for the designated time and analyzed by glpc.
[b] Based on moles of t-BuOOH converted.
[c] Based on moles of 2-cyclohexene-1-ol converted.
[d] Reference [392].

$$ROOH + CH_2{=}CHCH_2OH \rightleftharpoons \text{[cyclic complex: } CH_2{=}CH{-}CH_2{-}O(H){-}V^{n+} \cdots \overset{\delta+}{R O}OH \text{]}$$

$$\text{[complex]} \longrightarrow ROH + \underset{O}{CH_2{-}}CHCH_2OH + V^{n+}\text{cat} \qquad (265)$$

Such geometrical arrangements direct reactions of this type in a highly stereoselective manner [391, 394]. Thus, 2-hydroxypentene-3 gives erythro-2-hydroxy-3,4-epoxypentane exclusively and 2-cyclohexene-1-ol gives *cis*-1,2-epoxycyclohexane-3-ol when epoxidized by *tert*-butyl hydroperoxide in the presence of vanadylacetylacetonate.

It is interesting to note that although the epoxidation of 2-cyclohexene-1-ol by *tert*-butyl hydroperoxide in the presence of vanadium, molybdenum and tungsten complexes gives *cis*-1,2-epoxycyclohexane-3-ol predominantly, the stereoselectivity of the reaction under the conditions listed in Table 14 is dependent on the metal which is used [392]. When vanadium complexes are used, the reaction is almost completely stereoselective whereas complexes of molybdenum and tungsten produce both the *cis*- and *trans*-isomers in a ratio of 2/1, Table 14. Furthermore, the rate of reaction is faster and the epoxy alcohol yield is higher in the vanadium-catalyzed reaction than with molybdenum or tungsten complexes. Since both *cis*- and *trans*-1,2-epoxycyclohexane-3-ol completely retain their stereochemical integrity under reaction conditions, it would appear that the stereoselectivity of these reactions reflect mechanistic characteristics of the metal-catalyzed epoxidation process.

The mechanism suggested by Sheng and Zajacek [409] is consistent with stereochemical findings using vanadium complexes [392]. An intermediate in which oxygen enters from the side of the molecule to which the – OH group is attached will result in the formation of the *cis*-epoxy alcohol in a cyclic system, equation (266). If there is a weaker attraction between the metal center and the allylic OH group, as may be the case in these systems with molybdenum and tungsten complexes, some *trans*-epoxy alcohol could be formed *via* equation (267).

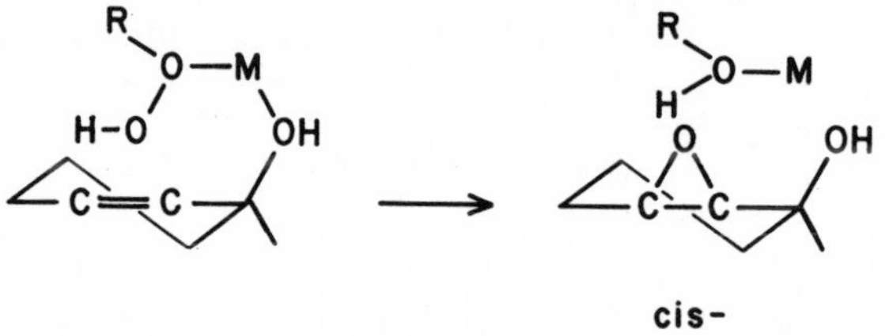

cis- (266)

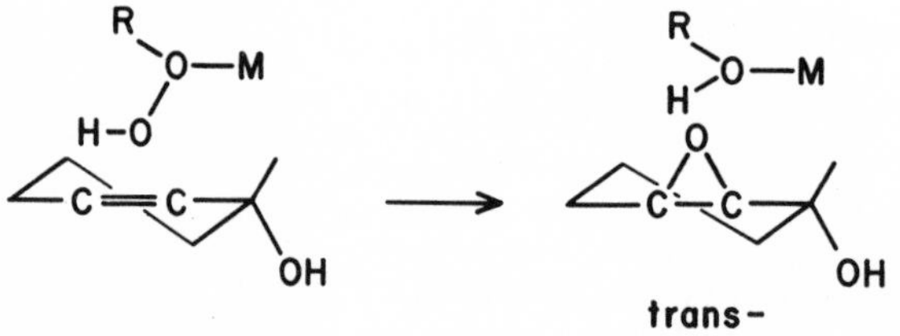

trans- (267)

16.2.3. SUMMARY

Reactions of hydroperoxides with olefins in the presence of a variety of other metal centers have also been investigated. Hydrogen peroxide epoxidizes olefins as well in the presence of oxy compounds of W, Mo, V, Os, Ti, Zr, Th, Nb, Ta, Cr and Ru [411–422]. Although CrO_3-oxidation of olefins has been shown to give epoxides [423–425], chromium complexes such as $[Cr(acac)_3]$ are not particularly effective epoxidation catalysts at elevated temperatures [426]. It has recently been shown [427] that OsO_4 is an effective catalyst for the hydroxylation of olefins by *tert*-butyl hydroperoxide in base equation (268).

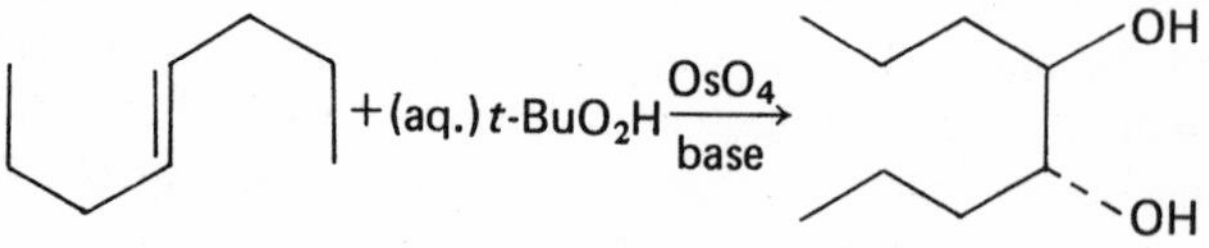

+ (aq.) t-BuO_2H $\xrightarrow[\text{base}]{OsO_4}$ (268)

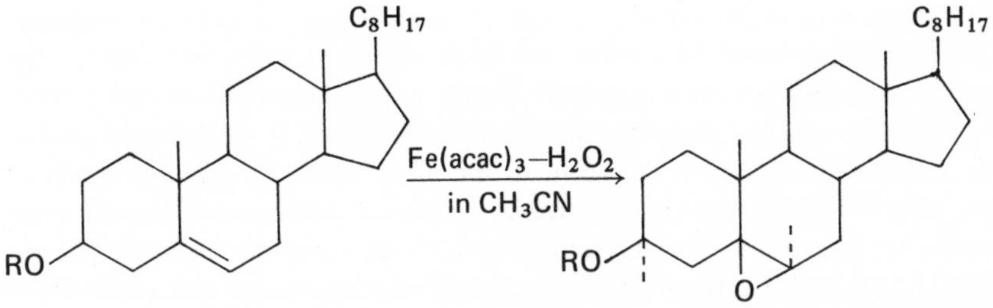

C_8H_{17} C_8H_{17}

RO $\xrightarrow[\text{in } CH_3CN]{Fe(acac)_3-H_2O_2}$ RO O (269)

Stereoselective *β*-epoxidation of cholesterol and its derivatives can be carried out in acetonitrile with $[Fe(acac)_3]$ and H_2O_2 in good yield under mild conditions [428] Eq. (269). Thus, a variety of interesting reactions between hydroperoxides and olefins of which we have mentioned here only a few, are catalyzed by metal complexes in solution.

The properties of a metal complex which make it an efficient catalyst for the epoxidation of an olefin by a hydroperoxide have been considered by Sheldon and Van Doorn [375] and by Gould, Hiatt and Irwin [408]. Such a metal complex will have a high charge, a relatively small size and have low lying *d* orbitals which are at least partly unoccupied [408]. Complexes of metals in low oxidation states (*e.g.* $V(CO)_6$, $Mo(CO)_6$, or $W(CO)_6$) are rapidly oxidized by hydroperoxides to their highest oxidation states [Mo(VI), W(VI), V(V), Ti(IV)] which are the active catalysts [375]. Since the major function of the catalyst is to withdraw electrons from the peroxidic oxygens, an active catalyst must be a good Lewis acid [375]. It is also necessary that the complex does not participate significantly in one electron transfer reactions under strongly oxidizing conditions [408]. Finally, in order for the catalyst to be active it must form complexes which are substitution labile [408].

The Lewis acidity of the transition metal oxides increases in the order CrO_3, $MoO_3 \gg WO_3 > TiO_2$, V_2O_5 [375]. Thus, it is apparent why Mo(VI) is the most effective epoxidation catalyst. Presumably Cr(VI) would also be expected to be a good catalyst, however, Cr(VI) is a strong oxidant and readily causes decomposition of the hydroperoxide [426]. Group VIII complexes in many instances exhibit the same difficulty.

16.3. Catalytic oxidation of olefins

16.3.1. COMPLEXES OF GROUPS VII, VIII AND IB AS CATALYSTS

It has long been known that metal salts and complexes promote the reaction of olefins with oxygen in the liquid phase. Early work ([429] and references cited therein) established that during olefin oxidation in the presence of various copper, cobalt and manganese salts, free radicals arise *via* decomposition of a catalyst-hydroperoxide complex formed from allylic hydroperoxide generated *in situ.* Although the metal modifies the nature of the observed products in many cases, most homogeneous metal-catalyzed oxidations exhibit characteristics of free radical initiated autoxidations.

The oxidation of cyclohexene in the presence of copper, cobalt and manganese carboxylates has continued to receive attention in recent years [430–441]. The stable monomeric products of reaction are largely 2-cyclohexene-1-one and 2-cyclohexene-1-ol with smaller amounts of cyclohexene oxide. Cyclohexenyl hydroperoxide formed by attack of dioxygen on the allylic radical produced by allylic-hydrogen abstraction, has been established to be the reaction intermediate. The product profile has been found to vary somewhat with the metal complex used. It was found [431] that with Co(II) or Mn(II) carboxylates reaction rate and selectivity to 2-cyclohexene-1-one were maximal at 46 °C.

Cyclohexene oxidations in the presence of a variety of acetylacetonates [442] were found to be free radical chain reactions having the same homogeneous propagation steps and yielding as the principle primary product, cyclohexenyl hydroperoxide. The metal catalyzed decomposition of the primary product appeared to give rise to varying amounts of the principle stable monomeric products of oxidation: 2-cyclohexene-1-one, 2-cyclohexene-1-ol and cyclohexene oxide.

Substituted cyclohexenes have also been oxidized in the presence of metal carboxylates and acetylacetonates [443–445]. The autoxidation of (+)-limonene in the presence of cobalt, nickel, manganese and copper acetates gave carvone in yields up to 49% as well as limonene epoxide, *cis*- and *trans*-2,8-*p*-menthadiene-1-ol, and (±)-carveol [444]. Similarly myrcene, β-pinene and α-phellandrene were reacted with oxygen in the presence of cobalt, copper, iron, manganese and nickel acetates and acetylacetonates to produce the expected mixtures of alcohols, aldehydes and ketones [445]. Percherskaya, Logus and Tsybul'ko [446] have reported several interesting examples of asymmetric synthesis during autoxidation using complexes containing optically active ligands. For example, dl-menthene was oxidized at 28 °C in the presence of Mn *d*(–)-mandelate to form (+)-*p*-menthene hydroperoxide whereas use of Mn 1 (+)-mandelate gave the (–)-hydroperoxide.

The use of cobalt and manganese carboxylates to initiate the oxidation of a large number of olefins such as the butenes [447, 448], propylene [449], oleic [450] linoleic [451], and stearic [452, 453] acids or their derivatives and α-methylstyrene [454, 455] is well known. The kinetics of oxidation of α-methylstyrene in the presence of cobaltous and manganous acetylacetonates as well as copper phthalocyanine have been investigated [454, 455]. The results of this study led Kamiya to postulate a mechanism involving formation of radical species by a metal dioxygen complex, equation (270), concurrent with radical generation by hydroperoxide decomposition.

$$MO_2 + \; >C{=}C< \; \longrightarrow MO_2{-}\overset{|}{\underset{|}{C}}{-}\overset{|}{\underset{|}{C}}\cdot \qquad (270)$$

Although the predominant products of oxidations of olefins in the presence of cobalt complexes are usually α, β-unsaturated aldehydes and ketones, a report in the patent literature [456] asserts that cobalt di-(salicylal)-arylene-diimines such as XXXVIII, are efficient catalysts for the direct epoxidation of olefins with molecular oxygen. Octene-1, for example, was reported to give 1,2-epoxyoctane in 57% selectivity at 50% conversion. It is interesting to note that reactions were carried out in acetonitrile, a solvent which has been reported [457] to enhance epoxide yields in reactions of olefins with dioxygen. It has long been known that hydrogen peroxide epoxidizes olefins in the presence of acetonitrile at slightly alkaline pH [458]. A peroxycarboximidic acid intermediate is postulated. Whether this type of chemistry is occurring during autoxidations in acetonitrile [456, 457] is uncertain.

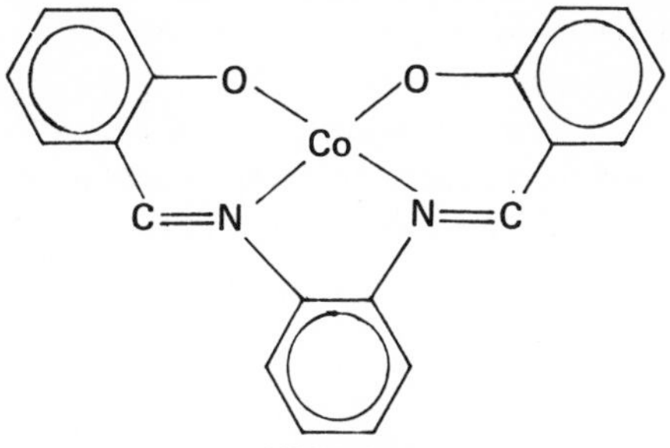

XXXVIII

A recent report by Budnik and Kochi [459] has shown that $[Co(acac)_3]$ promotes the oxidation of olefins such as *tert*-butylethylene, norbornene, and 1,1-dineopentylethylene, which are incapable of undergoing allylic hydrogen abstraction. High yields of epoxides result, however, the reactions exhibit all the characteristics of free radical reactions. These authors propose the pathway shown in equations (271)–(274) to account for their observations.

Initiation: $Co(acac)_3 + O_2 \longrightarrow Co(oxide) + CO_2 + R\cdot$ (271)

Propagation: $R\cdot + O_2 \longrightarrow RO_2^{\cdot}$ (272)

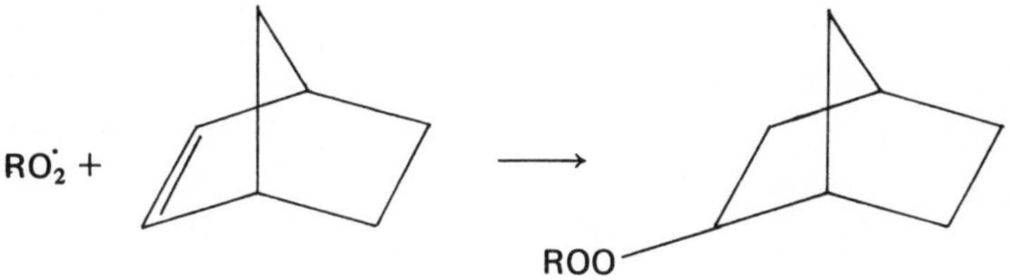

$RO_2^{\cdot}$ + ⟶ ROO (273)

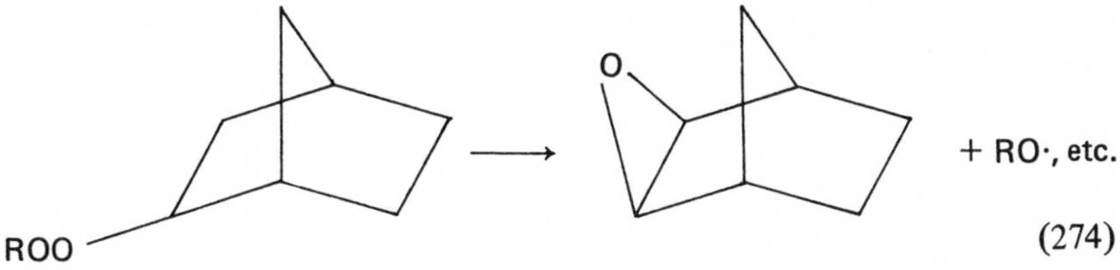

⟶ O + RO·, etc.

ROO (274)

Several years after the complexes: *trans*-$[IrX(CO)(PPh_3)_2]$, (X = Cl, Br, I) were shown to exhibit reversible oxygen carrying properties, Collman, Kubota and Hosking found [460] that compounds of this type as well as the rhodium(I) complex, $[RhCl(Ph_3P)_3]$, promoted the oxidation of cyclohexene to give 2-cyclohexene-1-one predominantly. These authors postulated that 3-cyclohexene hydroperoxide was a likely intermediate and suggested that coordination of oxygen to the metal center was probably involved as a necessary step in the oxidation reaction.

Shortly after Collman's observations it was found [461] that the oxidation of cyclohexene in the presence of $[RhCl(Ph_3P)_3]$ exhibits the characteristics of a radical chain process similar to reactions carried out in the presence of cobalt carboxylates. It was postulated that hydroperoxides formed *in situ* were decomposed by $[RhCl(Ph_3P)_3]$ to radical species in a Haber–Weiss manner, equations (222)–(225), and that these radical species initiated autoxidation.

Since metal dioxygen complexes are usually diamagnetic, it was of interest to determine whether the allylic hydroperoxide formed in autoxidation reactions catalyzed by $[RhCl(Ph_3P)_3]$ arose *via* an 'ene' addition pathway, equation (275), or *via* conventional radical pathways.

O=O M H ⟶ O—O M H (275)

To this end Baldwin and Swallow studied the oxidation of (+)-carvomenthene both photolytically and using $[RhCl(Ph_3P)_3]$ as the catalyst, equations (276) and (277) [462]. The photochemical reaction gave the isomeric carvotanacetols by way of the corresponding hydroperoxide. The rhodium catalyzed oxidation gave a mixture of racemic carvotanacetone, piperitone, and alcohols. Since the intermediate hydroperoxide was found to be optically stable under reaction conditions, these authors conclude that the bulk of the rhodium-catalyzed reaction proceeds *via* a symmetrized free radical intermediate.

HOO, $h\nu$, O_2, H ⟶ $-1/2\ O_2$ HO, H (276)

catalytic, H ⟶ O_2 ⟶ O + O (277)

Although the majority of the evidence [461–463] indicates that the oxidation of cyclohexene in the presence of d^8 and d^{10} complexes proceeds *via* decomposition of hydroperoxides which initiate autoxidation, there is some doubt as to mechanism of formation of initial quantities of the allylic hydroperoxide. Tsyskovskii, Fedorov and Moskovich [464] have suggested that a metal dioxygen complex is capable of interacting directly with the hydrocarbon with C–H bond cleavage to form a peroxy radical, ROO•.

James and Ochiai [472, 473] studied the oxidation of a cyclooctene rhodium(I) complex and found infrared evidence for cyclooctene hydroperoxide as an intermediate. These authors postulated oxygen transfer directly to the olefin within the co-ordination sphere, equation (278), and suggested that such a scheme may account for the formation of 2-cyclohexene-1-one during oxidation of cyclohexene catalyzed by Rh(I) complexes.

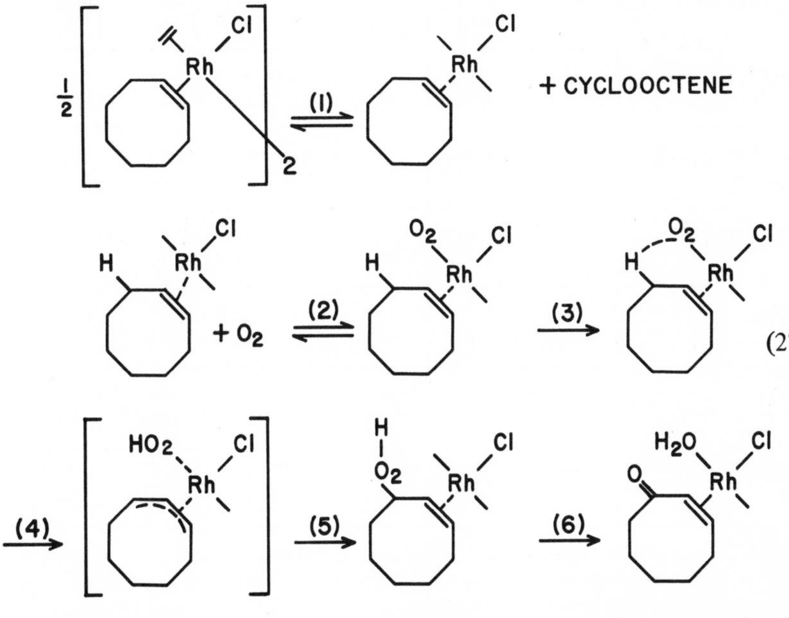

$\frac{1}{2}$ [Rh(Cl)(C₈H₁₄)₂]₂ ⇌(1) Rh(Cl)(C₈H₁₄) + CYCLOOCTENE

+ O_2 ⇌(2) →(3) (278)

→(4) →(5) →(6)

Holland and Milner [474] re-examined this reaction recently in some detail. It was found that reaction of $[(C_8H_{14})_2RhCl]_2$ with dioxygen in benzene at 74 °C gave equimolar amounts of 2-cyclooctene-1-one, cyclooctanone and water, equation (279). Reaction proceeds *via* a pathway which is independent of radical chains and does not involve a Wacker cycle. By contrast, cyclooctene was autoxidized by cobalt naphthenate by a radical pathway to give mainly cyclooctene oxide and this reaction was completely suppressed by the radical inhibitor 2,6-di-*t*-butyl-*p*-cresol. . This inhibitor had no effect on reaction (279). Labeling experiments with added H_2O^{18} showed that water did not take part in this reaction, ruling out a Wacker process.

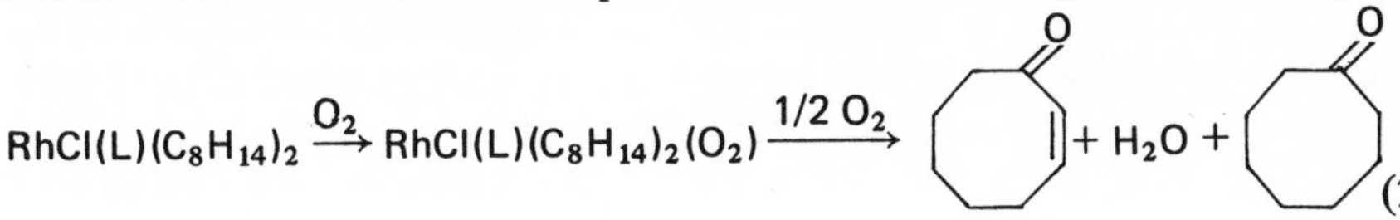

$$RhCl(L)(C_8H_{14})_2 \xrightarrow{O_2} RhCl(L)(C_8H_{14})_2(O_2) \xrightarrow{1/2\ O_2} \text{2-cyclooctene-1-one} + H_2O + \text{cyclooctanone} \qquad (279)$$

Table 15

CYCLOHEXENE OXIDATION[a]

Catalyst			Reaction time (min)	Conversion (%)	Product distribution (%)[b]		
Type	Electron. config.	Molarity ($\times 10^{-3}$)			Cyclo-hexene oxide	2-Cyclo-hexen-1-one	2-Cyclo-hexen-1-ol
$Ir(CO)(PPh_3)_2Cl(O_2)$	d^6	7.38	600	71.5[c], 72.2[d]	0.64	1.01	Traces
$Rh(PPh_3)_3Cl$	d^8	6.66	33	10.5 , 24.0	2.31	13.18	6.28
$Rh(PPh_3)_3Br$	d^8	6.67	540	79.2 , 80.9	0.10	19.68	7.10
		6.73	30	10.0 , 15.5	2.96	20.76	15.73
		6.66	540	30.2 , 45.9	2.16	17.56	7.97
$Rh(PPh_3)_3I$	d^8	9.04	37	10.0 , 23.3	2.28	21.56	17.78
		6.76	525	77.2 , 87.7	0.09	19.16	5.58
$Rh(CO)(PPh_3)_2Cl$	d^8	6.84	37	10.0 , 27.5	1.75	13.76	10.00
		6.84	555	83.6 , 85.5	Traces	17.83	5.98
$Ir(CO)(PPh_3)_2Cl$	d^8	6.64	30	10.0 , 25.0	2.82	15.19	11.05
		6.50	600	82.8 , 71.9	0.06	2.32	0.42
$Pt(PPh_3)_2O_2$	d^8	6.67	210	10.7 , 20.9	1.41	4.10	3.46
		6.65	570	28.1 , 39.2	2.52	5.01	2.53
$Pt(PPh_3)_3$	d^{10}	6.80	240	11.0 , 40.3	0.66	Traces	Traces
		6.73	600	29.7 , 38.7	0.80	0.50	Traces
$Au(PPh_3)Cl$	d^{10}	13.35	450	10.0 , 25.9	1.41	1.73	Traces
		52.07	540	18.2 , 43.5	1.63	1.30	0.55

[a] Table taken from reference 463. [b] Refers to reacted cyclohexene. [c] The first figures refer to O_2, assuming as 100% the formation of cyclohexene hydroperoxide only. [d] The second figures refer to initial cyclohexene.

DMA solutions of the analogous iridium(I) complex, $[IrCl(C_8H_{14})_2]_2$, absorb O_2 irreversibly with expulsion of free cyclooctene to the solution [480]. No evidence of any olefin oxidation products was found. Oxygen uptake measurements in benzene solution gave O_2 : Ir ratios of 0.6 to 1.1 depending on the conditions, and the reaction products exhibited IR bands at 830 cm^{-1} (coordinated dioxygen) and 1260 cm^{-1} (possibly the OH of hydroperoxide). This iridium system was not effective at ambient conditions for olefin oxidation.

Bonnaire and Fougeroux have found [481] that the oxidation of $[IrCl(cod)]_2$, (cod = 1,5-cyclooctadiene) in methylene chloride or benzene gave complexes having the formula: $[IrCl(OH)(cod) \cdot 1/2\ L]_n$ where L is a molecule of solvent. The IR spectra indicate the presence of an OH group in the products.

Fusi, Ugo, Fox, Pasini and Cenini [463] have investigated the oxidation of

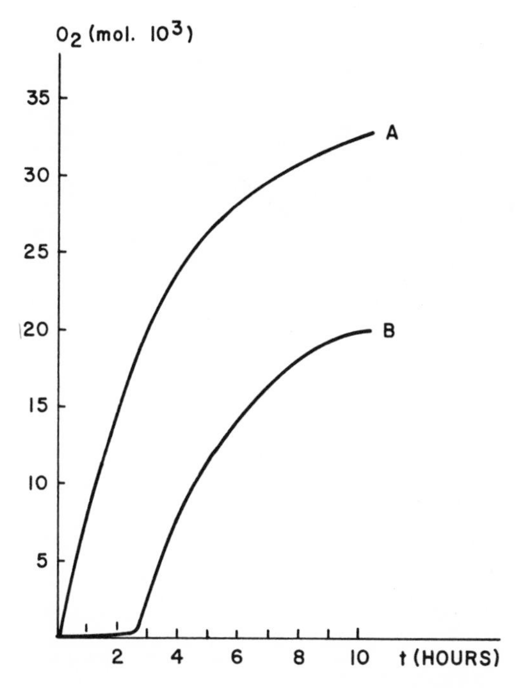

Figure 4. Effect of the presence of hydroperoxide on the $Ir(CO)(PPh_3)_2Cl$ catalysed cyclohexene (39.45×10^{-3} mol) oxidation at 65 ± 1°C. *A*, 6.50×10^{-3} mol/l in cyclohexene with traces of hydroperoxides; *B*, 6.69×10^{-3} mol/l in cyclohexene without hydroperoxides. Figure taken from reference [463].

cyclohexene in the presence of a number of metal complexes [463], Table 15. It was found that d^8 complexes were more active catalysts than d^{10} complexes. No relation was found between oxidation activity and the strength of metal–dioxygen bonding. Activity did not vary with the nature of the anionic ligand, X in complexes of the type $[MX(CO)(Ph_3P)_3]$. Ruthenium complexes such as $[RuCl_2(PPh_3)_3]$ also catalyze the oxidation of cyclohexene with predominant formation of 2-cyclohexene-1-one [463*b*, *c*, *d*].

A unique activity of $[RhCl(Ph_3P)_3]$ during early stages of reaction was noticed, however [463]. For example, minute traces of hydroperoxides totally eliminated the pronounced induction periods which were observed when hydroperoxide free cyclohexene was oxidized in the presence of $[IrCl(CO)(Ph_3P)_2]$, Figure 4. With $[RhCl(Ph_3P)_3]$ as the catalyst, elimination of hydroperoxides had little effect on initial reaction rates, Figure 5.

It can be concluded that in the case of the iridium-catalyzed reactions, radicals must be generated from which allylic hydroperoxides can form; but perhaps another pathway exists for hydroperoxide formation in the presence of the rhodium complex.

Fusi and co-workers [463] also found that dioxygen exhibits a rather unique

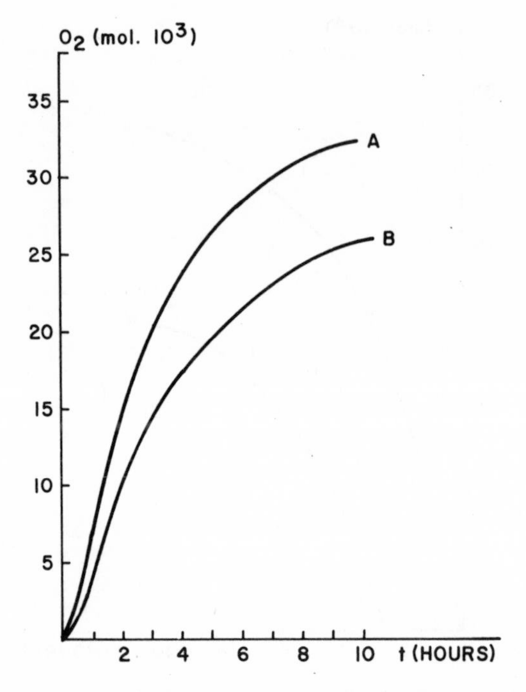

Figure 5. Effect of the presence of hydroperoxide on the $[Rh(PPh_3)_3Cl]$ catalysed cyclohexene (39.45×10^{-3} mol) oxidation at $65 \pm 1°C$. *A*, 6.60×10^{-3} mol/l in cyclohexene with traces of hydroperoxides; *B*, 6.84×10^{-3} mol/l in cyclohexene without hydroperoxides. Figure taken from reference [463].

reactivity in the presence of $[RhCl(Ph_3P)_3]$, equations (280) and (281) but not in the presence of $[IrCl(CO)(Ph_3P)_2]$.

$$\text{(tetraphenylcyclopentadienone)} + O_2 \xrightarrow{RhCl(Ph_3P)_3} PhC(O)C(Ph)=C(Ph)C(O)Ph \tag{280}$$

$$\text{(anthracene)} + O_2 \xrightarrow{RhCl(Ph_3P)_3} \text{(anthraquinone)} \tag{281}$$

Both $[RhCl(Ph_3P)_3]$ and $[IrCl(CO)(Ph_3P)_2]$ decompose cyclohexenyl hydroperoxide [464, 465] to oxygen and organic products, however, the product profile is somewhat different from that observed for reaction of cyclohexene with dioxygen. When cyclohexene is oxidized in the presence of $[RhCl(Ph_3P)_3]$, low yields of cyclohexene oxide are formed together with 2-cyclohexene-1-one and 2-cyclohexene-1-ol. Decomposition of cyclohexenyl hydroperoxide in the presence of $[RhCl(Ph_3P)_3]$ does not give rise to cyclohexene oxide even with added cyclohexene [465]. Thus, cyclohexene oxide forms *via* a different route than epoxidation of cyclohexene by cyclohexenyl hydroperoxide during oxidation. It may proceed *via* a pathway which does not involve fully formed hydroperoxide as suggested by Emanuel and coworkers [466, 467], equation (282).

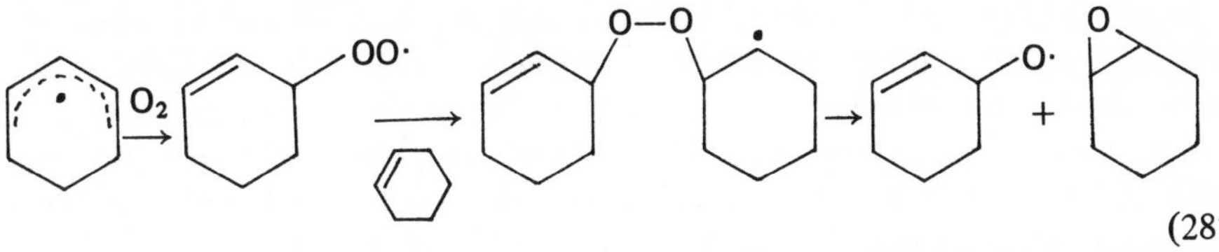

O_2 OO· O—O O· + O

(282)

Similar behavior may be occurring during the oxidative isomerization of nonene-1 in the presence of $[RhCl(CO)(Ph_3P)_2]$ and $[RhCl(Ph_3P)_3]$. Although nonene-1 is oxidized by dioxygen in the presence of $[RhCl(Ph_3P)_3]$ [468], when hydroperoxides are decomposed by rhodium(I) complexes in the presence of nonene-1, no epoxides were reported [469].

Dudley, Read and Walker [475, 123] have reported a novel rhodium(I)-catalyzed oxidation of 1-olefins. Hexene-1, heptene-1 and octene-1 were converted to the corresponding methyl ketones with dioxygen at ambient temperature and pressure in benzene solutions of the complexes: $[RhH(CO)(PPh_3)_3]$ and $[RhCl(PPh_3)_3]$. Methyl ketones are not normally produced in Haber–Weiss initiated radical reactions. Furthermore, radical inhibitors such as hydroquinone or 2,6-di-*t*-butyl-*p*-cresol do not retard methyl ketone formation. Since these authors are unable to detect radical chain processes, they suggest that reactions involve co-oxygenation of coordinated PPh_3 and olefin at the metal center, equation (283).

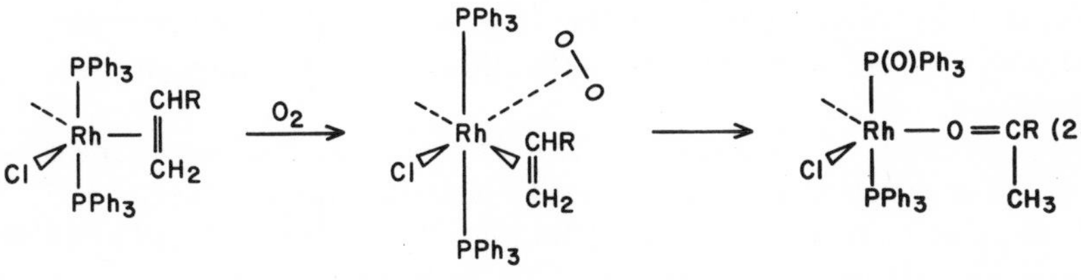

PPh_3 CHR Rh Cl CH2 PPh_3 O_2 → PPh_3 O O Rh CHR Cl CH2 PPh_3 → P(O)Ph3 Rh—O=CR (283) Cl PPh_3 CH_3

The oxidation of tetramethylethylene has also been studied in the presence of the complexes: *trans*-$[MCl(CO)(Ph_3P)_2]$ (M = Rh, Ir) [470]. The oxidation was found to be rapid and quite selective under mild conditions yielding 2,3-dimethyl-2,3-epoxybutane and 2,3-dimethyl-3-hydroxybutene-1 as the major oxidation products, equation (284).

$$2\ (CH_3)_2C{=}C(CH_3)_2 + O_2 \xrightarrow{MCl(CO)(Ph_3P)_2} (CH_3)_2\underset{O}{C{-}C}(CH_3)_2 + (CH_3)_2C(OH){-}C(CH_3){=}CH_2 \qquad (284)$$

The reactions were inhibited by hydroquinone which is consistent with a free radical initiated autoxidation. The reaction of tetramethylethylene was more rapid than was oxidation of less substituted olefins in the presence of the Rh(I) and Ir(I) complexes suggesting that initial coordinative interaction between the olefin and the metal center is not an important factor.

The metal complexes: $[MCl(CO)(Ph_3P)_2]$ (M = Rh, Ir) catalyze epoxidation of tetramethylethylene with *tert*-butyl hydroperoxide in good yield and the selectivity was better (~ 90%) with rhodium than iridium, equation (285). In this case a reasonable mechanism for epoxide formation involves epoxidation of unreacted olefin with the intermediate allylic hydroperoxide, XXXIX. The allylic hydroperoxide was found to reach levels in excess of 11% during the course of the metal catalyzed reactions of tetramethylethylene with oxygen [470], equation (284).

$$(CH_3)_3C{-}OOH + (CH_3)_2C{=}C(CH_3)_2 \xrightarrow{MCl(CO)(Ph_3P)_2} (CH_3)_3C{-}OH + (CH_3)_2\underset{O}{C{-}C}(CH_3)_2 \qquad (285)$$

$$(CH_3)_2C(OOH){-}C(CH_3){=}CH_2$$

XXXIX

Other highly substituted olefins such as 4-methylpentene-2, have been oxidized in the presence of $[IrCl(CO)(Ph_3P)_2]$ and $[RhCl(Ph_3P)_3]$ to give largely epoxide and allylic alcohol [471] and the pathway is probably similar to that for tetramethylethylene. It would seem that in the case of substituted olefins which form stable hydroperoxides, a much larger fraction of the epoxide which is formed arises *via* epoxidation by allylic hydroperoxides. Thus, when allylic hydrogens are present a variety of reactions are possible and often those proceeding through allylic radicals predominate.

The oxidation of styrene, an olefin having no allylic hydrogens for abstraction, has been the subject of studies by several groups of workers [476–479, 125]. Takao and co-workers found that both $[RhCl(Ph_3P)_3]$ [476] and $[IrCl(CO)(Ph_3P)_2]$ catalyze the oxidation of styrene at 110 °C. The iridium complex was active in acetic acid solution giving benzaldehyde and acetophenone in 8/1 ratio as principle products. Turnover numbers (moles product/gram atoms Ir) were ~ 10. Neither styrene oxide nor formaldehyde were reported. The iridium(I) complex was less active in toluene and dioxane but gave acetophenone exclusively in less than stoichiometric amounts (moles acetophenone/gram atoms Ir = 0.5–0.9). The rhodium complex was

Table 16

OXIDATION OF STYRENE IN THE PRESENCE OF Ir and Rh COMPLEXES[a]

Catalyst	Reaction temp, °C	Reaction time, hr	Solvent	Product and yield[b]		
				Benzaldehyde	Styrene oxide	Acetophenone
$IrCl(CO)(PPh_3)_2$	110	8	Acetic Acid	1280	0	152
$IrCl(CO)(PPh_3)_2$	110	8	Dioxane	0	0	55
$IrCl(CO)(PPh_3)_2$	110	8	Toluene	0	0	84
$RhCl(PPh_3)_3$	110	8	Acetic Acid	Trace	Trace	Trace
$RhCl(PPh_3)_3$	110	8	Dioxane	611	161	55
$RhCl(PPh_3)_3$	110	8	Toluene	148	Trace	348
$RhCl(PPh_3)_3$	80	8	Toluene	213	Trace	13

[a] Data from references [125] and [476]. [b] Yields as % based on moles of catalyst used.

active in several solvents but the product profile was highly dependent on the reaction medium. These workers considered this reaction as an example of coordination catalysis and postulated the schematic pathway shown in equation (286). The results of their studies are summarized in Table 16.

$$RhCl(PPh_3)_3 \xrightarrow[O_2]{STYRENE} [Rh(Cl)(PPh_3)_2(O_2)(CH_2{=}CHC_6H_5)] \quad \text{– OXIDATION PRODUCTS} \rightarrow RhCl(PPh_3)_3 \tag{286}$$

It is most interesting to note the change in product profile of the $[RhCl(PPh_3)_3]$-catalyzed oxidation of styrene as a function of temperature, Table 16. At 80 °C far more benzaldehyde is formed than at 110 °C but the yield of acetophenone has become extremely low. Nearly 30 times the amount of acetophenone is formed at 110 °C than is formed at 80 °C.

Takao and co-workers [125, 476] used rather dilute solutions of styrene in toluene (3–10 volume %) and a rather high catalyst/styrene ratio (2×10^{-2} to 6×10^{-3}) and observed turnover numbers rarely exceeding 20 in 8 hr. When more concentrated solutions of styrene (50 volume %) in toluene were used at 75 °C [477], rate and turnover numbers (500–1000) increased as expected, however, the product profile changed considerably. Substantial quantities of styrene oxide are formed which seem to arise *via* the co-oxidation of styrene and formaldehyde which is formed by oxidative cleavage of the double bond [477]. Reaction of styrene with dioxygen in the presence of $[RhCl(CO)(PPh_3)_2]$ and $[IrCl(CO)(PPh_3)_2]$ at 75 °C exhibit induction periods and are severely retarded by radical inhibitors.

The oxidation of styrene, α- and β-methylstyrene and *cis*- and *trans*-stilbene in the presence of $[MX(CO)(PPh_3)_2]$ (M = Rh, Ir; X = Cl, I) was examined further [478]. The catalytic activity of the complexes decreased in the order: RhCl > RhI > IrCl > IrI and the olefin reactivity was: $PhCH = CH_2 > PhCH = CHCH_3$, $PhC(CH_3){=}CH_2$ > *cis*- and *trans*-stilbene. Muto and Kamiya [478*a*] suggest that while cyclohexene oxidation follows radical pathways in the presence of group VIII complexes, α-methylstyrene oxidation depends on the coordinated oxygen molecule. Solvent effects were found to be particularly important, especially in the oxidation of α-methylstyrene in the presence of $[Ru(OH)(NO)(PPh_3)_2(CO)]$ which exhibited unusually high activity in acetic acid [478*a*].

Platinum and palladium phosphine complexes were effective for the oxidation of olefins to carbonyl compounds [478*b*]. Thus, propene at 1.5 l/hr and O_2 at 1.5 l/hr were fed into a flask containing 0.15 g $[Pd(PPh_3)_4]$ and 50 ml benzene at 60 °C for 3 hr to give 2.3 g of acetone. Cyclohexanone was reported to be formed from cyclohexene, and styrene gave a mixture of PhCOMe and $PhCH_2CHO$ [478*b*].

Ruthenium and osmium complexes which most readily cleave the C=C double bond form the highest yields of styrene oxide during oxygenation at 75 °C [477, 482–484], equation (287).

$$O_2 + PhCH = CH_2 \xrightarrow[75^\circ C,\ 3.5\,hr]{(RuCl_2(PPh_3)_3)} \underset{43\%}{PhCHO} + \underset{16\%}{CH_2O} + \underset{26\%}{PhCH\overset{O}{—}CH_2} + \underset{2\%}{PhC(=O)CH_3} \tag{287}$$

The initial step in the oxidation of styrene in the presence of $[RuCl_2(PPh_3)_3]$ appears to be radical initiated oxidative cleavage of the double bond. It was demonstrated that $[RuCl_2(PPh_3)_3]$ efficiently cleaved styrene polyperoxide, a likely intermediate, to give benzaldehyde and formaldehyde [477]. A possible reaction pathway was suggested [482], equations (288)–(292).

$$CH_2{=}CHPh + O_2 \longrightarrow [\text{-(}CH_2CH(Ph)OO\text{)-}]_n \rightarrow PhCHO + CH_2O \tag{288}$$

$$CH_2O + O_2 \rightarrow [HCO_3H] \tag{289}$$

$$[HCO_3H] + PhCH{=}CH_2 \rightarrow PhCH\overset{O}{—}CH_2 + HCOOH \tag{290}$$

$$HCOOH \rightarrow H_2 + CO_2 \tag{291}$$

$$H_2 + 1/2\ O_2 \rightarrow H_2O \tag{292}$$

Both CO_2 and water are by-products of the $[RuCl_2(PPh_3)_3]$-catalyzed oxidation of styrene and the yield of styrene oxide is approximately equal to the yield of benzaldehyde minus that of formaldehyde [482].

Rudel and Maizus [484] found that the nature of the solvent had a sizable effect on the oxidation of styrene catalyzed by $[RuCl_2(Ph_3P)_3]$. The initial rates of oxidation of styrene in various solvents diminish in the order: tetrachloroethane > chlorobenzene > toluene > anisole ≫ acetonitrile. It was postulated that coördinating solvents diminish the rate by preventing coordination of styrene polyperoxide to the ruthenium center and thus inhibit its decomposition. Reactions run in anisole, therefore, show a large buildup of styrene polyperoxide (~ 20%) over that of oxidations in toluene or chlorobenzene (~ 6%). Anisole slows the rate by competitive coordination and acetonitrile forms a stable complex $[RuCl_2(PPh_3)_2(CH_3CN)_2]$.

Rudel and co-workers have reported very interesting oxidation behavior of 1-nonene at 110 °C in the presence of ruthenium(II) complexes such as $[RuCl_2(PPh_3)_3]$ and $[RuCl_3(NO)(PPh_3)_2]$ [485]. In the presence of $[RuCl_2(PPh_3)_3]$ 1-nonene was oxidized to 1,2-epoxynonane in 52% yield and octanoic acid in 30% yield while in the presence of $[RuCl_3(NO)(PPh_3)_2]$ octanoic acid was formed in 39% and the epoxide in 43% yields, equation (293). Similar reactions were reported to occur [485*a*, *b*] when $RuCl_3$ was used as the catalyst.

$$CH_3(CH_2)_6CH=CH_2 + O_2 \xrightarrow{RuCl_2(PPh_3)_3} CH_3(CH_2)_6\underset{\diagdown O \diagup}{C-CH_2} + CH_3(CH_2)_6COOH \qquad (293)$$

Other oxidative cleavages reported to be catalyzed by metal complexes include the oxidation of cyclic ketenes to cyclic ketones in the presence of Co, Mn, and Ag carboxylates [485*c*], and the conversion of aldehydes having CHO attached to secondary carbon atoms to the corresponding ketones using $CuCl_2$ in pyridine [485*d*]. An example of the latter reaction is the conversion of $(CH_3)_2CHCHO$ to $(CH_3)_2C{=}O$ in 75% yield [485*d*].

Products of the oxidative cleavage of olefinic double bonds are more prevalent during Rh(I)-catalyzed oxidation of cinnamaldehyde [486] and styryl acetate [486] than with styrene itself, equations (294) and (295).

$$(H)(Ph)C{=}C(CHO)(H) + O_2 \xrightarrow{RhCl(CO)(PPh_3)_2} PhCHO + [OCHCHO] \qquad (294)$$

$$(H)(Ph)C{=}C(OCCH_3)(H) + O_2 \xrightarrow{RhCl(CO)(PPh_3)_2} PhCHO + CH_3COOH + CH_3OCH \qquad (295)$$

(In equation (295), the OCCH$_3$ group bears C=O, and CH$_3$OCH bears C=O.)

Takao and co-workers [487] propose a mechanism involving a four-center intermediate having olefin and dioxygen in the coordination sphere. Benzaldehyde and a cleavage product which decarbonylates to acetic acid and methyl formate are postulated, equation (296).

$$RhCl(PPh_3)_3 \xrightarrow{-PPh_3} RhCl(PPh_3)_2 \xrightarrow[\text{styryl acetate}]{O_2} \left[\begin{array}{c} \text{H(Ph)C=C(H)–OC(=O)CH}_3 \text{ with O–O–Rh bridge across C=C} \end{array} \right] \xrightarrow{\text{cleavage}} \tag{296}$$

$$\overset{O}{\overset{\|}{H C O C H_3}} + \overset{O}{\overset{\|}{H O C C H_3}} \xleftarrow{-CO} \left[\overset{O\ O}{\overset{\|\ \ \|}{HCOCCH_3}} \right] + PhCHO$$

Farrar, Holland and Milner [479] have found that dioxygen reacts with styrene in the presence of $[RhCl(C_2H_4)_2]_2$ at 110 °C in decalin solutions containing high concentrations of the radical inhibitor, 2,6-di-*t*-butyl-*p*-cresol. This severely inhibited system (~ 1:1 weight ratios of styrene : inhibitor) gave acetophenone and smaller amounts of benzaldehyde by a pathway which does not involve either radical chains or a Wacker cycle. It was shown that styrene oxide is not an intermediate in the oxidation reaction. Turnover numbers of 17 for actophenone and 3 for benzaldehyde were reported. Thus, the small amount of acetophenone formed at 75–80 °C [476–478] and larger quantities formed at 110 °C which are not accounted for by the mechanism postulated in equations (288)–(292) may arise *via* a mechanism involving coordination of both styrene and oxygen to the metal center [479].

The initial rate of acetophenone formation in this system conforms with equation (297). A sequential mechanism is proposed, equation (298), in which an initially formed catalyst-styrene adduct reacts with oxygen.

$$\text{rate} = \frac{C[\text{styrene}][\text{catalyst}][O_2]}{1 + C[O_2]} \tag{297}$$

$$\begin{array}{l} \text{catalyst} + \text{styrene} \rightleftharpoons \text{catalyst-styrene} \\ \text{catalyst-styrene} + O_2 \longrightarrow \text{products} + \text{catalyst} \end{array} \tag{298}$$

In oxygen-saturated solution, the activation energy was found to be 70 kJ mole^{-1}. In contrast to metal-assisted, radical-initiated autoxidations, styrenes having methyl or phenyl substituents at the olefinic positions are resistant to oxidation under these conditions.

Copper(II) triazine complexes promote cyclohexene oxidation *via* free radical pathways involving cyclohexenyl hydroperoxide [488] in a manner not dissimilar to copper salts. Copper(I) complexes, on the other hand have been reported [489] to be capable of promoting direct epoxidation of an olefin to an epoxide in either the liquid or the vapor phase. The Cu(I) compound is thought to form a complex with the olefin. This complex reacts with oxygen to form Cu(II) and the epoxide. Copper(II) may be converted to Cu(I) by reduction in a separate step. Greater than normal yields of epoxides are produced during oxidation of olefins such as cyclohexene in the presence of the gold(I) complex $[AuCl(PPh_3)_3]$ [463, 490]. Acetylene may be stoichiometrically oxidized in water in the presence of a gold(III) triphenylphoshine complex [491]. Reaction products are glyoxal (80% yield based on Au) and a gold(I) triphenylphosphine complex. The gold(III) complex can be regenerated by halogen giving yields up to 250% based on Au [491]. Silver chelates have been shown to be effective catalysts for the liquid phase oxidation of propylene to propylene oxide [491*b*]. Yields of propylene oxide as high as 65% at 7.5% olefin conversion are reported [491*b*]. Thus, the limited information available on complexes of the coinage metals indicates that these complexes are capable of catalyzing interesting oxidation reactions of unsaturated hydrocarbons.

Another class of complexes whose oxidation activity has been investigated is the metalporphyrins [491*a*]. Compounds such as [Co(II)(TPP)], [Rh(III)(CO)(TPP)Cl] and [Rh(III)(TPP)Cl] have been found to catalyze the oxidation of cyclohexene. These catalysts produce substantial amounts of cyclohexenyl hydroperoxide and in the case of [Rh(III)(CO)(TPP)Cl], the hydroperoxide was found to be the sole product of oxidation [491*a*]. The oxidation of olefins in the presence of iron(III) *meso*-tetraphenylporphin has been reported in some detail [491*a*]. Cyclohexene gives the expected products: 2-cyclohexene-1-one (73%), 2-cyclohexene-1-ol (24%) and cyclohexene oxide (3%). Tetramethylethylene gives equimolar amounts of the allylic alcohol and epoxide, equation (284) (45 and 41% selectivity, respectively) and minor amounts of acetone. α-Methylstyrene gives acetophenone in 95% selectivity. Free radical traps instantly quench these reactions and no evidence could be found for complex formation between [Fe(III)(TPP)Cl] and either oxygen or olefin.

In summary, it appears that in most cases of olefin oxidation in the presence of salts or complexes of groups VII, VIII and IB, free radical pathways play a large role. In many cases, this is the only pathway which occurs. The pathway depends both on the nature of the olefin and that of the metal complex. Substituted olefins having allylic hydrogens capable of easy abstraction generally prefer radical pathways. First row metals generally initiate radical pathways. Even in the absence of allylic hydrogens radical pathways are possible and cobalt salts are involved in free radical initiated autoxidation. Rhodium complexes appear to be capable of entering into some unusual reactions which have been described as involving coordination of O_2 and olefin in a catalytic cycle, however, iridum complexes do not react in this manner. Ruthenium complexes are capable of catalyzing oxidative cleavage of double bonds and as a result give high yields of epoxides as a co-product. The nature of this catalysis is not clearly defined. It seems reasonable that coordination catalysis is occurring in some of these systems and it is expected that future work in this area will clarify the situation.

16.3.2. COMPLEXES OF GROUPS IVB, VB AND VIB AS CATALYSTS

We have shown that complexes of groups IVB, VB and VIB taken as a class, tend to differ from complexes of VIIB, VIII and IB regarding reactions of relevance to olefin oxidation. Group VIII complexes tend to react far more readily with molecular oxygen to form dioxygen complexes. Some group VIB dioxygen complexes, however, react smoothly with olefins to give epoxides while group VIII dioxygen complexes do not. Hydroperoxides are decomposed rapidly in the presence of many complexes of groups VIIB, VIII and IB whereas hydroperoxides are considerably more stable in the presence of many of the higher valent complexes of groups IVB, VB and VIB. Olefins are far more easily and selectively epoxidized in the presence of complexes of groups IVB, VB and VIB than in the presence of complexes of groups VIIB, VIII and IB. Thus, it is not surprising that the product profile of oxidations of olefins using the two groups of complexes differs considerably in many instances.

Of these complexes, those most widely used in olefin oxidation have been complexes of molybdenum, vanadium, tungsten and chromium. Because of the well-known epoxidation activity of these complexes, a great effort has been made to selectively epoxidize propylene to the industrially important monomer, propylene oxide [492–504]. Table 17 summarizes the results of several recent studies of this reaction. Both the metal center and the ligand system seem to have an effect on reaction selectivity. Reaction conditions such as oxygen, partial pressure, and temperature as well as solvents are also important. It is generally accepted [492–504] that the group VB and VIB complexes serve to catalyze epoxidation of propylene with peroxidic double bond cleavage products as the epoxidizing agents, equations (299)–(306).

$$CH_2{=}CHCH_2O_2H \longrightarrow CH_3CHO + CH_2O \tag{299}$$

$$CH_2{=}CHCH_2O_2^{\cdot} \longrightarrow CH_3\dot{C}O + CH_2O \tag{300}$$

$$CH_3CHO + O_2 \longrightarrow CH_3\dot{C}O + {\cdot}O_2H \tag{301}$$

$$CH_3\dot{C}O + O_2 \longrightarrow CH_3CO_3^{\cdot} \tag{302}$$

$$CH_3CO_3^{\cdot} + CH_2{=}CHCH_3 \longrightarrow CH_3CO_2^{\cdot} + \underset{\backslash\,O\,/}{CH_2{-}CHCH_3} \tag{303}$$

$$CH_3CO_2^{\cdot} + CH_2{=}CHCH_3 \longrightarrow CH_3CO_2H + CH_2{=}\dot{C}HCH_2^{\cdot} \tag{304}$$

$$CH_3CO_3^{\cdot} + CH_2{=}CHCH_3 \longrightarrow CH_3CO_3H + CH_2{=}CHCH_2^{\cdot} \tag{305}$$

$$CH_3CO_3H + CH_2{=}CHCH_3 \longrightarrow CH_3CO_2H + \underset{\backslash\,O\,/}{CH_2{-}CH{-}CH_3} \tag{306}$$

As expected, minor reaction products are methanol, acetaldehyde, methyl formate, CO_2, CO, and polymeric residue. No allyl alcohol was detected [501].

It is interesting to note that $[Mo(acac)_3]$ catalyzes the oxidative cleavage of

Table 17

EPOXIDATION OF PROPYLENE BY REACTION WITH DIOXYGEN IN THE PRESENCE OF METAL COMPLEXES

Catalyst	Solvent	Temp. °C	PP_{O_2} Atm.	Conversion, %	Epoxide selectivity, %	Ref.
None	Benzene	150	10	9	21	492
AIBN	Benzene	150	10	9	21.5	492
MoO_5(DMF)	Benzene	150	10	8.5	44	492
MoO_5(HMPA)	Benzene	150	10	8	45	492, 493
MoO_5(HMPA)	Heptane	150	15	8	53	493
$Mo(acac)_3$	Benzene	145	3	15	58	492
$MoO_2(acac)_2$	Benzene	165	10	8	53	501
$MoO_2(acac)_2$	Acetonitrile	165	10	8	57	501
$MoO_2(acac)_2$	CH_2Cl_2	165	10	8	70	501
MoO_2(azo 11)[a]	Benzene	150	12	7.5	70	494
MoCl(azo)$_2$[b]	Benzene	150	12	7.5	85	495
WO_5(HMPA)	Benzene	150	10	8.5	44	492
WO_2(azo 11)[a]	Benzene	150	12	7.5	71	494
Cr(azo 11)[a]	Benzene	150	12	7.5	45	494
Co(azo 11)[a]	Benzene	150	12	7.5	46	494
Cu(azo 11)[a]	Benzene	150	12	7.5	31	494

[a] (Azo 11) refers to the azo complex prepared by addition of 2-anisidine-4-sulfone-diethylamine and 6,8-disulfonaphthol(2) acid to the corresponding metal complex. Selectivities may be enhanced in these cases because small amounts of acetaldehyde were added.
[b] (Azo) refers to amino-2-phenolsulfonic acid plus 6,8-disulfonaphthol(2) acid.

cyclopentene smoothly to glutaraldehyde and *trans*-1,2-cyclopentane diol when hydrogen peroxide is used as the oxidant, equation (307).

$$\text{cyclopentene} + H_2O_2 \xrightarrow{Mo(acac)_3} OHC(CH_2)_3CHO + \textit{trans}\text{-1,2-cyclopentanediol} \qquad (307)$$

Heating 32% H_2O_2 in hexamethyphosphoramide, 15.5 g; cyclopentene, 5.1 g; and [$Mo(acac)_3$], 0.8 g; at 60 °C in an autoclave gave 3.36 g glutaraldehyde and 1.10 g *trans*-1,2-cyclopentane diol [501*a*].

A radical pathway going through the allylic hydroperoxide to give allyl alcohol and propylene oxide cannot be the only pathway involved since the maximum epoxide selectivity would be only 50% by this route. DeRuiter [501] concludes that such a pathway is a minor one since allyl alcohol was stable under reaction conditions and its incorporation into oxidation reactions led to no increase in epoxide yields. Even a route such as (299)–(306) would not appear to give selectivities as high as are currently being reported in some systems [504]. Soviet workers describe the direct liquid phase oxidation of propylene to propylene oxide in 89% selectivity at 15% conversion [504], equation (308). If indeed radical pathways are involved, perhaps reactions such as those proposed by Budnick and Kochi [459] might be occurring.

$$CH_3CH{=}CH_2 + O_2 \xrightarrow[170\,°C,\ 50\,atm.]{LaO/SiO_2} CH_3{-}\overset{O}{C}{-}CH_2 \qquad (308)$$

The direct oxidation of a number of olefins other than propylene using Group V and VIB metal complexes have also been reported. Cyclohexene reacts with molecular oxygen in the presence of oxodiperoxo-*bis*(dimethylformamido) molybdenum(V) to give cyclohexene oxide in 57% selectivity at 12% olefin conversion [492]. Ethylene may be oxidized with molecular oxygen to ethylene oxide in the presence of trialkoxychromium(III) complexes [505], hexene-1 to 1,2-epoxyhexane using the same complexes [505], and octene-1 to 1,2-epoxy octane using chromium acetylacetonate [506].

Whereas the major products of the $[C_5H_5Mo(CO)_3]$-catalyzed oxidation of substituted olefins are epoxides and allylic alcohols [390, 392], oxidation of substituted olefins in the presence of vanadium complexes gives rise to epoxy alcohols as the major products [390, 392, 507–511]. When cyclohexene is the olefin used, reaction is observed to occur with a high degree of stereoselectivity [511], equation (309).

$$\text{cyclohexene} + O_2 \xrightarrow{[C_5H_5V(CO)_4]} \text{2,3-epoxycyclohexanol (OH)} \qquad (309)$$

SOLVENT : CH_2ClCH_2Cl (6.0 ml)
CYCLOHEXENE : 5 M
CATALYST : 2×10^{-2} M
OXYGEN : 1.0 L/Hr.

YIELD : 65 %
CONVERSION : 10 %
STEREOSELECTIVITY : ~99 %

It has been noted [390, 392, 512] that the intermediate allylic hydroperoxide is stereoselectively converted to the *cis*-epoxy alcohol in the presence of vanadium complexes. Cross-product experiments [390, 392], equations (310) and (311), experiments measuring relative rates of epoxidation of cyclohexene and 2-cyclohexene-1-ol [390, 392], effects of added 2-cyclohexene-1-ol, and other data [390, 392] indicate that in the case of vanadium-complex catalyzed oxidation of olefins, epoxy alcohols are formed *via* intermolecular epoxidation of allylic alcohols rather than by intramolecular rearrangement of allylic hydroperoxides, equation (312).

$$\text{(allylic alcohol, OH)} + \text{cyclohexenyl–OOH} \xrightarrow{\text{V-complex}} \text{(epoxy alcohol, O, OH)} + \text{cyclohexenyl–OH} \qquad (310)$$

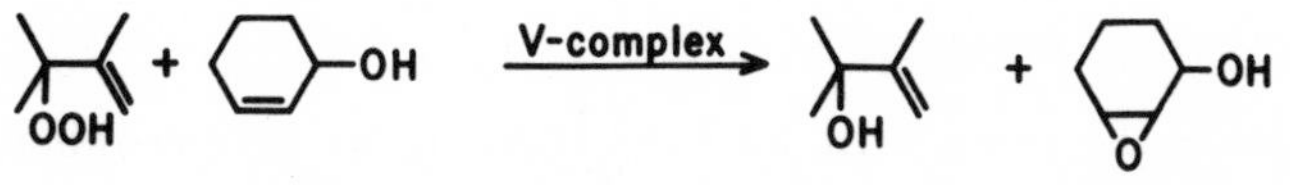

V-complex (311)

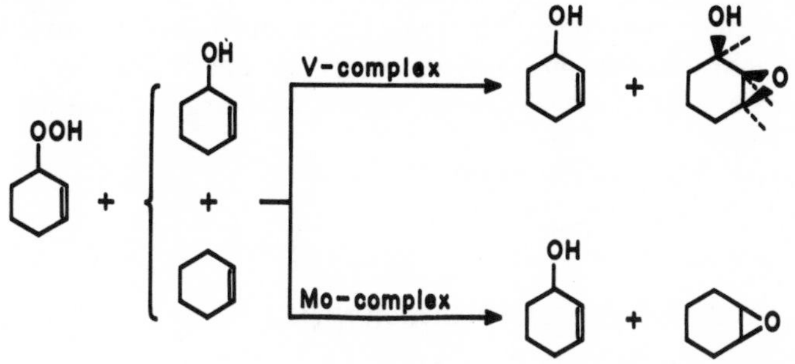

V-complex
Mo-complex (312)

Thus, vanadium complexes preferentially epoxidize small amounts of allylic oxygen species formed *in situ* to give epoxy alcohols whereas molybdenum complexes catalyze epoxidation of the excess of unreacted olefin to give epoxides. The mechanism of vanadium catalyzed epoxidation of allylic alcohols has been discussed in an earlier section.

Titanium tetrachloride has been found to catalyze a unique reaction between triplet oxygen and ergosteryl acetate [512*a*] to first form the 5,8-peroxide which reacts in a facile manner with unreacted olefin to give 6α-chloro-5-hydroxy-5α-ergosta-7,22-dien-3β-yl acetate, equations (313) and (314).

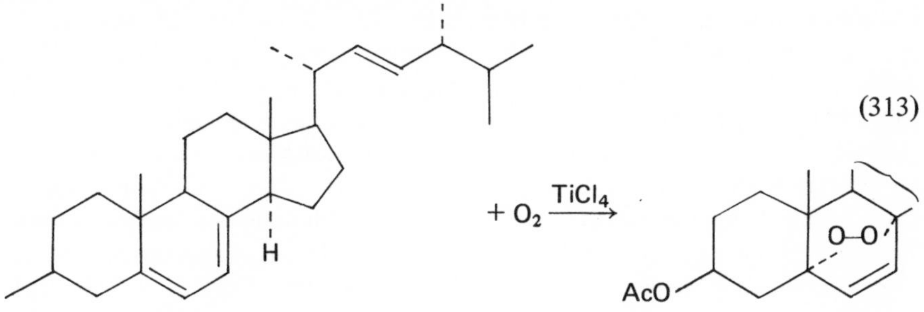

$+ O_2 \xrightarrow{TiCl_4}$ (313)

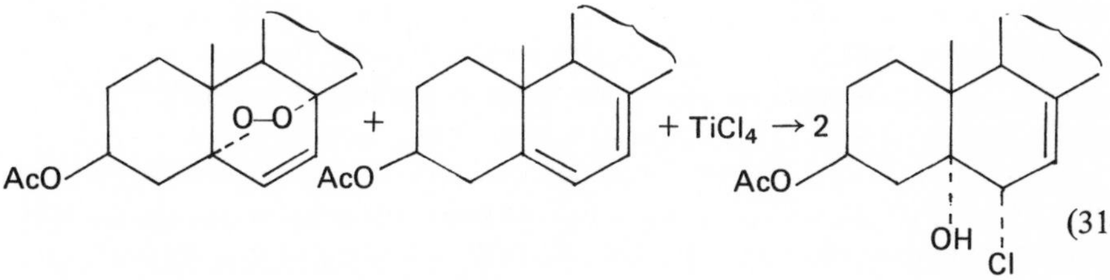

$+ TiCl_4 \rightarrow 2$ (314)

This novel type of peroxide-diene oxygen transfer is most interesting and it is hoped that it will be extended to simpler systems where its general utility could be examined. Reaction (313) is also unusual since the formation of the *endo* peroxide is a characteristic reaction of singlet oxygen.

16.3.3. MIXTURES OF METAL COMPLEXES

Since complexes of the group VIIB, VIII and IB metals are capable of initiating rapid oxidation of olefins by several routes, and since complexes of IVB, VB and VIB are capable of selectively converting initial oxidation products (hydroperoxides and peroxidic species) into desirable products, studies of the use of mixtures of these metal complexes have been undertaken in the hope of obtaining reactions which are both facile and selective [513–517].

Shin and Kehoe have reported [513] a method for the epoxidation of allyl alcohol using $[(CN)_5CoOOCo(CN)_5]^{6-}$ and tungstic acid in acidic aqueous medium. The authors suggest the reaction sequence, equations (315)–(318) for this reaction.

$$2K_3Co(CN)_5 + O_2 \longrightarrow K_3Co(CN)_5OOCo(CN)_5K_3 \quad (315)$$

$$K_3Co(CN)_5OOCo(CN)_5K_3 + H_2O + H^+ \longrightarrow K_3Co(CN)_5OOH + K_2Co(CN)_5.H_2O + K^+ \quad (316)$$

$$K_3Co(CN)_5OOH + H^+ + H_2O \longrightarrow K_2Co(CN)_5{\cdot}H_2O + H_2O_2 + K^+ \quad (317)$$

$$CH_2{=}CHCH_2OH + H_2O_2 \xrightarrow{H_2WO_4} \underset{\diagdown\ O\ \diagup}{CH_2{-}CHCH_2OH} + H_2O \quad (318)$$

A convenient synthetic approach to the selective oxidation of olefins to epoxides uses mixtures of two types of metal complexes classed type A and type B. Complexes of metals of type A (groups VIIB, VIII, IB) tend to efficiently catalyze the oxidation of an olefin to an allylic hydroperoxide or peroxy species but then slowly and unselectively decompose it. Complexes of metals of type B (groups IVB, VB, VIB) are inefficient initiators of oxidation but catalyze rapid epoxidation of olefins by hydroperoxides and alklylperoxy species. Reaction products of oxidations carried out in this way are roughly 1:1 mixtures of allylic alcohols and epoxides. Table 18 lists some representive catalyst systems which have been applied to the oxidation of cyclohexene. In addition, it has been found that mixtures of molybdenum complexes and copper phthalocyanine catalyze oxidation of olefins such as octene-2 or heptene-1 to mixtures of allylic alcohols and epoxides [514]. Fusi, Ugo and Zanderighi noted that the ratio of epoxide/alcohol inexplicably rose much higher than 1:1 (>2:1) in some instances raising the possibility of a different type of epoxide-forming reaction [515] occurring concurrently.

Mixed-metal catalyst systems containing vanadium(IV) or vanadium(III) such as $[RuCl_2(PPh_3)_3]/[(C_5H_5)_2VCl_2]$ give rise to epoxy alcohols as well as epoxides, allylic alcohols and ketones but epoxy alcohol selectivity is inferior to that obtained with low valent vanadium complexes alone, i.e., $[C_5H_5V(CO)_4]$ [392]. This is the result of competing decompostion of the intermediate allylic hydroperoxide by the group VIII metal complex.

In summary, a wide spectrum of reaction chemistry is presented by the metal catalyzed oxidation of olefins. Because of the breadth of the subject matter, some

Table 18

DUAL METAL COMPLEX CATALYSTS FOR THE OXIDATION OF CYCLOHEXENE[a]

				Products, %					
Catalyst A	Catalyst B	Time hr	Conversion %	OOH	OH	O	O	Mole ratio epoxide/ alcohol	Ref.
$RhCl(PPh_3)_3$	–	4	20	76	2	15	7		516
$Pt(PPh_3)_2O_2$	–	0.3	16	39	18	21	4		515
–	$MoO_2(acac)_2$	0.3	4	tr	47	0	53		515
–	$MoO_5(HMPA)(H_2O)$	4	–	–	–	–	–		516
$RhCl(PPh_3)_3$	$MoO_5(HMPA)(H_2O)$	4	9	28	27	12	33	1.2	516
$RhCl(PPh_3)_3$	$MoO_2(acac)_2$	3.3	16	4	37	11	48	1.3	515
$Pt(PPh_3)_2O_2$	$MoO_2(acac)_2$	12	6		46	tr	54	1.2	515
$RuCl_2(PPh_3)_3$	$MoO_2(acac)_2$	1.3	23	6	37	27	30	0.8	515
$IrCl(CO)(PPh_3)_2$	$MoO_5(HMPA)(H_2O)$	4	19	31	27	22	20	0.7	516

[a] Oxidations carried out using a 1:1 molar ratio of A/B.

reactions were not treated and others, I am sure, were inadvertantly overlooked by the author. For this reason, there is listed in the bibliography [327*b*, 521–524] a number of additional reviews of this subject which may aid the reader.

When one is discussing oxidation of olefins with dual metal catalysts it is difficult to avoid the subject of the Wacker reaction. This extremely important synthetic method using various mixtures of metal salts (Pd/Cu, Pd/Fe, Rh/Fe, etc.) allows the preparation of acetaldehyde or vinyl acetate from ethylene equations (1) and (2), as well as many other useful synthetic procedures. Reactions (1) and (2) occur, however, *via* nucleophilic attack of coordinated water or acetate, equations (319)–(321), followed by β-hydrogen elimination.

$$[Pd(OAc)_3]^- + C_2H_4 \longrightarrow [Pd(OAc)_3C_2H_4]^- \tag{319}$$

$$[Pd(OAc)_3C_2H_4]^- \longrightarrow AcOCH_2CH_2PdOAc + AcO^- \tag{320}$$

$$\underset{\substack{|\\ H}}{AcOCH}-CH_2-Pd-OAc \longrightarrow AcOCH{=}CH_2 + HPdOAc \xrightarrow{-HOAc} Pd \tag{321}$$

Palladium is maintained in the active higher valence state by added copper salts which in turn are oxidized by dioxygen. Since this reaction does not involve transfer of dioxygen to the organic substrate we will not consider this chemistry here. Many excellent reviews have considered this subject, however, and may be referred to by the interested reader [327*b*, 516–520].

17. CONCLUDING REMARKS

The oxidation of organic substrates will continue to play a major role in furnishing commercially important and synthetically interesting compounds. As work progresses in this area, metal catalyzed oxidations will continue to replace costly and cumbersome stoichiometric oxidizing agents which were once necessary for effecting selective reactions. This review has dealt mainly with one aspect of metal catalyzed oxidation – the continuing search for homogeneous catalyst systems which will activate molecular oxygen toward reaction with organic substrates. Clearly dioxygen interacts with the majority of metal complexes during homogeneous catalytic oxidations and often forms dioxygen complexes. Ample evidence has been presented for the high reactivity of many dioxygen complexes and thus during catalytic processes metal must transfer coordinated dioxygen to the organic molecule – especially unsaturated substrates. In many instances, however, the extent to which catalytic oxygen transfer proceeds is extremely slight because rapid radical reactions dominate. Clear indications exist, however, that new pathways are possible under conditions wherein radical reactions are disfavored. Indeed with sufficiently active metal complexes, selective oxidations will occur in the presence of large amounts of radical inhibitor. The challenge to the chemists interested in this area is clearly to find the highly active metal complexes which will achieve selective transformations whether they be by radical processes or *via* coordination catalysis. The intensive efforts of many groups of workers in this field insure that this goal will ultimately be achieved for large numbers of new catalytic oxidation reactions.

18. REFERENCES

[1] T. Dumas and W. Bulani, *Oxidation of Petrochemicals: Chemistry and Technology*, John Wiley & Sons, New York (1974).
[2] A. Aguilo, *Advan. Chem. Ser.*, **5**, 321, (1967).
[3] L. Vaska, *Accounts of Chemical Research*, **9**, 175, (1976).
[4] G. Henrici-Olivé and S. Olivé, *Angew. Chem. Internat. Edit.*, **13**, 29, (1974).
[5] J. Valentine, *Chem. Rev.*, **73**, 235, (1973).
[6] L. Klevan, J. Peone and S. Madan, *J. Chem. Ed.*, **50**, 670, (1973).
[7] V. Choy and C. O'Connor, *Coord. Chem. Rev.*, **9**, 145, (1972/73).
[8] R. Stomberg, *Arkiv Kemi*, **24**, 283 (1965), and references cited therin.
[9] J. McGinnety, N. Payne and J. Ibers, *J. Amer. Chem. Soc.*, **91**, 6301 (1969).
[10] J. McGinnety and J. Ibers, *Chem. Commun.*, **235**, (1968).
[11] D. Schwarzenbach, *Helv. Chim. Acta*, **55**, 2990, (1972).
[12] D. Schwarzenbach, *Inorg. Chem.*, **9**, 2391, (1970).
[13] R. Drew and F. Einstein, *Inorg. Chem.*, **12**, 829, (1973).
[14] R. Stomberg and I. Ainalem, *Acta Chem. Scand.*, **22**, 1439, (1968).
[15] R. Stomberg, *Acta Chem. Scand.*, **23**, 2755, (1969).
[16] J. Ibers and S. LaPlaca, *Science*, **145**, 920, (1964).
[17] S. LaPlaca and J. Ibers, *J. Amer. Chem. Soc.*, **87**, 2581, (1965).
[18] W. Spofford and E. Amma, Unpublished results quoted in [3].
[19] J. McGinnety, R. Doedens and J. Ibers, *Inorg. Chem.*, **6**, 2243, (1967).
[20] J. McGinnety, R. Doedens and J. Ibers, *Science*, **155**, 709, (1967).
[21] M. Weininger, I. Taylor and E. Amma, *Chem. Commun.*, **1172**, (1971).
[22] N. Terry, E. Amma and L. Vaska, *J. Amer. Chem. Soc.*, **94**, 653, (1972).
[23] G. Cook, P. Cheng and S. Nyburg, *J. Amer. Chem. Soc.*, **91**, 2123, (1969).
[24] P. Cheng, C. Cook, S. Nyburg and K. Wan, *Can. J. Chem.*, **49**, 3772, (1971).
[25] T. Kashiwagi, N. Yasuoka, N. Kasai, M. Kakudo, S. Takahashi and N. Hagihara, *Chem. Commun.*, **743**, (1969).
[26] W. Schaefer, *Inorg. Chem.*, **7**, 725, (1968).
[27] J. Fritch, G. Christoph and W. Schaefer, *Inorg. Chem.*, **12**, 2170, (1973).
[28] M. Galligaris, G. Nardin, L. Randaccio and A. Ripamonti, *J. Chem. Soc. (A)*, 1069, (1970).
[29] L. Lindblom, W. Schaefer and R. March, *Acta Cryst.*, **B27**, 1461, (1971).
[30] F. Fronczek and W. Schaefer, *Inorg. Chim. Acta*, **9**, 143, (1974).
[31] M. Bennett and R. Donaldson, *J. Amer. Chem. Soc.*, **93**, 3307, (1971).
[32] G. Rodley and W. Robinson, *Nature*, **235**, 438, (1972).
[33] L. Brown and K. Rámond, *Chem. Commun.*, 470, (1974).
[34] R. Marsh and W. Schaefer, *Acta Cryst.*, **B24**, 246, (1968).
[35] W. Schaefer and R. Marsh, *Acta Cryst.*, **B21**, 735, (1966).
[36] G. Christoph, R. Marsh and W. Schaefer, *Inorg. Chem.*; **8**, 291, (1969).
[37] U. Thewalt and R. Marsh, *J. Amer. Chem. Soc.*, **89**, 6364, (1967).
[38] B. Bosnich, W. Jackson, S. Lo and J. McLaren, *Inorg. Chem.*, **13**, 2605 (1974).
[39] M. Calvin, R. Bailes and W. Wilmarth, *J. Amer. Chem. Soc.*, **68**, 2254, (1946).
[40] C. Floriani and F. Caldarazzo, *J. Chem. Soc. A.*, 946, (1969).
[41] C. Busetto, C. Neri, N. Palladino and E. Perrotti, *Inorg. Chim. Acta*, **5**, 129, (1971).
[42] D. Diemente, B. Hoffman and F. Basolo, *Chem. Commun.*, 467, (1970).
[43] H. Diehl, *Iowa State Coll. J. Sci.*, **22**, 271, (1946).
[44] A. Crumblis and F. Basolo, *J. Amer. Chem. Soc.*, **92**, 55, (1970).

[45] A. Sykes and J. Weil, *Progr. Inorg. Chem.*, **13**, 1, (1970).
[46] J. Simplicio and R. Wilkins, *J. Amer. Chem. Soc.*, **89**, 6092, (1967).
[47] F. Miller and R. Wilkins, *J. Amer. Chem. Soc.*, **92**, 2687, (1970).
[48] F. Miller, J. Simplicio and R. Wilkins, *J. Amer. Chem. Soc.*, **91**, 1962, (1969).
[49] J. Bayston, F. Looney and M. Winfield, *Australian J. Chem.*, **16**, 557, (1963).
[50] J. Ellis, J. Pratt and M. Green, *Chem. Commun.*, 781, (1973).
[51] E. Ochai, *J. Inorg. Nucl. Chem.*, **35**, 3375, (1973).
[52] E. Ochiai, *Inorg. Nucl. Chem. Letters*, **10**, 453, (1974).
[53] J. Baldwin and J. Huff, *J. Amer. Chem. Soc.*, **95**, 5757, (1973).
[54] J. Collman and C. Reed, *J. Amer. Chem. Soc.*, **95**, 2048, (1973).
[55] J. Collman, R. Gagne, T. Hubert, J. Marchon and C. Reed, *J. Amer. Chem. Soc.*, **95**, 7868, (1973).
[56] F. Calderazzo, C. Floriani, R. Henzi and F. L'Eplattenier, *J. Chem. Soc. A*, 1378, (1969).
[57] C. Chang and T. Traylor, *J. Amer. Chem. Soc.*, **95**, 5810, (1973).
[58] H. Stynes and J. Ibers, *J. Amer. Chem. Soc.*, **94**, 5125, (1972).
[59] D. Anderson, C. Weschler and F. Basolo, *J. Amer. Chem. Soc.*, **96**, 5599, (1974).
[60] W. Brinigar, C. Chang and T. Traylor, *J. Amer. Chem. Soc.*, **96**, 5597, (1974).
[61] J. Almog, J. Baldwin, R. Dyer, J. Juff and C. Wilkerson, *J. Amer. Chem. Soc.*, **96**, 5600, (1974).
[62] G. Schrauzer and L. Lee, *J. Amer. Chem. Soc.*, **92**, 1151, (1970).
[63] J. Drake and R. Williams, *Nature*, **182**, 1084, (1958).
[64] M. Cowan, J. Drake and R. Williams, *Discuss. Faraday Soc.*, **27**, 217, (1959).
[65] J. Swinehart, *Chem. Commun.*, 1443, (1971).
[66] J. Chien, W. Kruse, D. Bradley and C. Newing, *Chem. Commun.*, 1177, (1970).
[67] S. Kim and T. Takizawa, *Chem. Commun.*, 356, (1974).
[68] C. Weschler, B. Hoffman and F. Basolo, *J. Amer. Chem. Soc.*, **97**, 5280, (1975).
[69] L. Vaska, *Science*, **140**, 809, (1963).
[70] L. Vaska, L. Chen and W. Miller, *J. Amer. Chem. Soc.*, **93**, 6671, (1971).
[71] L. Vaska and L. Chen, *Chem. Commun.*, 1080, (1971).
[72] G. Clark, C. Reed, W. Roper, B. Skelton and T. Waters, *Chem. Commun.*, 758, (1971).
[73] L. Vaska, *Inorg. Chim. Acta*, **5**, 295, (1971).
[74] A. Addison and R. Gillard, *J. Chem. Soc. A*, 2523, (1970).
[75] B. James, F. Ng and E. Ochiai, *Can. J. Chem.*, **50**, 590, (1972).
[76] B. Graham, K. Laing, C. O'Connor and W. Roper, *Chem. Commun.*, 1272, (1970).
[77] K. Laing and W. Roper, *Chem. Commun.*, 1556, 1558, (1968).
[78] D. Christian and W. Roper, *Chem. Commun.*, 1271, (1971).
[79] B. Graham, K. Laing, C. O'Connor, and W. Roper, *J. Chem. Soc. Dalton*, 1237, (1972).
[80] B. Cavit, K. Grundy and W. Roper, *Chem. Commun.*, 60, (1972).
[81] T. Kahn, R. Andal and P. Manoharan, *Chem. Commun.*, 561, (1971).
[82] S. Otsuka, A. Nakamura and Y. Tatsuno, *Chem. Commun.*, 836, (1967).
[83] S. Otsuka, A. Nakamura and Y. Tatsuno, *J. Amer. Chem. Soc.*, **91**, 6994, (1969).
[84] S. Takahashi, S. Sonogashira and N. Hagihara, *J. Chem. Soc. Jap.*, **87**, 610, (1966).
[85] G. Wilke, H. Shott and P. Heimbach, *Angew, Chem. Int. Ed.*, **6**, 92, (1967).
[86] C. Cook and G. Jauhal, *Inorg. Nucl. Chem. Lett.*, **3**, 31, (1967).
[87] J. Birk, J. Halpern and A. Pickard, *J. Amer. Chem. Soc.*, **90**, 4491, (1968).
[88] J. Halpern and A. Pickard, *Inorg. Chem.*, **9**, 2798, (1970).
[89] S. Otsuka, A. Nakamura, Y. Tatsuo and M. Miki, *J. Amer. Chem. Soc.*, **94**, 3761, (1972).
[90] J. Connor and E. Ebsworth, *Adv. Inorg. Chem. and Radiochem.*, **6**, 279, (1964).
[91] M. Orhanovic and R. Wilkins, *J. Amer. Chem. Soc.*, **89**, 278, (1967).
[92] A. Samuni, D. Meisel and G. Czapski, *J. Chem. Soc. Dalton*, 1274, (1972).
[93] H. Chan, *Chem. Commun.*, 1550, (1970).
[94] J. Baldwin, J. Swallow and H. Chan, *Chem. Commun.*, 1407, (1971).
[95] H. Mimoun, I. Seree DeRoch and L. Sajus, *Tetrahedron*, **26**, 37, (1970).
[96] K. Garbett and R. Gillard, *J. Chem. Soc. A*, 1725, (1968).
[97] J. Levinson and S. Robinson, *J. Chem. Soc. A.*, 762, (1971).
[98] T. Nappier and D. Meek, *J. Amer. Chem. Soc.*, **94**, 306, (1972).

[99] M. Stiddard and R. Townsend, *Chem. Commun.*, 1372, (1969).
[100] D. Christian, G. Clark, W. Roper, J. Waters and K. Whittle, *Chem. Commun.*, 458, (1972).
[101] P. Hayward, D. Blake, C. Nyman and G. Wilkinson, *Chem. Commun.*, 987, (1967).
[102] F. Cariati, R. Mason, G. Robertson and R. Ugo, *Chem. Commun.*, 408, (1967).
[103] P. Hayward and C. Nymon, *J. Amer. Chem. Soc.*, **93**, 617, (1971).
[104] R. Ugo, F. Conti, S. Cenini, R. Mason and G. Robertson, *Chem. Commun.*, 1498, (1968).
[105] P. Hayward, S. Saftich and C. Nyman, *Inorg. Chem.*, **10**, 1311, (1971).
[106] K. Sharpless, J. Townsend and D. Williams, *J. Amer. Chem. Soc.*, **94**, 295, (1972).
[107] C. Brown and G. Wilkinson, *Chem. Commun.*, 70, (1971).
[108] C. Brown, D. Georgiou and G. Wilkinson, *J. Chem. Soc., A*, 3120, (1971).
[109] W. Siegl, S. Lapporte and J. Collman, *Inorg. Chem.*, **10**, 2158, (1971).
[110] Y. Iwashita and A. Hayata, *J. Amer. Chem. Soc.*, **91**, 2525, (1969).
[111] K. Grundy, K. Laing and W. Roper, *Chem. Commun.*, 1500, (1970).
[112] D. Schmidt and J. Yoke, *J. Amer. Chem. Soc.*, **93**, 637, (1971).
[113] L. Johnson and J. Page, *Can. J. Chem.*, **47**, 4241, (1969).
[114] C. Giannotti, A. Gaudenier and C. Fontaine, *Tetrahedron Lett.*, **37**, 3209, (1970).
[115] B. Booth, R. Hazeldine and G. Neuss, *Chem. Commun.*, 1074, (1972).
[116] M. Nolte, E. Singleton and M. Laing, *J. Amer. Chem. Soc.*, **97**, 6396, (1975).
[117] J. Birk, J. Halpern and A. Pickard, *Inorg. Chem.*, **7**, 2672, (1968).
[118] S. Takahashi, K. Sonogashira and N. Hagihara, *Nippon Kagaku Zasshi*, **87**, 610, (1966).
[118*a*] E. Stern, *Transition Metals in Homogeneous Catalysis*, Marcel Dekker, Inc., New York, p. 138, (1971).
[118*b*] R. Ugo, *Coord. Chem. Rev.*, **3**, 319, (1968).
[118*c*] L. Malatesta and C. Cariello, *J. Chem. Soc.*, 2323, (1958).
[118*d*] L. Malatesta and M. Angoletta, *J. Chem. Soc.* 1187, (1957).
[119] S. Cenini, A. Fusi and G. Capparella, *J. Inorg. Nucl. Chem.*, **33**, 3576, (1971).
[120] R. Poddar and U. Agarwala, *Inorg. Nucl. Chem. Letters*, **9**, 785, (1973).
[121] B. VanVugt, N. Koole, W. Drenth and F. Kuijpers, *Rec. Trav. Chim.*, **92**, 1321, (1973).
[122] R. Augustine and J. van Peppen, *Chem. Commun.*, 497, (1970).
[123] C. Dudley and G. Read, *Tetrahedron Lett.*, 5273, (1972).
[124] H. Arai and J. Halpern, *Chem. Commun.*, 1571, (1971).
[125] K. Takao, Y. Fujiwara, T. Imanaka and S. Teranishi, *Bull. Chem. Soc. Japan*, **43**, 1153, (1970).
[126] D. Schmidt and J. Yoke, *J. Amer. Chem. Soc.*, **13**, 637, (1971).
[127] J. Drapier and A. Hubert, *J. Organometal. Chem.*, **64**, 385, (1974).
[128] J. Halpern, B. Goodall, G. Khare, H. Lim and J. Pluth, *J. Amer. Chem. Soc.*, **97**, 2301, (1975).
[129] R. Barral, C. Bocard, I. Seree deRoch and L. Sajus, *Tetrahedron Lett.*, 1693, (1972).
[130] R. Barral, C. Bocard, I. Seree deRoch and L. Sajus, *Fr., 2,115,598,* (1972).
[131] R. Barral, C. Bocard, I. Seree deRoch and L. Sajus, *Kinet. Katal.*, **14**, 164, (1973).
[132] F. Moore and M. Larson, *Inorg. Chem.*, **6**, 998, (1967).
[133] N. Sutin and J. Yandell, *J. Amer. Chem. Soc.*, **95**, 4847, (1973).
[134] Y. Doyfman, T. Rakitskaya and T. Sapova, *Tr. Inst. Org. Katal. Elektrokhim., Akad. Nauk Kaz. SSR*, **8**, 59, (1974); *Chem. Abs.*, **82**, 160744s.
[135] D. Sokolskii, Y. Dorfman and I. Kazantseva, *Izv. Akad. Nauk Kaz. SSR, Ser. Khim.*, **22**, 36, (1972); *Chem. Abs.*, **77**, 10091j.
[136] H. Mimoun, C. Bocard and I. Seree deRoch, *Ger. Offen. 2,348,261,* (1974).
[137] S. Takase and S. Tomonori, *Japan, 72 30,162,* (1972).
[137*a*] M. Ledlie and I. Howell, *Tetrahedron Lett.*, 785, (1976).
[138] L. Avdeeva and A. Mashkina, *Neftekhimiya*, **14**, 461, (1974).
[139] J. Trocha-Grimshaw and H. Henbest, *Chem. Commun.*, 1035, (1968).
[140] H. Henbest and J. Trocha-Grimshaw, *J. Chem. Soc., Perkin Trans.*, **1**, 607, (1974).
[141] N. Connon, *Fr., 2,095,718,* (1972).
[142] N. Connon, *Org. Chem. Bull.*, **44**, 1, (1972).
[143] I. Seree deRoch, *Fr., 1,540,284,* (1968).
[144] L. Kuhnen, *Angew. Chem., Int. Ed.*, **5**, 893, (1966).

[145] V. List and L. Kuhnen, *Erdöl* und Kohle, **20**, 192, (1967).
[146] R. Curci, F. Di Furia, R. Testi and G. Modena, *J. Chem. Soc., Perkin II*, 752, (1974).
[147] S. Otsuka and M. Tatsuno, *Japan, 70 19,884*, (1970).
[148] V. Van Rheenen, *Chem. Commun.*, 314, (1969).
[149] T. Ho, *Synth. Commun.*, **4**, 135, (1974).
[149*a*] T. Itoh, K. Kaneda, I. Watanabe, S. Ikeda and S. Teranishi, *Chem. Lett.*, 227, (1976).
[149*b*] E. Balogh-Hergovich and G. Speier, *Reaction Kinetics and Catalysis Letters*, **3**, 139, (1975).
[149*c*] M. Barker and S. Perumal, *Tetrahedron Lett.*, 349, (1976).
[150] T. Itoh, K. Kaneda and S. Teranishi, *Tetrahedron Lett.*, 2801, (1975).
[151] H. Bach, *U.S. 3,719,701*, (1973).
[152] H. Takahashi, T. Kajimoto and J. Tsuji, *Synth Commun.*, **2**, 181, (1972).
[152*a*] T. Kajimoto, H. Takahashi, and J. Tsuji, *J. Org. Chem.*, **41**, 1389, (1976).
[152*b*] J. Tsuji, H. Takajanagi and I. Sakai, *Tetrahedron Lett.*, 1245, (1975).
[153] J. Tsuji, H. Takahashi and T. Kajimoto, *Tetrahedron Lett.*, 4573, (1973).
[154] C. Kramer, G. Davies, R. Davis and R. Slaven, *Chem. Commun.*, 606, (1975).
[155] J. Tsuji, S. Hayakawa and H. Takayanagi, *Chem. Lett.*, 437, (1975).
[156] D. Wagnerova, E. Schwertnerova and J. Veprek-Siska, *Coll. Czek. Commun.*, **38**, 756, (1973).
[157] L. Dohnal and J. Zyka, *Microchem. J.*, **19**, 63, (1974); *Chem. Abs.*, **80**, 108091w.
[158] J. Elsworth and M. Lamchen, *J. S. Afr. Chem. Inst.*, **25**, 85, (1972); *Chem. Abs.*, **77**, 14155x.
[159] R. Kreher and K. Deckardt, *Z. Naturforsch, Teil B*, **29**, 234, (1974); *Chem. Abs.*, **81**, 63548c.
[160] K. Maeda, I. Moritani, T. Hosokawa and S. Murahashi, *Tetrahedron Lett.*, 797, (1974).
[161] N. Faleev, Y. Belokon, V. Belikov and L. Mel'nikova, *Chem. Commun.*, 85, (1975).
[162] F. Benfield and M. Green, *Chem. Commun.*, 1274, (1971).
[163] F. Benfield and M. Green, *J. Chem. Soc. Dalton*, 1245, (1974).
[164] W. McWhinnie, J. Miller, J. Watts and D. Waddan, *Chem. Commun.*, 629, (1971).
[165] W. McWhinnie, J. Miller, J. Watts and D. Waddan, *Inorg. Chim. Acta*, **7**, 461, (1973).
[166] J. Miller, D. Watts and D. Waddan, *Inorg. Chim. Acta*, **12**, 267, (1975).
[167] E. Paniago, D. Weatherburn and D. Margerum, *Chem. Commun.*, 1427, (1971).
[168] J. Vasilevskis, D. Olson and K. Loos, *Chem. Commun.*, 1718, (1970).
[169] Reviewed by W. Nigh in *Oxidation in Organic Chemistry, Part B*, W. Trahanovski, ed., Academic Press, New York (1973).
[170] E. Derouane, J. Braham and R. Hubin, *J. Catal.*, **35**, 196, (1974).
[171] A. Terentev and Y. Mogilyansky, *Dokl. Akad. Nauk, SSSR*, **103**, 91, (1955); *Chem. Abs.*, **50**, 4807, (1956).
[172] K. Kinoshita, *Bull. Chem. Soc. Japan*, **32**, 780, (1956).
[173] K. Kinoshita, *Bull. Chem. Soc. Japan*, **32**, 777, (1956).
[174] A. Terentev and Y. Mogilyansky, *J. Gen. Chem. USSR*, **31**, 298, (1961).
[175] K. Pausacker, *J. Chem. Soc.*, 1989, (1953).
[176] K. Pausacker, *et al.*, *J. Chem. Soc.*, 4003, (1954).
[177] K. Wurthrich and S. Fallab, *Helv. Chim. Acta.*, **47**, 1440, (1970).
[178] K. Wurthrich and S. Fallab, *Helv. Chim. Acta.*, **47**, 1609, (1964).
[179] L. Denisova and E. Denisova, *Isz. Akad. Nauk SSR, Ser. Khim*, **12**, 2220, (1966).
[180] E. Ochiai, *Inorg. Nucl. Chem. Lett.*, **9**, 987, (1973).
[181] L. Kuhnen, *Chem. Ber.*, **99**, 3384, (1966).
[182] M. Sheng and J. Zajacek, *J. Org. Chem.*, **33**, 588, (1968).
[183] G. Tolstikov, U. Dzhemilev, V. Yur'ev, A. Pozdeeva and F. Gerchikova, *Zh. Obshchei Khim.*, **43**, 1360, (1973).
[184] G. Tolstikov, U. Jemilev, V. Jurjev, F. Gershanov and S. Rafikov, *Tetrahedron Lett.*, 2807, (1971).
[185] G. Tolstikov, U. Dzhemilev and V. Yur'ev, *Zh. Obshchei Khim*, **8**, 2204, (1972).
[186] K. Kosswig, *Liebigs Ann. Chem.*, **749**, 206, (1971).
[187] G. Howe and R. Hiatt, *J. Org. Chem.*, **35**, 4007, (1970).
[188] H. DeLaMare, *J. Org. Chem.*, **25**, 2114, (1960).
[189] Halcon Int. Inc., *Belg. Patents 661,500*, (1965) and *668,811*, (1965).

[190] N. Johnson and E. Gould, *J. Amer. Chem. Soc.*, **95**, 5198, (1973).
[191] N. Johnson and E. Gould, *J. Org. Chem.*, **39**, 407, (1974).
[192] I. Kolomnikov, T. Belopotapova, T. Lysyak and M. Vol'pin, *J. Organometal. Chem.*, **67**, C25, (1974).
[193] J. Kiji and J. Furukawa, *Chem. Commun.*, 977, (1970).
[194] G. Mercer, J. Shu, T. Rauchfuss and D. Roundhill, *J. Amer. Chem. Soc.*, **97**, 1967, (1975).
[195] J. Byerly and J. Lee, *Can. J. Chem.*, **45**, 3025 (1967).
[196] R. McAndrew and E. Peters, *Can. Met. Quart.*, **3**, 153, (1964).
[197] S. Nakamura and J. Halpern, *J. Amer. Chem. Soc.*, **83**, 4102, (1961).
[198] C. Costa, G. Mestroni, G. Pellizer and T. Licari, *Inorg. Nucl. Chem. Letters*, **5**, 515, (1969).
[199] E. Hirsch and E. Peters, *Can. Met. Quart.*, **3**, 137, (1964).
[200] A. C. Harkness and J. Halpern, *J. Amer. Chem. Soc.*, **83**, 1258, (1961).
[201] B. R. James and G. L. Rempel, *Chem. Commun.*, 158, (1967).
[202] B. R. James and G. L. Rempel, *J. Chem. Soc. (A)*, 78, (1969).
[203] G. Rosenberg, PhD. thesis, University of British Columbia, (1974).
[204] J. Stanko, G. Petrov and C. Thomas, *Chem. Commun.*, 1100, (1969).
[205] J. Bayston and M. Winfield, *J. Catal.*, **9**, 217, (1967).
[206] L. Lee and G. Schrauzer, *J. Amer. Chem.*, **90**, 5274, (1968).
[207] J. Bercaw, L. Goh and J. Halpern, *J. Amer. Chem. Soc.*, **94**, 6535, (1972).
[208] T. Kruck and M. Noach, *Chem. Ber.*, **97**, 1693, (1964).
[208*a*] D. Darensbourg and D. Drew, *J. Amer. Chem. Soc.*, **98**, 275, (1976).
[209] W. Friedrich, *Z. Naturforsch, B.*, **25**, 1431, (1970).
[210] J. Nicholson, J. Powell and B. Shaw, *Chem. Commun.*, 174, (1966).
[211] J. Powell and B. Shaw, *J. Chem. Soc. (A)*, 583, (1968).
[212] W. Dent, R. Long and A. Wilkinson, *J. Chem. Soc.*, 1585, (1964).
[213] V. Likholobov, V. Zudin, N. Eremenko and Y. Ermakov, *Kinet. Katal.*, **15**, 1613, (1974).
[214] W. Lloyd and D. Rowe, *U.S. 3,849, 336*, (1975).
[215] T. Imanaka, S. Matsumoto, S. Wakimura and S. Teranishi, *Kogyo Kagaku Zasshi*, **74**, 1071, (1971); *Chem. Abs.*, **75**, 67955g (1972).
[216] T. Imanaka, S. Matsumoto, S. Wakimura and S. Teranishi, *Kogyo Kagaku Zasshi*, **74**, 2222, (1971).
[217] V. Avdeev, L. Kozhevina, K. Matveev, I. Ovsyannikova, N. Eremenko and L. Rachkovskaya, *Kinet. Katal.*, **15**, 935, (1974).
[218] N. Eremenko, K Matveev, L. Rachkovskaya, S. Katanakhova and S. Mel'gunova, *Kinet. Katal.*, **12**, 147, (1971).
[219] N. Eremenko and K. Matveev, *Kinet. Katal.*, **8**, 538, (1967).
[220] V. Markov, V. Golodov and A. Fasman, *Izd. Sib. Otd. Akad Nauk SSSR, Ser. Khim. Nauk*, 36, (1968); *Chem. Abs.*, **70**, 6864b, (1969).
[221] D. Fenton and P. Steinwand, *J. Org. Chem.*, **39**, 701, (1974).
[222] W. Gaenzler, K. Klaus and G. Schroeder, *Ger. Offen. 2,213,435*, (1973).
[223] H. Verter, *The Chemistry of the Carbonyl Group*, Chapter 2, Vol. 2, Interscience Publishers, New York, (1970).
[224] J. McNesby and C. Heller, Jr., *Chem. Rev.*, **54**, 325, (1954).
[225] C. Bawn and J. Jolley, *Proc. Roy. Soc. (London)*, **A237**, 297, (1956).
[226] F. Marta, E. Boga and M. Matok, *Discuss. Faraday Soc.*, **46**, 173, (1968).
[227] E. Boda and F. Marta, *Acta Chim. (Budapest)*, **78**, 105, (1973).
[228] E. Boga, I. Kiricsi, A. Deer and F. Marta, *Acta Chim. (Budapest)*, **78**, 89, (1973).
[229] E. Boga and F. Marta, *Acta Chim. (Budapest)*, **80**, 333, (1974).
[230] E. Boga and F. Marta, *Kem. Kozlem.*, **41**, 297, (1974); *Chem. Abstr.*, **81**, 497374m, (1974).
[231] A. Ivanov, T. Grimalovskaya and L. Ivanova, *Zh. Fiz. Khim.*, **49**, 893, (1975); *Chem. Abstr.*, **83**, 42967q, (1975).
[231*b*] A. Ivanov and O. Bagira, *Zh. Prikl. Khim. (Leningrad)*, **48**, 243–7; *Chem Abstr.*, **82**, 111302n, (1975).
[232] N. Digurov and V. Sedlyarov, *Tr. Mosk. Khim.-Technol. Inst.*, **66**, 76, (1970); *Chem.*

Abstr., **75**, 76334f, (1971).
[233] S. Imamura and Y. Takegami, *Kogyo Kagaku Zasshi,* **74**, 2490, (1971); *Chem. Abstr.*, **76**, 45483v, (1972).
[234] A. Ivanov and V. Antonyuk, *Zh. Org. Khim.*, **11**, 756, (1975); *Chem. Abstr.*, **83**, 9377z, (1975).
[235] N. Ota and T. Imamura, *Japan, 71 05,687,* (1971).
[236] N. Ota and T. Imamura, *Japan, 71 05,688,* (1971).
[237] Y. Fujita and S. Higasa, *Japan, 73 30,255,* (1973).
[238] E. Hesse and H. Steger, *Ger. (East), 60,553,* (1968).
[239] T. Matsuzaki, J. Imamura and N. Ohta, *Kogyo Kagaku Zasshi,* **71**, 706, (1968).
[239*a*] C. Bawn, F. Hobin and L. Raphael, *Proc. Roy. Soc. (London),* **A237**, 297, (1956).
[239*b*] M. Vinogradov, R. Kereselidze and G. Nikishin, *Izv. Akad. Nauk SSR, Ser. Khim.*, **6**, 1195, (1971).
[240] T. Motoda, T. Koyama and K. Sakai, *Japan, 73 01,366,* (1973).
[241] S. Sakai, N. Kasuag and Y. Ishii, *Kogyo Kagaku Zasshi,* **72**, 1687, (1969).
[242] Y. Ishii, S. Sakai and N. Kasuga, *Japan, 71, 19,931,* (1971).
[243] M. Manakov, A. Litovka and N. Lebedev, *Izv. Vyssh. Ucheb. Zaved., Khim. Khim. Tikhnol.*, **16**, 975, (1973); *Chem. Abstr.*, **79**, 77793z, (1973).
[244] N. Lebedev, M. Manakov and A. Litovka, *Kinet. Katal.*, **15.**, 15, 791, (1974).
[245] M. Ladhabhoy and M. Sharma, *J. Appl. Chem.*, **19**, 267, (1969).
[246] M. Ladhabhoy and M. Sharma, *J. Appl. Chem.*, **20**, 274, (1970).
[247] S. Yamaguchi, H. Nakajima, H. Kimura and H. Nishimaru, *Japan, 74 35,317,* (1974).
[248] S. Yamaguchi, H. Nakajima, H. Kimura and H. Nishimaru, *Japan, 73 97,809,* (1973).
[249] M. Gay, *Ger. Offen., 2,136,481,* (1972).
[250] Soc. Des Usines Chim. Rhone-Poulenc, *Fr. Demande 2,141,441,* (1973).
[251] H. Inoue, Y. Kida and E. Imoto, *Bull. Chem. Soc. Jap.*, **40**, 184, (1967).
[252] H. Inoue, Y. Kida and E. Imoto, *Bull. Chem. Soc. Jap.*, **41**, 692, (1968).
[253] K. Takagi and I. Takaharu, *Japan, 75 19,717,* (1975).
[254] E. Brasset, *Fr. Demande, 2,069836,* (1971).
[255] T. Ishikawa, T. Tezuka, K. Kaneko, T. Hayakawa and Y. Kin, *Yukagaku,* **15**, 16, (1966); *Chem. Abstr.*, **66**, 10338r, (1966).
[256] I. Sulima and A. Klyuchkivs'kii, *Visn. L'viv. Politekh. Inst.*, **10**, 37, (1966); *Chem. Abstr.*, **69**, 3055b, (1967).
[257] T. Takeuchi, H. Shimono and A. Tanabe, *Yukugaku,* **20**, 25, (1971).
[258] A. Misono, T. Osa, Y. Ohkatsu and M. Takeda, *Kogyo Kagaku Zasshi,* **69**, 2129, (1966); *Chem. Abstr.*, **67**, 1093h, (1966).
[259] Y. Ohkatsu, T. Osa and A. Misono, *Bull. Chem. Soc. Jap.*, **40**, 2116, (1967).
[260] H. Volger and W. Brackman, *Rec. Trav. Chim. Pays-Bas,* **85**, 817, (1966).
[261] H. Volger and W. Brackman, *Ibid.*, **84**, 1203, (1965).
[262] H. Volger and W. Brackman, *Ibid.*, **84**, 1233, (1965).
[263] H. Volger and W. Brackman, *Ibid.*, **84**, 1017, (1965).
[264] A. Liu, *Brit., 1,119,728,* (1968).
[265] H. Jun-Ichi, S. Yuasa, N. Yamazo, I. Mochida and T. Seiyama, *J. Catal.*, **36**, 93, (1975).
[266] H. Jun-Ichi, M. Aramaki, N. Yamazo and T. Seiyama, *Kyushu Daigaku Kogaku Shuho,* **47**, 332, [1974]; *Chem. Abstr.*, **81**, 135279r, (1974).
[267] J. Byerly and W. Teo, *Can. J. Chem.*, **47**, 3355, (1969).
[268] A. Vaskelis and J. Kulsyte, *Liet. TSR Moksulu Akad. Darb. Ser. B,* **2**, 3, (1968); *Chem. Abstr.*, **70**, 28155y, (1969).
[269] V. Shanker and M. Singh, *Indian J. Chem.*, **6**, 702, (1968); *Chem. Abstr.*, **70**, 114353t, (1969).
[270] V. Singh, M. Gangwar, B. Saxena and M. Singh, *Can. J. Chem.*, **47**, 1051, (1969).
[271] V. Komissarov and E. Denisov, *Zh. Fiz. Khim.*, **43**, 769, (1969).
[272] V. Komissarov and E. Denisov, *Zh. Fiz. Khim.*, **44**, 390, (1970).
[273] V. Komissarov and E. Denisov, *Neftekhimiya,* **8**, 595, (1968).
[274] V. Komissarov and E. Denisov, *Neftekhimiya,* **7**, 420, (1967).
[275] M. Saitova and V. Komissarov, *Kinet. Katal.*, **13**, 496, (1972).
[276] R. Kozlenkova, V. Kamzolkim and A. Bashkirov, *Neftekhimiya,* **10**, 707, (1970).
[277] H. Den Hertog, Jr. and E. Kooyman, *J. Catal.*, **6**, 357, (1966).

[278] S. Suzuki, T. Tokumaru, E. Ando and Y. Watanabe, *Japan Kokai, 74, 00,235,* (1974).
[279] P. Sukhopar, B. Zubko and K. Chervinskii, *Zh. Prikl. Khim.*, **47**, 115, (1974).
[280] S. Kamath and S. Chandalia, *J. Appl. Chem. Bio-technol.*, **23**, 469, (1973).
[281] E. Baranova and K. Chervinskii, *Khim. Technol.*, **19**, 151, (1971).
[282] K. Chervinski and V. Mal'tsev, *Neftekhimiya,* **7**, 264, (1967).
[283] N. Sakai, M. Ogawa and M. Kitabatake, *Japan, 69 12,128* (1969).
[284] M. Kusunoki and M. Ogawa, *Japan, 69 05,858,* (1969).
[285] M. Ogawa, M. Kusunoki and M. Kitabatake, *Japan, 69 26,283* (1969).
[286] R. Barker, *et al., U.S. 3,234,271,* (1968).
[287] E. Yasui, *et al., Brit. 1,169,777,* (1969).
[288] R. Lidov, *U.S. 3,361,806,* (1968).
[289] Y. Ishimoto, H. Togawa and S. Nakahachi, *Japan, 72 26,768,* (1972).
[290] H. Charman, *Brit. 1,114,885,* (1968).
[291] K. Takagi and T. Ishida, *Ger. Offen., 2,124,712,* (1971).
[292] Ashai Chemical Industry Co., *Brit. 1,103,885,* (1968).
[292*a*] V. Serov, A. Seroezhko, V. Proskuryakov and V. Potekhin, *Zh. Prikl. Khim.*, **48**, 252, (1975).
[293] I. Korsak, V. Agabekov, and N. Mitskevich, *Neftekhimiya,* **15**, 130, (1975).
[294] Y. Yamiya, *Kogyo Kagaku Zasshi,* **74**, 1811, (1971).
[295] A. Semenchenko, V. Solyanikov and E. Denisov, *Kinet. Katal.*, **13**, 1153, (1972).
[296] C. Gardner, H. Gilbert and W. Morris, *Brit. 1,250,192,* (1971).
[297] W. Morris, *Ger. Offen., 1,912,878,* (1969).
[298] W. Morris, *Ger. Offen., 2,037,189,* (1971).
[299] Z. Zamchuk and O. Kuznetzova, *Zh. Prikl. Khim.*, **48**, 859, (1975).
[300] K. Tanaka, *et al., Kogyo Kagaku Zasshi,* **73**, 938, (1970).
[301] G. Koshel and M. Farberov, *Uch Zap. Yaroslavskogo Technol. Inst. Khim. i Khim. Tekhnolog.*, **27**, 39, (1971).
[302] V. Belov and L. Kheifits, *Usp Khim.*, **25**, 969, (1956).
[303] D. Welton, *Neth. Pat. 7,309,271,* (1974).
[304] H. Kojima, S. Takahashi and N. Hagigara, *Tetrahedron Lett.*, 1991, (1973).
[305] H. Kojima, S. Takahashi, H. Yamayaki and N. Hagihara, *Bull. Chem. Soc. Japan,* **43**, 272, (1970).
[306] H. Kojima, S. Takahashi and N. Hagihara, *Chem. Commun.*, 230, (1973).
[307] C. Cullis and A. Fish, *Chemistry of the Carbonyl Group, Chapter 2, Carbonyl-Forming Oxidations,* S. Patai, Ed., John Wiley & Sons, New York, (1966).
[308] W. Brackman, *U.S., 2,883,426,* (1959).
[309] F. Steinbach and H. Schmidt, *J. Catal.*, **29**, 515, (1973).
[310] A. Savitskii, S. Skachilova, R. Vlaskina, A. Gaevskii and G. Ratushenko, *USSR, 240,696,* (1969).
[311] M. M. Cocivera, C. Fyfe, S. Vaish and H. Chen, *J. Amer. Chem. Soc.*, **96**, 1611, (1974).
[312] P. Camerman and J. Hanotier, *Fr., 2,094,808,* (1972).
[313] P. Camerman and J. Hanotier, *Fr., 2,095,160,* (1972).
[314] S. Matsuda, A. Uchida and T. Yamazi, *Nippon Kagaku Kaishi,* 296, (1973).
[315] V. Sapunov, E. Selyutina, O. Tolchinskaya and N. Libedev, *Kinet. Kata.*, **15**, 605, (1974).
[316] A. Savatskii, *Zh. Obsch. Khim.*, **44**, 1548, (1974).
[317] J. Groves and M. Van Der Puy, *J. Amer, Chem. Soc.*, **96**, 5274, (1974).
[318] J. Groves and M. Van Der Puy, *Tetrahedron Lett.*, 1949, (1975).
[319] J. Groves and M. Van Der Puy, *J. Amer. Chem. Soc.*, **97**, 7118, (1975).
[320] G. Schreyer, W. Schwarze and W. Weigert, *Ger. Offen., 2,052,815.*
[321] U. Zeidler and H. Lepper, *Ger. Offen. 2,256,888.*
[322] U. Zeidler, H. Lepper and W. Stein, *Fette, Seifen. Anstrichm.*, **76**, 260, (1974); *Chem. Abs.*, **81**, 91007u, (1974).
[323] H. Connon, B. Sheldon and K. Harding, *J. Org. Chem.*, **38**, 2020, (1973).
[324] A. Van den Besselaer, J. Lubach and W. Drenth, *Rec. Trav. Chim. Pays-Baz,* **93**, 108, (1974).
[325] T. Rutledge, *Ger. Offen. 2,360,269,* (1974).
[326] G. Sosnovsky and D. Rawlinson, *Organic Peroxides,* Vol. II, D. Swern, Ed., John Wiley

& Sons, Inc., New York, p. 153, (1971).
[327] I. Skibida, *Uspkhi Khimii,* **44,** 1729, (1975).
[327a] R. Sheldon and J. Kochi, *Advan. Catal.,* **25,** 272, (1976).
[328] R. Hiatt, K. Irwin and C. Gould, *J. Org. Chem.,* **33,** 1430, (1968).
[329] N. Indictor and T. Jochsberger, *J. Org. Chem.,* **31,** 4271, (1966).
[330] M. Dean and G. Skirrow, *Trans. Faraday Soc.,* **54,** 349, (1958).
[331] W. Richardson, *J. Amer. Chem. Soc.,* **87,** 247, (1965).
[332] W. Richardson, *J. Amer. Chem. Soc.,* **87,** 1096, (1965).
[333] W. Richardson, *J. Org. Chem.,* **30,** 2804, (1965).
[334] W. Richardson, *J. Amer. Chem. Soc.,* **88,** 973, (1966).
[335] K. Kirschke and H. Oberender, *Z. Chem.,* **9,** 105, (1969).
[336] B. Booth, R. Haszeldine and G. Neus, *Chem. Commun.,* 1074, (1972).
[337] R. A. Sheldon, *Chem. Commun.,* 788, (1971).
[338] N. Emanuel, E. Denisov and Z. Maizus, *Liquid Phase Oxidation of Hydrocarbons,* English Translation, Plenum Press, New York, p. 51 (1967).
[339] R. Hiatt, T. Mill, K. Irwin and J. Castleman, *J. Org. Chem.,* **33,** 1421, (1968).
[340] R. Hiatt, J. Clipsham and T. Visser, *Can. J. Chem.,* **42,** 2754, (1964).
[341] J. Kochi and F. Rust, *J. Amer. Chem. Soc.,* **83,** 2018, (1961).
[342] J. Kochi, *J. Amer. Chem. Soc.,.* **84,** 1193, (1962).
[343] J. Kochi and P. Mocadlo, *J. Org. Chem.,* **30,** 1134, (1965).
[344] W. deKlein and E. Kooyman, *J. Catal.,* **4,** 626, (1965).
[345] B. Booth, R. Haszeldine and G. Neuss, *J. Chem. Soc., Perkin I,* 209, (1975).
[346] H. Berger and A. Bickel, *Trans. Faraday Soc.,* **57,** 1325, (1961).
[347] L. Matienko, I. Skibida and Z. Maizus, *Izv. Akad. Nauk SSR, Ser. Khim.,* 1322, (1975).
[348] L. Tyutchenkova, L. Privalova, Z. Maizus and I. Emanuel, *Izv. Akad. Nauk SSSR, Ser. Khim,* 48, (1975).
[349] D. Lindsay, J. Howard, E. Horswill, K. Ingold and T. Cobbley, *Can. J. Chem.,* **51,** 870, (1973).
[350] J. Bennett and J. Howard, *J. Amer. Chem. Soc.,* **95,** 4008, (1973).
[351] A. Factor, C. Russel and T. Traylor, *J. Amer. Chem. Soc.,* **87,** 3692, (1965).
[352] R. Hiatt and T. Traylor, *J. Amer. Chem. Soc.,* **87,** 3766, (1965).
[353] W. Pritzkow and K. Müller, *Ber.,* **89,** 2321, (1956).
[354] I. Berezin, E. Denisov and N. Emanuel, *The Oxidation of Cyclohexane,* Moscow University Press, 1962, Translated by K. Allen, Pergamon Press, Oxford (1966).
[355] G. Pustarnakova and V. Solyanikov, *Izd. Akad. Nauk SSSR, Ser. Khim.,* 2191, (1974).
[356] G. Pustarnakova, V. Solyanikov and E. Denisov, *Izv. Akad, Nuak SSSR, Ser. Khim.,* 547, (1975).
[357] I. Skibida, M. Brodskii, M. Gervits, L. Goldena and Z. Maizus, *Kinet. Katal.,* **14,** 885, (1973).
[358] L. Matienko, I. Skibida and Z. Maizus, *Izv. Akad. Nauk SSSR, Ser. Khim.,* 2834, (1974).
[359] L. Matienko and Z. Maizus, *Izv. Akad. Nauk SSSR, Ser. Khim.,* 1207, (1971).
[360] L. Matienko and Z. Maizus, *Izv. Akad. Nauk SSSR, Ser. Khim.,* 1524, (1972).
[361] L. Matienko, I. Skibida and Z. Maizus, *Kinet. Katal.,* **12,** 95, (1971).
[362] V. Vinogradova and Z. Maizus, *Kinet. Katal.,* **12,** 1322, (1971).
[363] G. Bulgakova, I. Skibida and Z. Maizus, *Kinet. Katal.,* **12,** 76, (1971).
[364] N. Emanuel, Z. Maizus and I. Skibida, *Angew. Chem. Internat. Edit.,* **8,** 97, (1969).
[365] B. Balkov, I. Skibida and Z. Maizus, *Izv. Akad. Nauk SSSR, Ser. Khim.,* 1780, (1970).
[366] H. Arzoumanian, A. Blanc, J. Metzger and J. Vincent, *J. Organometal. Chem.,* **82,** 161, (1974).
[367] J. Kollar, *U.S., 3,350,422,* (1967); *U.S., 3,351,635,* (1967); *U.S., 3,360,584,* (1967).
[368] N. Sheng and J. Zjacek, *Can., 799,502,* (1968); *Can., 799,503,* (1968); *Can., 799,504,* (1968).
[369] I. DeRoch and P. Menguy, *Fr., 1,505,337,* (1967); *Fr., 1,505,332,* (1967).
[370] R. Hiatt, in *Oxidation Techniques and Applications in Organic Synthesis,* (R. Augustine, Ed.), Vol. 2, pp. 133–138, Marcel Dekker, Inc., New York (1971).
[371] D. Metelitsa, *Russian Chemical Reviews,* (Engl. Transl.), **41,** 807, (1972).
[372] F. Mashio and S. Kaito, *Yuki Gosei Kagaku Kyokaishii,* **26,** 367, (1968).

[373] A. Doumaux, *Oxidation*, Vol. 2, R. Augustine, Ed., Marcel Dekker, Inc., New York, pp. 141–185 (1971).
[374] N. Indictor and W. Brill, *J. Org. Chem.*, **30**, 2074, (1964).
[375] M. Sheng and J. Zajacek, *Advan. Chem. Ser.*, **76**, 418, (1968).
[376] R. Sheldon and J. Van Doorn, *J. Catal.*, **31**, 427, (1973).
[377] M. Sheng, J. Zajacek and T. Baker, Symposium on New Olefin Chemistry, Houston, Texas, February (1970).
[378] G. Howe and R. Hiatt, *J. Org. Chem.*, **36**, 2493, (1971).
[379] R. Sheldon, *Recl. Trav. Chim. Pays-Bas*, **92**, 253, (1973).
[380] M. Farberov, G. Stozhkova, A. Bondarenko and T. Kirik, *International Chemical Engineering*, **12**, 634, (1972).
[381] M. Farberov, G. Stozhkova, A. Bondarenko and T. Kirik, *Kinet. Katal*, **13**, 291, [1972].
[382] T. Baker, G. Mains, M. Sheng and J. Zajacek, *J. Org. Chem.*, **38**, 1145, (1973).
[383] M. Farberov, G. Stozhkova and A. Bondarenko, *Neftekhimiya*, **11**, 1578, (1971).
[384] V. Gavrilenko, E. Evzerikhin, V. Kolosov, G. Larin and I. Moiseev, *Izv. Akad. Nauk SSR, Ser. Khim.*, 1954, (1974).
[385] V. Sapunov, I. Margitfal'vi and N. Lebedev, *Kinet, Katal.*, **15**, 1442, (1974).
[386] C. Su, J. Reed and E. Gould, *Inorganic Chemistry*, **12**, 337, (1973).
[387] R. Sheldon, J. VanDoorn, C. Schram and A. DeJong, *J. Catal.*, **31**, 438, (1973).
[388] V. Yur'ev, I. Gailyunas, Z. Isaeva and G. Tolstikov, *Izv. Akad. Nauk SSSR, Ser. Khim.*, 919, (1974).
[389] V. Yur'ev, I. Gailyunas, L. Spirikhin and G. Tolstikov, *Zhur. Obshch. Khim.*, **45**, 2312, (1975).
[390] J. Lyons, Homogeneous Catalysis-II, Joint Symposium of the Division of Industrial and Engineering Chemistry and Petroleum Chemistry, 166th Meeting, ACS, Chicago, Illinois, August (1973).
[391] K. Sharples and R. Michaelson, *J. Amer. Chem. Soc.*, **95**, 6136, (1973).
[392] J. Lyons, *Catalysis in Organic Synthesis*, P. Rylander and H. Greenfield, Eds., Academic Press, New York, pp. 235–255, (1976).
[393] Y. Paushkin, I. Kolesnikov, B. Sherbanenko, S. Nizova and L. Vilenskii, *Kinet. Katal.*, **13**, 493, (1972).
[394] S. Tanaka, H. Yamamoto, H. Nozaki, K. Sharples, R. Michaelson and J. Cutting, *J. Amer. Chem. Soc.*, **96**, 5254, (1974).
[395] U. Dzhemilev, V. Yur'ev, G. Tolstikov, F. Gershanov and A. Rafikov, *Dokl. Akad. Nauk SSSR*, **196**, 588, (1971).
[396] R. Sheldon, *Recl. Trav. Chim. Pays-Bas*, **92**, 367, (1973).
[397] J. Kaloustian, L. Lena and J. Metzger, *Tetrahedron Lett.*, 599, (1975).
[398] H. Mimoun, I. Seree DeRoch, L. Sajus and P. Menguy, *Fr., 1,549,184,* (1968).
[399] H. Mimoun, I. Seree DeRoch, P. Menguy and L. Sajus, *Ger. Offen., 1,815,998,* (1969).
[400] H. Mimoun, I. Seree DeRoch, P. Menguy and L. Sajus, *Ger. Offen., 1,817,717,* (1970).
[401] H. Arakawa, Y. Moro-oka and A. Ozaki, *Bull. Chem. Soc. Japan*, **47**, 2958, (1974).
[402] A. Achrem, T. Timoschtschuk and D. Metelitza, *Tetrahedron*, **30**, 3165, (1974).
[403] K. Sharples, J. Townsend and D. Williams, *J. Amer. Chem. Soc.*, **94**, 296, (1972).
[404] K. Khcheyan, L. Samter and A. Sokolov, *Neftekhimiya*, **15**, 415, (1975).
[405] J. Kaloustian, L. Lena and J. Metzger, *Bull. Soc. Chim. France*, 4415, (1971).
[406] H. Arakawa and A. Ozaki, *Chem. Lett.* 1245, (1975).
[407] G. Tolstikov, V. Yur'ev, I. Gailyunas and U. Dzhemilev, *Zh. Obshch. Khim.*, **44**, 215, (1974).
[408] E. Gould, R. Hiatt and K. Irwin, *J. Amer. Chem. Soc.*, **90**, 4573, (1968).
[409] M. Sheng and J. Zajacek, *J. Org. Chem.*, **35**, 1839, (1970).
[410] M. Farberov, L. Mel'nik, B. Bobylev and V. Podgornova, *Kinet. Katal.*, **12**, 1144, (1971).
[411] M. Beg and I. Ahmad, *J. Catal.*, **39**, 260, (1975).
[412] M. Mugdan and D. P. Young, *J. Chem. Soc.*, 2988, (1949).
[413] G. Payne and P. Williams, *J. Org. Chem.*, **24**, 54, (1959).
[414] K. A. Saegebarth, *J. Org. Chem.*, **24**, 1212, (1959).
[415] G. G. Allan, *U.S. Patent 3156709*, November 10, (1964).
[416] G. G. Allan and A. N. Neogi, *J. Catal.*, **19**, 256, (1970).

[417] G. G. Allan and A. N. Neogi, *J. Phys. Chem.*, **73**, 2093, (1969).
[418] O. L. Lebedev and S. N. Kazarnovskii, *J. Gen. Chem. USSR*, **30**, 1629, 3079, (1960.
[419] Z. Raciszewski, *J. Amer. Chem. Soc.*, **82**, 1267, (1960).
[420] H. C. Stevens and A. J. Kaman, *J. Amer. Chem. Soc.*, **87**, 734, (1965).
[421] V. N. Sapunov and N. N. Lebedev, *Zh. Org. Khim.* (Eng. Transl.) **2**, 273, (1966).
[422] G. G. Allan and A. N. Neogi, *J. Catal.*, **16**, 197, (1970).
[423] P. Kalsi, K. Kumar and M. Wadia, *Chem. & Ind.*, 31, (1971).
[424] G. Aulakh, M. Wadia and P. Kalsi, *Chem. & Ind.*, 802, (1970).
[425] E. Glotter, S. Greenfield and D. Lavie, *J. Chem. Soc. C*, 1646, (1968).
[426] T. Jochsberger, D. Miller, F. Herman and N. Indictor, *J. Org. Chem.*, **36**, 4078, (1971).
[427] K. Sharpless and K. Akashi, *J. Amer. Chem. Soc.*, **98**, 1986, (1976).
[428] M. Tohma, T. Tomita and M. Kimura, *Tetrahedron Lett.*, 4359, (1973).
[429] A. Chalk and J. Smith, *Trans. Faraday Soc.*, **53**, 1214, (1957).
[430] E. Bordier, *Bull. Soc. Chim. Fr.*, 2621, (1973).
[431] E. Bordier, *Bull. Soc. Chim. Fr.*, 3291, (1973).
[432] S. Manole, I. Reibel and A. Sandu, *Issled. Khim. Khelatin. Soedin.*, 17, (1971). *Chem. Abstr.* **78**, 57613b, (1972).
[433] I. Reibel and A. Sandu, *Issled. Khim. Koord. Soedin. Fiz. Khim. Metod. Anal.*, 38, (1969); *Chem. Abstr.*, **75**, 129130v, (1971).
[434] N. Ariko and B. Erofeev, *Usp. Khim. Org. Perekiznykh Soedin. Autookisleniya, Dokl. Vses. Konf. 3rd 1965*, 354, (1969); *Chem. Abstr.*, **72**, 42520, (1969).
[435] N. Ariko and B. Erofeev, *Zh. Fiz. Khim.*, **43**, 2057, (1969).
[436] M. Prevost-Gangneux, G. Clement and J. Balaceanu, *Bull. Soc. Chim. Fr.*, 2085, (1966).
[437] M. Prevost-Gangneux, G. Clement and J. Balaceanu, *Bull. Soc. Chim. Fr.*, 2905, (1966).
[438] S. Imamura, T. Banba and Y. Takegami, *Bull. Chem. Soc. Jap.*, **46**, 856, (1973).
[439] S. Imamura, T. Otani and H. Teranishi, *Bull. Chem. Soc. Jap.*, **48**, 1245, (1975).
[440] Y. Nakamura, K. Kato, S. Tanaka and T. Asano, *Kogyo Kagaku Zasshi*, **69**, 444, (1966).
[441] C. Sharma, S. Sethi and S. Dev, *Synthesis*, 45, (1974).
[442] E. Gould and M. Rado, *J. Catal.*, **13**, 238, (1969).
[443] A. Akhrem and A. Moiseenov, *Izv. Akad. Nauk SSSR, Ser. Khim.*, 1801, (1969).
[444] F. Matsubara, *Yukagaku*, **22**, 724, (1973).
[445] M. Garnier and A. Gaiffe, *C. R. Acad. Sci., Paris, Ser. C*, **264**, 1065, (1967).
[446] K. Pecherskaya, M. Logus and A. Tsybul'ko, *Geterogennyl Reakts. Reakts. Sposobnost*, 234, (1964); *Chem. Abstr.*, **66**, 37248a, (1967).
[447] G. Yakimova and O. Levanevskii, *Izv. Akad. Nauk Kirg. SSR*, 68, (1972); *Chem. Abstr.*, 77, 151379w, (1972).
[448] J. Imamura, T. Saito and N. Ohta, *Kogyo Kagaku Zasshi*, **71**, 1642, (1968).
[449] M. Naylor, *U.S. 3,271,447*, (1966).
[450] S. Nan'ya and K. Fukuzumi, *Kogyo Kagaku Zasshi*, **72**, 589, (1969).
[451] T. Labuza, J. Maloney and M. Karel, *J. Food Sci.*, **31**, 885, (1966).
[452] A. Schwab, *J. Amer. Oil Chem. Soc.*, **50**, 74, (1973).
[453] A. Schwab, E. Frankel, E. Dufek and J. Cowan, *J. Amer. Oil Chem. Soc.*, **49**, 75, (1972).
[454] Y. Kamiya, *J. Catal.*, **24**, 69, (1972).
[455] Y. Kamiya, *Tetrahedron Lett.*, 4965, (1968).
[456] H. Klein, C. Bell and J. Coyle, *Canadian Patent, 888221*, (1971).
[457] G. Gadelle and I. Seree de Roch, *Fr., 2,070,406*, (1971).
[458] G. Payne, P. Deming and P. Williams, *J. Org. Chem.*, **26**, 659, (1961).
[459] R. Budnik and J. Kochi, *J. Org. Chem.*, **41**, 1384, (1976).
[460] J. Collman, M. Kubota and J. Hosking, *J. Amer. Chem. Soc.*, **89**, 4809, (1967).
[461] V. Kurkov, J. Pasky and J. Lavigne, *J. Amer. Chem. Soc.*, **90**, 4744, (1968).
[462] J. Baldwin and J. Swallow, *Angew. Chem., Int. Ed.*, 8, 601, (1969).
[463] A. Fusi, R. Ugo, F. Fox, A. Pasini and S. Cenini, *J. Organometal. Chem.*, **26**, 417, (1971).
[463*a*] W. Strohmeier and E. Eder, *J. Organometal. Chem.*, **94**, C14, (1975).
[463*b*] D. Dowden and M. Spencer, *Brit. Pat., 928,443*, (1963).
[463*c*] S. Cenini, A. Fusi and G. Capparella, *Inorg. Nucl. Chem. Lett.*, 8, 127, (1972).

[464] V. Tsykovskii, V. Fedorov and Y. Maskovich, *Zh. Obshch. Khim.*, **45**, 248, (1975).
[465] K. Kaneda, T. Itoh, Y. Fujiwara and S. Teranishi, *Bull. Chem. Soc. Japan*, **46**, 3810, (1973).
[466] E. Blyumberg, M. Bulygin and N. Emanuel, *Dokl. Akad. Nauk SSSR*, **166**, 353, (1956).
[467] V. Rubailo, A. Gararina and N. Emanuel, *Kinet. Katal.*, **15**, 891, (1974).
[468] L. Tyutchenkova, L. Privalova, Z. Maizus, N. Gol'dshleger, M. Khidekel, I. Kalechits and N. Emanuel, *Dokl. Akad. Nauk SSSR*, **199**, 872, (1971).
[469] L. Tyutchenkova, L. Privalova, Z. Maizus and I. Emanuel, *Izv. Akad. Nauk SSSR, Ser. Khim.*, 48, (1975).
[470] J. Lyons and J. Turner, *J. Org. Chem.*, **37**, 2882, (1972).
[471] K. Allison, M. Chambers and G. Foster, *Brit. Pat., 1,206,166*, (1970).
[472] B. James and E. Ochiai, *Can. J. Chem.*, **49**, 975, (1971).
[473] B. James and F. Ng, *Chem. Commun.*, 908, (1970).
[474] D. Holland and D. Milner, *J. Chem. Soc. Dalton*, 2440, (1975).
[475] C. Dudley, G. Read and P. Walker, *J. Chem. Soc. Dalton*, 1926, (1974).
[476] K. Takao, M. Wayaku, Y. Fujiwara, T. Imanaka and S. Teranishi, **43**, 3898, (1970).
[477] J. Lyons and J. Turner, *Tetrahedron Letters*, 2903, (1972).
[478] H. Hasegawa, K. Yoshida and C. Kawashima, *Asahi Garasu Kogyo Gijitsu Shoreikai Kenkyu Hokoku*, **20**, 119, (1972); *Chem. Abstr.*, **78**, 124207c.
[478*a*] S. Muto and Y. Kamiya, *J. Catal.*, **41**, 148, (1976).
[479] J. Farrar, D. Holland and D. Milner, *J. Chem. Soc. Dalton*, 815, (1975).
[479*a*] M. Tsutsumi and S. Masayasu, *Japan, 74 24,451*, (1974).
[480] C. Chan and B. James, *Inorg. Nucl. Chem. Letters*, **9**, 135, (1973).
[481] R. Bonnaire and P. Fougeroux, *C. R. Hebd. Seances Acad. Sci. Ser. C*, **280**, 767, (1975); *Chem. Abstr.*, **83**, 70751h, (1975).
[482] J. Lyons, *Adv. Chem. Ser.*, **132**, 64, (1974).
[483] M. Pudel and Z. Maizus, *Neftekhimiya*, **14**, 412, (1974).
[484] M. Pudel and Z. Maizus, *Izv. Akad. Nauk SSSR., Ser. Khim.*, 43, (1975).
[485] M. Pudel, L. Privalova, Z. Maizus, L. Revenko, M. Khidekel and I. Kalechits, *Neftekhimiya*, **13**, 64, (1973).
[485*a*] M. Sheng, *U.S., 3,839,375, 3,839,376*, (1974).
[485*b*] P. Washechek, *U.S., 3,658,896*, (1972).
[485*c*] *Fr. Demande, 2,175,523*, (1972).
[485*d*] H. Charman, *Brit., 1,133,882*, (1968).
[486] K. Takao, M. Wayaku, Y. Fujiwara, T. Imanaka and S. Teranishi, *Bull. Chem. Soc. Japan*, **45**, 1505, (1972).
[487] K. Takao, H. Azuma, Y. Fujiwara, T. Imanaka and S. Teranishi, *Bull. Chem. Soc. Japan*, **45**, 2003, (1972).
[488] A. Salemov, A. Medzhidov, V. Aliev and T. Kutovaya, *Kinet. Katal.*, **16**, 413, (1975).
[489] A. Davies and G. Hobson, *Ger. Offen., 2,044,875*, (1971); *Fr., 2,061,168*, (1971); *Belgian, 775,119*, (1971).
[490] K. Masada and Y. Akio, *Jap., 71 09,691*, (1971).
[491] R. Rennie, *Brit., 1,071,902*, (1967).
[491*a*] D. Paulson, R. Ullman, R. Sloan and G. Closs, *Chem. Commun.*, 186, (1974).
[491*b*] J. Rouchaud and M. DePauw, *J. Catal.*, **14**, 114, (1969).
[492] C. Bocard, C. Gadelle, H. Mimoun and I. Seree deRoch, *French, 2,044,007*, (1971).
[493] J. Alagy, C. Busson, C. Gadelle, I. Seree deRoch and L. Sajus, *French, 2,052,068*, (1971).
[494] J. Rouchaud and J. Mawaka, *J. Catal.*, **19**, 172, (1970).
[495] J. Rouchaud and M. DePauw, *Bull. Soc. Chim. Fr.*, 2905, (1970), *ibid.*, 2914, (1970).
[496] J. Rouchaud and A. Mumbieni, *Bull. Soc. Chim. Fr.*, 2907, (1970).
[497] J. Rouchaud and F. Mingiedi, *Bull. Soc. Chim. Fr.*, 2909, (1970); *ibid.*, 2912, (1970).
[498] J. Rouchaud and F. Mingiedi, *Bull. Chim. Soc. Belg.*, **78**, 285, (1969).
[499] J. Rouchaud and P. Nsumba, *Bull. Chim. Soc. Belg.*, **77**, 551, (1968).
[500] J. Rouchaud, *Bull. Soc. Chim. Fr.*, 1189, (1971).
[501] E. deRuiter, *Erdol und Kohle*, 510, (1972).
[501*a*] H. Waldmann, W. Schwerdtel and W. Swodenk, *Ger. Offen., 2,201,456*, (1973).

[502] S. Cavitt, *U.S., 3,856,826*, (1974).
[503] S. Cavitt, *U.S., 3,856,827*, (1974).
[504] *Ger. Offen., 2,313,023*, (1973).
[505] V. Tulupov and T. Zakhar'eva, *Zh. Fiz. Khim.*, **49**, 272, (1975).
[506] P. Hayden, *Brit. 1,209,321*, (1970).
[507] K. Allison, G. Foster, P. Johnson and M. Sparke, presented at the Am. Chem. Soc. Nat'l Meeting, Detroit (1965).
[508] K. Allison, P Johnson, G. Foster and M. Sparke, *Ind. Eng. Chem., Prod. Res. Develop.*, **5**, 166, (1966).
[509] K. Allison, *Belg., 640,204*, (1964).
[510] K. Allison, *U.S., 3,505,360*, (1970).
[511] J. Lyons, *Tetrahedron Lett.*, 2737, (1974).
[512] T. Itoh, K. Kaneda and S. Teranishi, *Bull. Chem. Soc. Japan*, **48**, 1337, (1975).
[513] K. Shin and L. Kehoe, *J. Org. Chem.*, **36**, 2717, (1971).
[514] J. Sun, *Ger. Offen., 2,057,192*, (1971).
[515] A. Fusi, R. Ugo and G. Zanderighi, *J. Catal.*, **34**, 175, (1974).
[516] H. Arzoumanian, A. Blanc, U. Hartig and J. Metzger, *Tetrahedron Lett.*, 11011, (1974).
[517] H. Hashimoto, *Kagaku to Kogyo,* **20**, 1077, (1967).
[518] A. Aguilo, *Advan. Organometal. Chem.*, **5**, 321, (1967).
[519] G. Szonyi, *Advan. Chem. Ser.*, **70**, 53, (1968).
[520] I. Moiseev, *Kinet. Katal.*, **11**, 342, (1970).
[521] Y. Kamiya, *Yuki Gosei Kagaku Shi*, **26**, 957, (1968).
[522] N. Emanuel, *Neftekhimiya*, **13**, 323, (1973).
[523] A. Shilov and A. Shteinman, *Kinet. Katal.*, **14**, 149, (1973).
[524] E. Stern, *Transition Metals in Homogeneous Catalysis*, G. Schrauzer, Ed., Marcel Dekker, Inc., New York, pp. 93–146 (1971).

Chapter 2

Structure and Electronic Relations Between Molecular Clusters and Small Particles: An Essay to the Understanding of Very Dispersed Metals

J.M. BASSET

Institut de Recherches sur la Catalyse, 79, Boulevard du 11 novembre 1918, 69626 Villeurbanne, Cedex, France

R. UGO

Istituto di Chimica Generale e Inorganica dell'Universita CNR Centre, Via G. Venezian, 21, 20133 Milano, Italy

1. INTRODUCTION

Heterogeneous catalysis over supported metals requires, in general, the largest metallic surface for the minimum content of the metal itself. This implies that the surface/volume ratio of the metallic particles must be the highest in order to achieve the highest efficiency of the catalyst. Consequently, researches done in this field were mainly oriented to produce more and more dispersed catalysts. Progressively the range of particle size has been decreased from very large particles of 1000 Å to about 100 Å and finally down to about 10 Å and below. As early as 1964 Rabo *et al.* [1] proposed the now well accepted concept of 'atomic dispersion' in the case of zeolite supported platinum.

In the meantime, for about ten years, there has been growing interest in the field of the preparation and characterisation of molecular complexes of transition metals with metal–metal bonds [2]. Starting from a single metal with its ligands, it is now possible to prepare molecular clusters with 2, 3, 4, 5, 6, 7, 8, 9, 11, 13 or even 15 transition metal atoms bonded together in different geometries and surrounded by ligands. The size of these molecules can approach that of particles of very finely dispersed metallic catalysts. In one case we are dealing with a kind of very small crystallites, which are stabilized by appropriate ligands, whereas in the other case the very small crystallites are stabilized by interaction with the surface of an oxide, or are trapped in the frame of a zeolite or in a matrix.

Is there a gap in properties or a kind of progressive change from the molecular state to the metallic state? The definition of precise relationships between the two cases is interesting; as a consequence we feel that a useful approach could be that of comparing what is known about properties of large clusters and small metallic crystallites. Therefore this review, which is not extensive, deals with the field of small metallic particles, including the related solid state physics, and describes a comparative approach to pertinent characteristics of molecular clusters, giving special relevance to transition metals.

2. AN ANALYSIS OF SIZE RANGE: A METHOD OF FINDING AN OVERLAP BETWEEN VERY SMALL PARTICLES AND MOLECULAR CLUSTERS

The simplest parameter that can be first considered is the size range which means the effect of the size of a crystal or of an aggregate on the number of atoms present in the aggregate and on the fraction of atoms located on the surface and finally on the geometry of the surface. Although the determination of these geometric properties would require an exact direct picture of the structure of very small particles (VSP) a rough calculation has been made by Poltorak *et al.* [3] in the case of platinum. In his approach a microcrystal of 5.5 Å in 'diameter' possesses 6 platimum atoms which are all located at the surface of the crystallite. When the size of the microcrystal reaches a value of 11 Å, 44 platinum atoms are included in the aggregate, and 87% are located at the surface of the crystallite. A particle of 20 Å has about 231 platinum atoms and 63% of them are present on the surface. Particles of 80 Å contain about 18×10^3 metal atoms, 20% of which are located on the surface. This description gives us the order of magnitude to predict the critical region at the border line with molecular clusters where the majority of atoms are on the surface or better on the external shell of the metallic cluster. Very recently a similar kind of approach has been carried out by Bond [4] with a simple uniform sphere model, whilst Poltorak *et al.* [3] have considered a cubic structure for the small particles. However the main features are very similar.

Table 1 gives the relative number of atoms in different states of the surface, that is vertices, edges and faces. The predominent structural changes are concentrated in the narrow interval of 5–50 Å. It is again in this range of size that the average coordination number of platinum is varying the most.

A different and probably more sophisticated type of classification of small particles has been introduced by Friedel [5] on the basis of various studies for crystal growth: four different steps must be considered.

(*a*) In the case of aggregates of a few atoms up to about 10–20 atoms, the shape will probably depend on the method of preparation; as a consequence electronic structure and properties will be quite dependent on particle size and shape.

(*b*) In the case of poly-tetrahedral aggregates of 10–10^2 atoms of a typical f.c.c. metal, one observes a regular non-crystalline arrangement with five fold symmetry axes as in an icosahedron. Various studies of molecular dynamics have shown that this kind of structure is more stable than a f.c.c. or h.c. structure.

(*c*) In the case of crystals having about 10^2 to 10^3 atoms the icosahedral symmetry is preserved, but with the appearance of a f.c.c. structure.

(*d*) Above 10^3 atoms there is a well defined f.c.c. structure typical of monocrystals.

As a consequence of the above considerations it is possible to define a very small particle as an unusual particle having less than about 10^2 atoms, the most unusual results being expected below 20 atoms. However the geometry of the surface [3, 4] or of the bulk [5] suggests to extend the domain up to about 10^3 atoms from which the normal expected f.c.c. structure is the most stable one.

Cluster compounds, contrary to very small particles, have a molecular character and already belong to a well defined group of molecules in the field of inorganic chemistry: therefore we shall designate them later on as 'molecular clusters' (MC). According to the definition given by Cotton [6] a cluster compound is a 'finite group of metal atoms which are held together entirely, mainly, or at least to a significant extent by bonds directly between the metal atoms even though some non-metal atoms may be associated intimately with the cluster'. This definition has been

Table 1

SIZE AND SURFACE PROPERTIES OF STABLE PLATINUM CRYSTALS ASSUMING A CUBIC GEOMETRY [3]

Edge length		Total number of atoms in the crystal	Fraction of atoms on surface	Relative number of atoms in different states on surface			Average coordination number of atoms on the surface
In number of atoms	In Å			In vertices	On edges	On faces	
2	5.5	6	1	1	0	0	4
3	8.25	19	0.95	0.33	0.67	0	6.0
4	11	44	0.87	0.16	0.63	0.21	6.94
5	13.75	85	0.78	0.09	0.55	0.36	7.46
6	16.50	146	0.7	0.06	0.47	0.47	7.76
7	19.25	231	0.63	0.04	0.41	0.55	7.97
10	27.5	670	0.49	0.02	0.29	0.69	8.31
15	41.25	2255	0.01	0.01	0.20	0.79	8.56
18	49.5	3894	0.3	0.005	0.17	0.83	8.64
30	82.5	18010	0.19	0.002	0.10	0.90	8.78
50	137.5	36484	0.13	0.006	0.06	0.94	8.87
∞	∞	∞	0	0	0	1	9

extended to the case of metal carbonyl derivatives which represent the most numerous cluster compounds discovered so far; Chini [2] makes a distinction between non-cluster compounds with no significant metal–metal bonds and cluster compounds with a significant metal–metal bond:

Each group is divided in two classes as follows:

Metal Carbonyl derivatives	Non-cluster compounds (no significant metal–metal bonds)	A. Monomeric compounds as $Cr(CO)_6$, $Ni(CO)_3PR_3$ B. Polymeric compounds as $[Ru(CO)_3Cl_2]_2$, $[Mn(CO)_3SR]_4$
	Cluster Compounds (significant metal–metal bonds)	C. Open clusters as $Mn_2(CO)_{10}$ D. Closed clusters as $Fe_3(CO)_{12}$, $Rh_6(CO)_{16}$

Although these classifications are not always realistic, since many compounds can be at a border line between two classes, they are very meaningful to estimate the type of compounds which present an analogy with VSP. Obviously compounds of class D, with 3 or more transition metal atoms are interesting in this respect. It must be said that up to now the number of transition metal atoms in this class may vary from 3 to 15. Nevertheless the most numerous examples are found with 3, 4, 5 and 6 transition metal atoms; for instance $Rh_6(CO)_{16}$ which has a well known octahedral arrangement with an approximate size of 7 Å [7]. This value is not far from that obtained with VSP of platinum included in the frame of zeolites. In this latter case values as low as 8 Å have been observed from X-ray low angle scattering data [8].

In conclusion, concerning the number of atoms involved, the most meaningful comparison between VSP and MC would require cluster compounds having more than 6 transition metals and VSP with less than about 20 atoms. It is in fact in this range of size that VSP differ from the atomic arrangement of metallic monocrystals, therefore approaching in some way the atomic frame of molecular clusters. Nevertheless, before going further, one has to point out that no comparison would be entirely satisfactory since the cluster compounds are stabilized by ligands coordinated to their 'surface' whereas aggregates usually present an entirely unsaturated surface provided they are suitably activated. In fact with MC other ligands are required to satisfy the valence requirements; they must, however, conform to the stereochemical restrictions imposed by the distances and angles of the dominating metal cluster frame [7]. These other ligands are incorporated into the total system in two different ways. First they may act as bridges spanning edges or faces of a polyhedral fragment, in which case they will form bonds with two or more metal atoms. Secondly they may be associated with vertices of the metal polyhedra, usually forming a bond with one metal atom only.

Therefore a correct picture of the comparison between VSP and MC would require 'poisoning' of the surface of a VSP by a kind of ligand such as CO, easily coordinated to a low valent transition metal cluster frame. It could be interesting if a thermal treatment would regenerate the VSP according to the following equilibrium:

$$x\,\mathrm{CO} + \mathrm{VSP} \rightleftharpoons \mathrm{MC}$$

Such a decomposition of MC has been observed in mass spectra of polynuclear metal carbonyl [9] but up to now inorganic chemists could not prepare MC with a sufficient number of transition metal atoms so that they can be stable enough either with or without extra ligands. In the latter case the molecular cluster would obviously have to be supported on a carrier and it would loose its ligands by a thermal treatment under vacuum [10]. A few works have recently appeared in this field [10–12] but it is not proved so far that the frame of the cluster is kept upon thermal treatment; however, molecular clusters seem to be obtained by treatment of the finely dispersed particles with the appropriate ligands [10–12].

It is obvious that the investigation of the above field is of primary importance in order to develop clear correlations between VSP and MC. Actually it is a field just in progress, where very often the evidence offered are not completely satisfactory, although it is quite clear that reversible equilibria between VSP and MC can occur under very well-defined conditions.

3. CHARACTERISATION OF VERY SMALL PARTICLES AND OF MOLECULAR CLUSTERS

Characterisation of size range and electronic properties of VSP and MC are the basis of any comparison. Since molecular clusters and very small particles belong to three different fields of research, namely inorganic chemistry, solid state physics and heterogeneous catalysis, it is not surprising to observe that each science has its own technique of characterisation which depend primarily on their own interest. Inorganic chemistry deals with the geometry and molecular nature of the cluster, that is mainly with the chemical and structural definition of the multi-metallic compounds [2, 7]. Catalysis research is more concerned with the geometric and electronic properties of the surface [4]. Solid state physics deals with physical properties of the bulk [5].

A great variety of experimental methods are available to give an estimate of the average size of very small particles [13–17]. The most direct method seems to be transmission microscopy, since it is the only method which is able to yield also the geometry through the images of the particle or even of the atoms [18, 19] in a visible form. However, in the case of very small particles the electron microscope is almost at the limit of its possibilities since the resolution power is equal to about 2–3 Å corresponding to the size of a single atom. Let us mention that Yates *et al.* [19] were able, by means of this technique, to obtain a picture of very small particles of rhodium deposited on silica. In their picture one can see an aggregate of five rhodium atoms at the surface of silica.

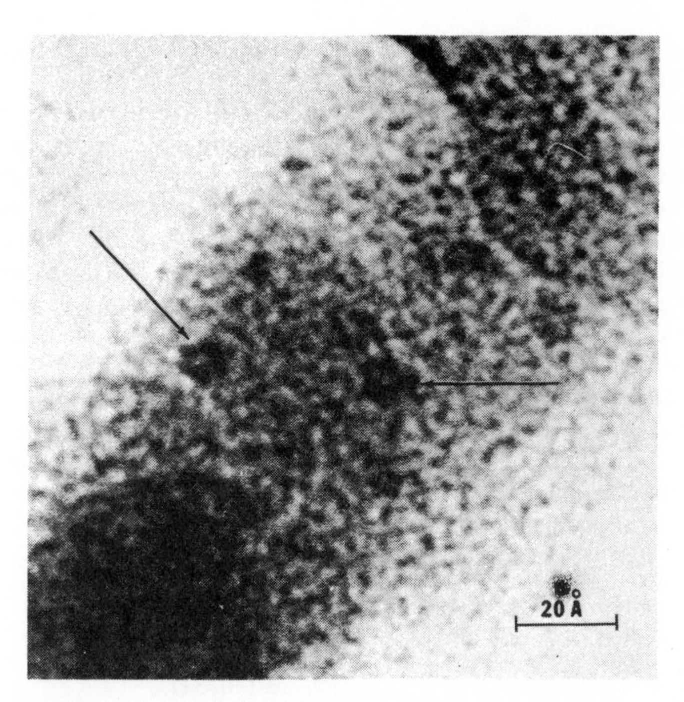

Figure 1. The high resolution electron microscopy of rhodium dispersed on silica (from reference [19]).

In conclusion, this technique could give us not only the size of the particles but also its geometry, at least when the particle is very small (Figure 1).

X-ray methods are also widely used in the field of heterogeneous catalysis. However X-ray line broadening [20] has a limited field for very small particles since it does not seem to be useful for particles smaller than about 20 Å and for metal content less than a few percent. Low angle X-ray scattering seems to be a more promising technique applicable to particles of 10 Å or less [8, 21].

Magnetic methods are also very useful in the case of ferromagnetic materials such as Fe, Co, Ni or their alloys. When the particles are very highly dispersed, the resulting superparamagnetic system obeys the Langevin law which is directly related to the number of atoms in the particle. The main interest of this technique lies in the fact that there is neither theoretical nor experimental limitation for very small particles [22] and that it can also be a useful technique in the case of paramagnetic cluster compounds. This technique may also be applied to diamagnetic transition metals aggregates. In this latter case the magnetic susceptibility is not related to the number of unpaired electrons but to the state density at the Fermi level. It seems that when the size of the particle is very small, this susceptibility increases considerably, when the number of atoms in the particle is even [23].

Very recently other spectroscopic methods have been developed and used to characterize supported metals as Mössbauer spectroscopy [25], Auger spectroscopy [25] and ESCA spectroscopy [26].

Chemisorption methods have also been widely used to determine the percentage of metal atoms present on the surface of the crystallite and indirectly its size. The method consists in the use of a gas molecule which chemisorbs selectively on the metal and not on the support. Assuming a given stoichiometry for this surface reaction, it is theoretically possible to obtain an estimate of the surface area and of the average particle size. In the case of platinum particles of 20 Å or above there is generally a good agreement between the number of surface atoms determined by chemisorption of O_2, H_2 or CO and by physical methods such as electron microscopy, magnetic methods or X-ray low angle scattering [21] or ESCA spectroscopy [26]. This indicates that the chemical methods can be very useful for the determination of the size of relatively large particles. However for very small particles of 15 Å or below the situation is rather more complicated and the simple stoichiometries of the adsorption are no longer observed; for instance in the case of very small particles of platinum or palladium of about 8 Å it seems that the simple stoichiometry of the O_2–H_2 titration [27–29]

$$\underset{\text{(surface)}}{Pt-O} + H_2 \longrightarrow \underset{\text{(surface)}}{Pt-H} + \underset{\text{(ads)}}{H_2O}$$

is no longer observed for the same type of catalysts [30–31].

An interesting new technique as dissolution of hydrogen in small particles of Pd seems to be a useful approach to the determination of the size of VSP of this latter metal [32].

Techniques available to solid state physicists are also useful in the case of very small particles; we will limit ourselves to the most frequently used techniques applied to this field. In most cases theoretical and experimental data are very closely connected, though a comparison, however, which involves an extrapolation from microcrystals of diameter less than 20 Å, the field of the theorist, to microcrystals of diameter greater than 50 Å, the field of the experimentalist. X-ray or electron diffraction are the most commonly used techniques [33]. In these cases the very small particles are obtained by condensation of the vapor phase on a support which is maintained at liquid nitrogen or liquid helium temperature. In some cases, electron microscopy and electron diffraction are simultaneously applied to the same sample [33, 34].

Besides the lattice parameters, which can be determined by diffraction methods, the lattice vibrations can be characterised by superconductive tunneling [35]. Atoms in crystals vibrate about their equilibrium position and the root-mean square amplitude of vibration is a good measure of the atomic motion. These vibration properties can be probed by superconductive tunneling and can be quite significantly modified when a disordered structure of a very small particles changes to an ordered structure typical of a well crystallised particle. Obviously this latter technique can also give some information on the geometry of the small particle.

Another approach of very small particles consists in determining the effect of the size on the calorimetric and thermodynamic properties of the particle itself, namely the melting temperature (phase diagram), the latent heat of melting and the pre-melting temperature. However in this case there are more theoretical calculations [36] than experimental evidence [37].

Besides size and geometric properties, it is also possible to study the electronic properties of very small particles through their ionisation potential or electron affinity. These parameters have been evaluated by mass spectrometry in the case of very small aggregates of selenium [38].

In conclusion, all the above techniques give us more or less indirect information about the size and the geometry of small crystallites.

The techniques available for the characterisation of molecular clusters belong to the field of inorganic and organo-metallic chemistry; besides microanalysis which represents the first step, it must be pointed out that X-ray is by far the best method of characterisation of these compounds [39]. Complementary techniques are infrared and Raman spectroscopy, which can supply information on the nature of the ligands and on the symmetry of the cluster molecules as well as on the vibrations of the metallic skeleton of the cluster [40]. Additional information can be obtained from mass spectrometry, magnetism and NMR. This latter technique gives information on the location and mobility of ligands bonded to the metal framework, a point which is important in the field of theory of chemisorption [41]. It should be noted here that a molecular cluster, due to its chemical and structural characterisation is certainly known with much more detail as to its size and geometry than a very small particle. On the contrary the actual state of their theoretical description and of the knowledge of their physico-chemical properties is rather far from satisfactory.

In conclusion the two fields, VSP and MC, are actually complementary, the first being rather advanced from the theoretical point of view, the second giving rise to a large series of information about real geometries.

4. GEOMETRIC ASPECTS OF THE RELATION BETWEEN VERY SMALL PARTICLES AND MOLECULAR CLUSTERS

It might seem arbitrary to make a clear distinction between electronic and geometric properties of VSP or MC, when they are considered separately, because often geometric factors can be rationalised from electronic considerations and *vice-versa.* Nevertheless we found it more suitable to consider these two factors separately for sake of simplicity.

Due to the technical difficulties in the preparation and characterisation of very small particles, experimental results concerning the structure and geometry of very small particles are very rare. Nevertheless this constitutes the subject of an intense field of research on the theory of crystal growth. The pioneer work in this field is due to Hoare and Pal [42] who have examined the growth sequence of a microcrystal starting with one atom and steadily adding one atom at a time. The starting hypothesis is based on a growth sequence in which atoms are interacting only by pair potentials, under vacuum. If the atoms are growing on a surface instead of under vacuum, the sequence will not be modified unless the atoms strongly interact with this surface.

The growth sequence to form an aggregate of 13 atoms [13] is shown in Figure 2. Two atoms come together to form a dimer. The third atom is added to form a trimer with a trangular configuration and the fourth atom gives a tetrahedron. It is at this stage that the growth deviates from f.c.c. packing since the fifth atom should go in the same plane as one of the triangular face of the tetrahedron. The sixth and seventh atoms are then placed in the middle of two triangles of the trigonal bipyramid so as to form a stable pentagonal bipyramid with a fivefold axis. The addition of five atoms on the five upper faces of the pentagonal bipyramid plus one further atom on the fivefold symmetry axis produces the 13 atoms icosahedron. It has been shown that if the 13 atoms cluster were formed in the f.c.c. packing or h.c.p. packing it would be unstable and deform spontaneously into icosahedral packing. Addition of 42 more atoms would produce a 55 atoms icosahedron [13].

In agreement with the previous study, theoretical calculations were performed by Fripiat *et al.* [43] in the case of lithium clusters using the recently developed self-consistent field Xα Scattered Wave method [44]. Besides electronic considerations on the effect of the size on the stability of the clusters, they find that the icosahedral Li_{13} aggregate is the most stable particle of all those considered, especially compared to the cubo-octahedron. This result is of interest in heterogeneous catalysis since up to now the cubo-octahedron which has a rather spherical shape, was assumed so far to be the most likely crystallite for dispersed metallic catalysts [4], although among the other ideal particles thoroughly treated by van Hardeveld and Hartog [45] are the tetrahedron, the cube, the octahedron, the dodecahedron, the hexagonal bipyramid and the truncated bipyramid. Recent calculations on Li clusters by an *ab initio* method have also shown that icosahedral clusters become more stable than others when the number of lithium atoms is more than 4 [46].

Direct experimental evidence in favor of very small particle having a fivefold symmetry axis was first given by Yates [19] in the case of a rhodium catalyst supported on silica. The size of the aggregate is equal to 12 Å, however the spacing between atoms is quite large (3.5 Å) compared to interatomic distance of rhodium in a f.c.c. structure (2.690 Å). Strangely enough this would indicate the absence of strong metal–metal bond in this cluster. Recent studies by Gillet *et al.* [33] seem to confirm the validity of the icosahedron packing in the case of gold aggregates evaporated on a support of NaCl. The gold aggregates seem to keep a fivefold symmetry axis from 30 Å up to 100 Å. Above this size there is a defect of closing around C_5 axis corresponding to the transition to a f.c.c. structure. Similar results were obtained by Alpress *et al.* [47] in the case of gold crystals on mica and by Mader [48] in the case of germanium crystals on NaCl. Polytetrahedral aggregates of argon were also observed by Farges *et al.* [49, 50] by looking at the very beginning of the condensation directly in the gas jet. These polytetrahedra recrystallise in the f.c.c. system when their size reaches an approximate value of 15 Å.

In comparison with the growth sequency of very small particles, we have a rather

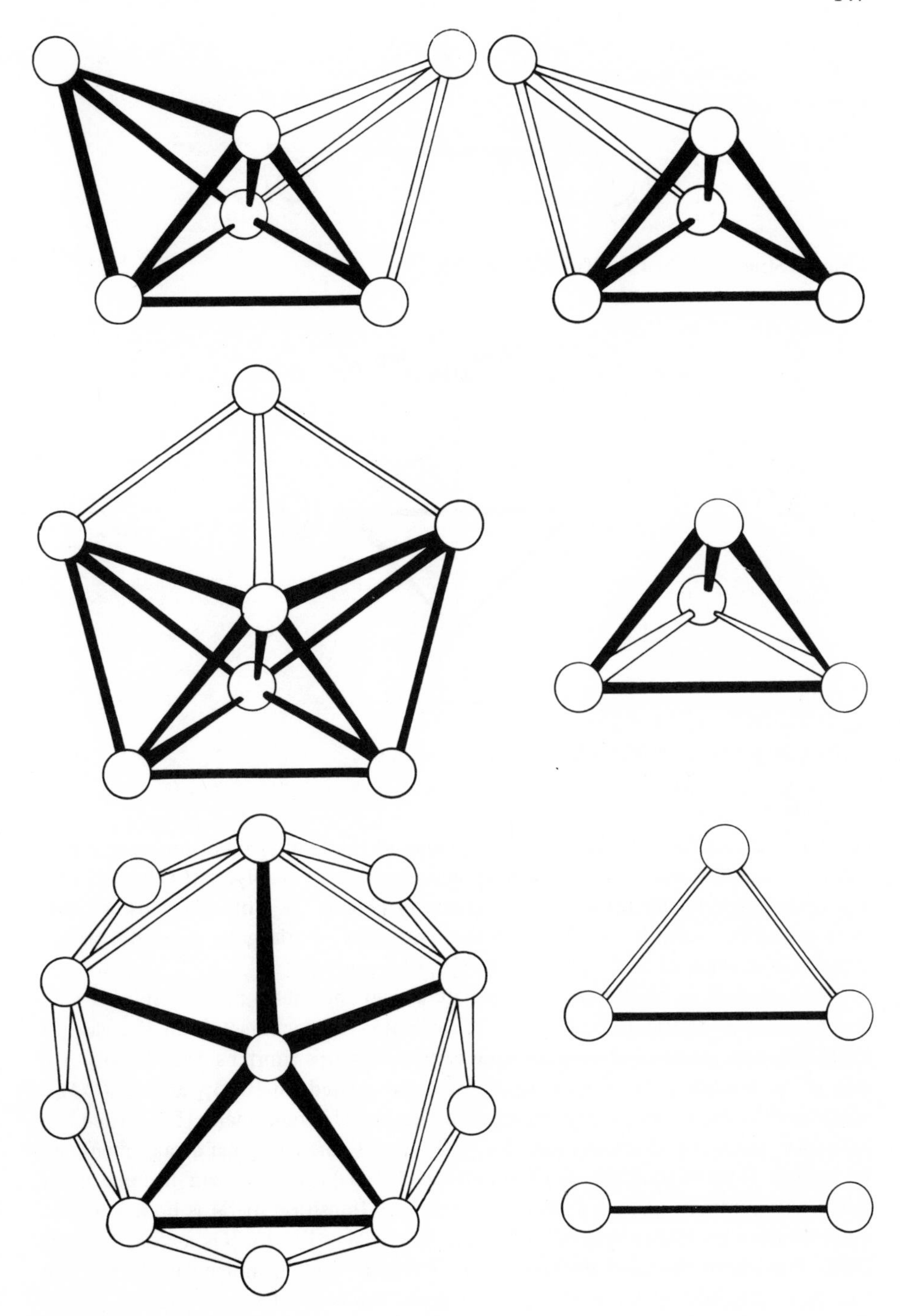

Figure 2. A growth sequence of a metallic aggregate [13].

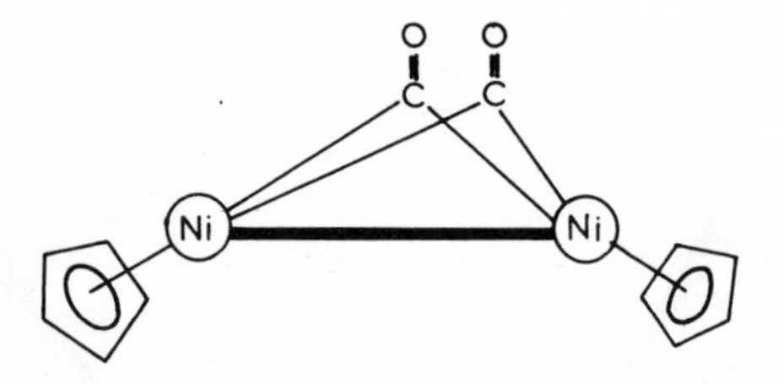

Figure 3. Structure of $Ni_2Cp_2(CO)_2$ ($Cp = \eta^5\text{-}C_5H_5$).

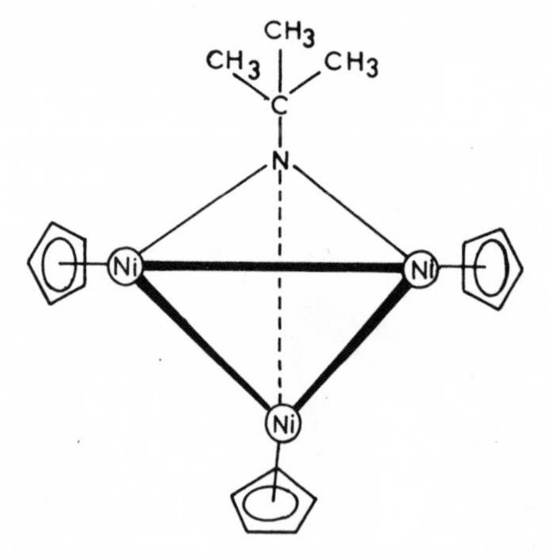

Figure 4. Structure of $Ni_3Cp_3N(t\text{-}C_4H_9)_3$ ($Cp = \eta^5\text{-}C_5H_5$).

complex but not too different kind of growing in the family of cluster compounds, which possess the same sizes. Evidently any dimer, such as $Ni_2Cp_2(CO)_2$ [51] (Cp means cyclopentadienyl), has a linear geometry (Figure 3). Molecular clusters with three transition metal atoms have, most of the time, a triangular geometry as for instance in the case of $Ni_3Cp_3N(-t\text{-}C_4H_9)$ [52] (Figure 4).

The triangle is not necessarily equilateral and the metal–metal distances may depend on the nature of the ligands coordinated to each metal: in the case of $Fe_3(CO)_{12}$ the presence of two bridging carbonyl groups shortens considerably one side of the triangle [53] (Figure 5), whilst in the related $Ru_3(CO)_{12}$ and $Os_3(CO)_{12}$ compounds a higher symmetry achieves an equilateral arrangement [54]. When the molecular symmetry decreases as in $Ru_3(CO)_{10}(NO)_2$ the triangular arrangement can be broken (Figure 6) [55]. Such non-triangular arrangements can become quite stable and common in mixed clusters, where one transition metal is bound to two non-transition metals as in $M(CO)_4(HgX)_2$ where M = Fe, Ru, Os and X = Cl, Br [56]. The above reported examples would suggest that in molecular clusters the triangular structure, which is the most stable on the basis of simple packing, can be disturbed and finally destroyed by the effect of ligands and of the total symmetry.

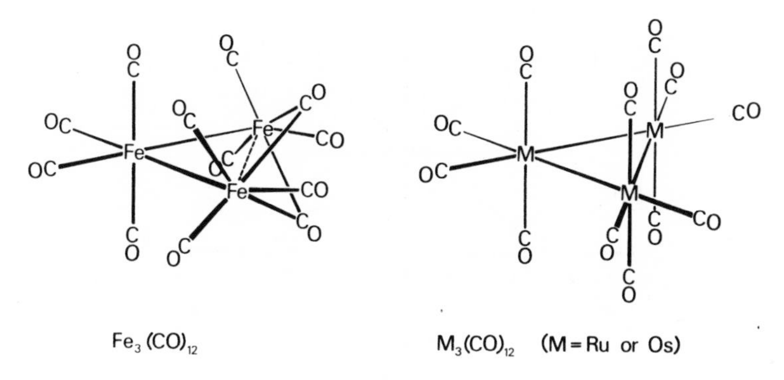

Figure 5. Structures of $M_3(CO)_{12}$ species (M = Fe, Ru, Os).

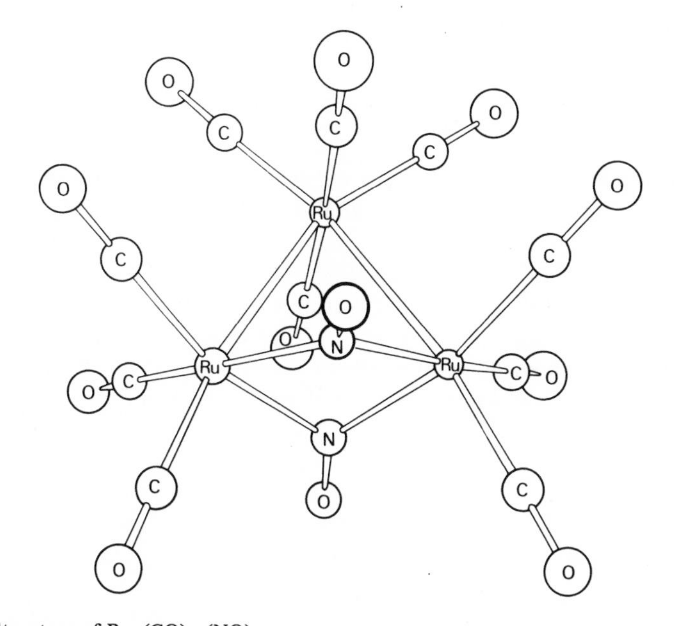

Figure 6. Structure of $Ru_3(CO)_{10}(NO)_2$.

Most of the molecular clusters with 4 transition metal atoms have a tetrahedral arrangement, a typical example being $Ir_4(CO)_{12}$ [57] (Figure 7).

Here also the tetrahedron, which can be a perfect one when all the ligands are equivalent, may be distorted along an axis if the ligands are different or differently coordinated to the 4 metal atoms as in $Ni_4Cp_4(\mu_3\text{-}H)_3$ [58] (Figure 8). In this latter case the face of the tetrahedron (characterised by Ni_1, Ni_2, Ni_3 atoms) is quite different from the other three faces, which have a bridging hydride ligand. As above, the lowering of total symmetry introduced by ligands in appropriate positions of the

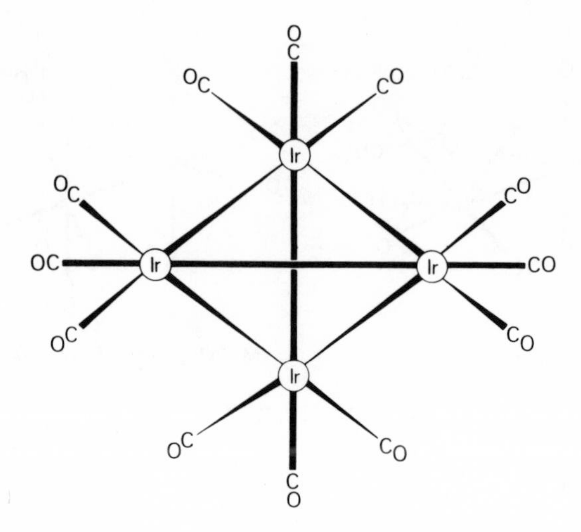

Figure 7. Structure of $Ir_4(CO)_{12}$.

metallic frame produce distortion of the regular tetrahedron. In a few extreme cases the distortion is such that one metal–metal bond is broken to produce a butterfly kind of structure. This effect can be produced in mixed clusters [59] (Figure 9) or in clusters with special organic ligands [60, 61] (Figures 10 and 11).

The butterfly kind of arrangement is still based on triangles, as it is in the pure tetrahedron. Larger distortions to produce either a nearly triangular star structure [62] (Figure 12) or a pseudo square-planar arrangement [63] (Figure 13) can be obtained in isolated cases where the original structure of the pure tetrahedron has been very much destroyed.

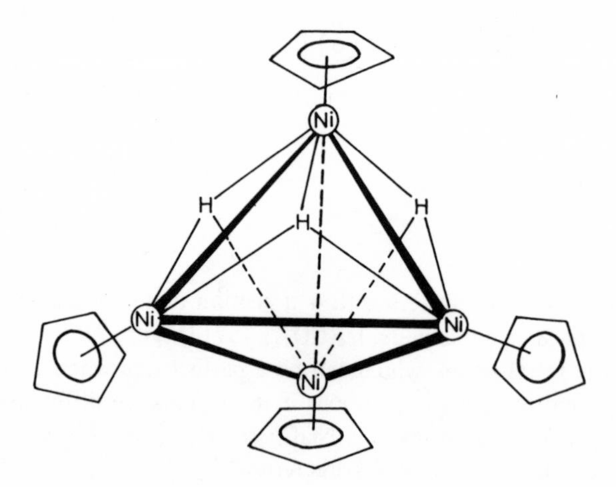

Figure 8. Structure of $Ni_4Cp_4(\mu_3\text{-}H)_3$ ($Cp = \eta^5\text{-}C_5H_5$).

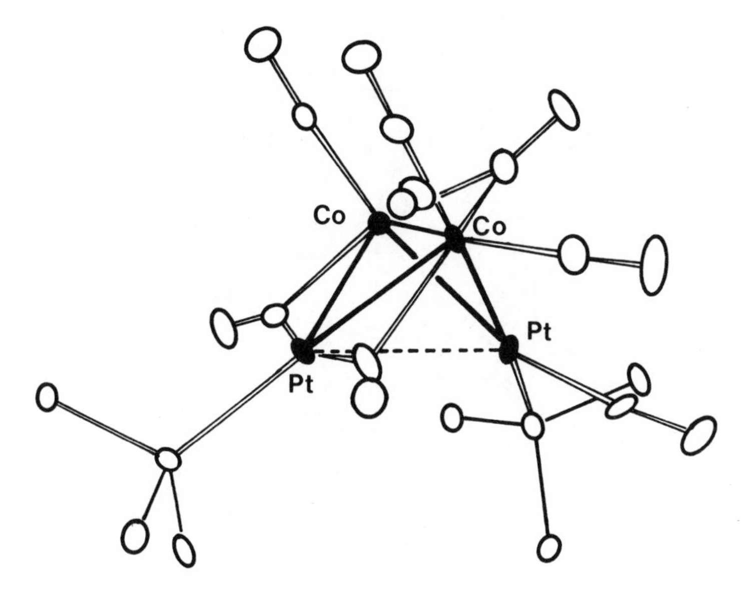

Figure 9. Structure of $Pt_2Co_2(PPh_3)_2(CO)_6$.

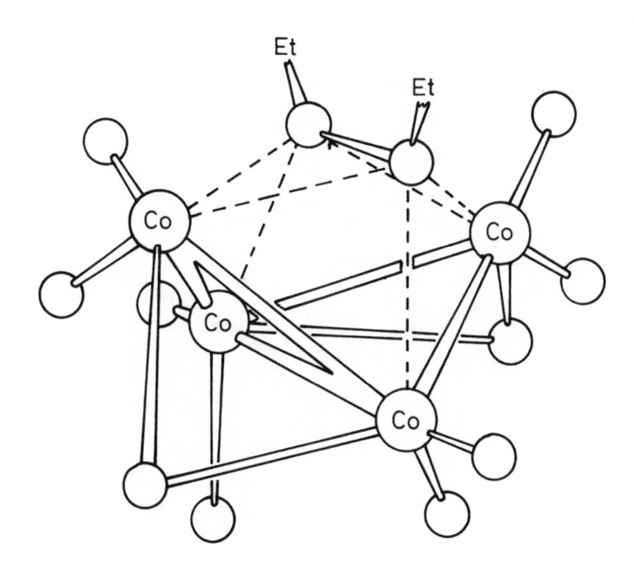

Figure 10. Structure of $Co_4(CO)_{10}$ (Et–C ≡ C–Et).

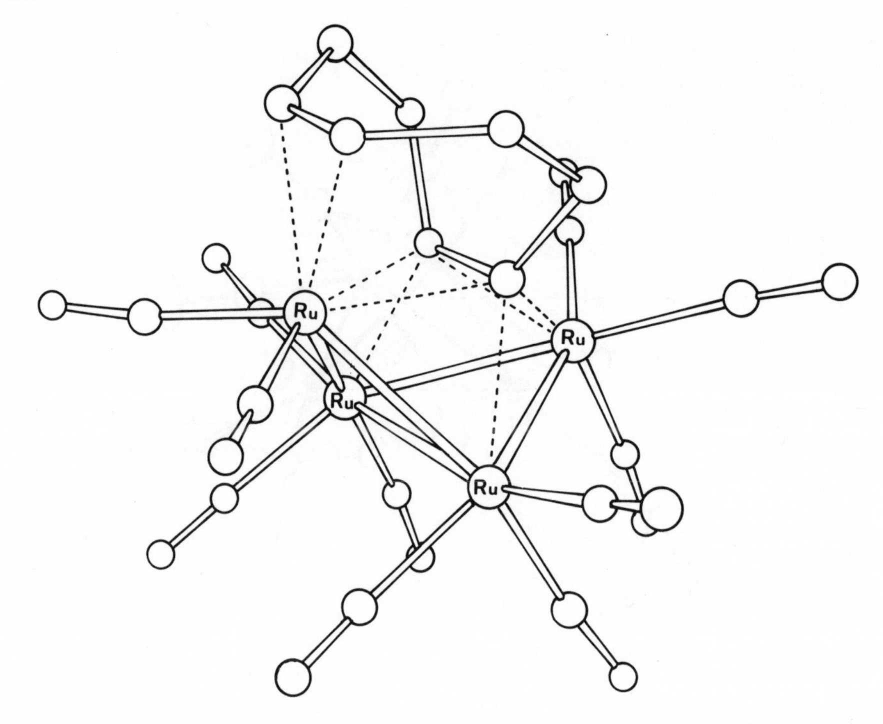

Figure 11. Structure of $Ru_4(CO)_{11}(C_8H_{10})$.

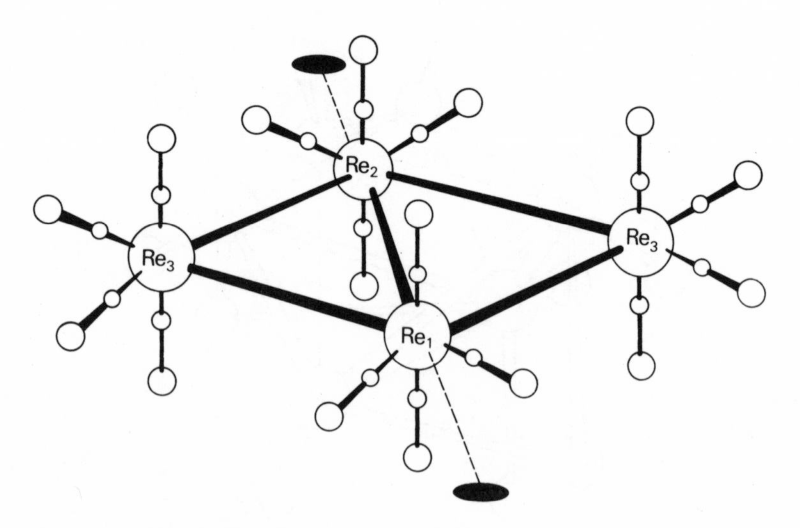

Figure 12. Structure of $[Re_4(CO)_{16}]^{2-}$.

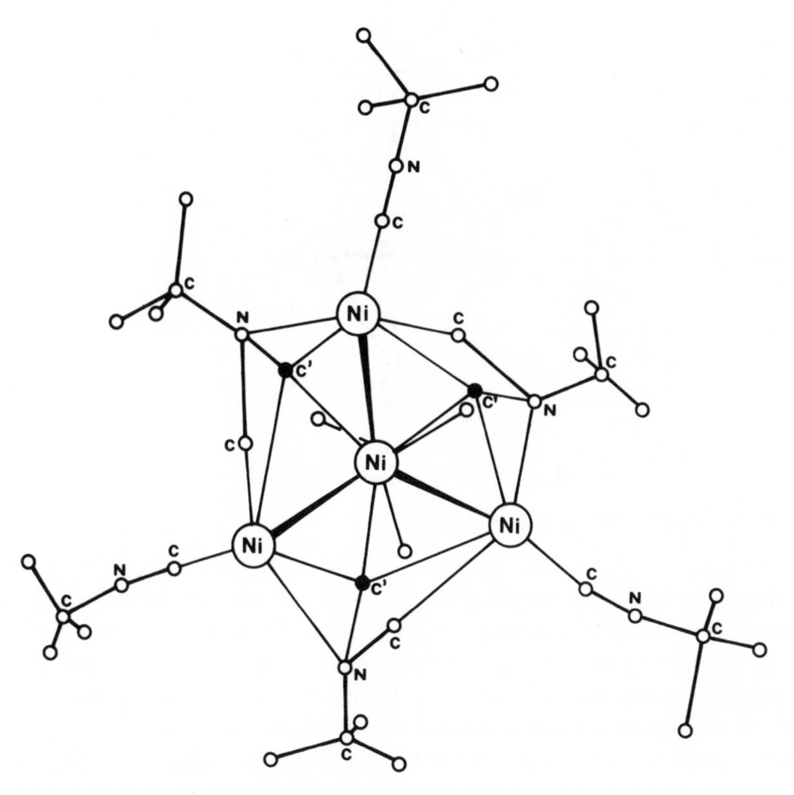

Figure 13. Structure of $Ni_4(CN\ t\text{-}C_4H_9)_7$ from reference [63].

A linear geometry has been first postulated in the case of the cluster

Cl–Os–Os--Os–Au–PPh_3
(CO)$_4$ (CO)$_4$ (CO)$_4$

which has been directly proved later by X-ray crystallography to be triangular [64].

A linear geometry does not seem to be frequently encountered. Let us mention however, that theoretical calculations have shown that for Ag, Li and Cd, a low energy form of very small particles could be linear chains [65]. Clusters with five metal atoms are less common. However some interesting examples, in perfect agreement with the proposed packing of very small particles, have been reported.

In the anionic cluster $[Ni_5(CO)_9(\mu_2\text{-}CO)_3]^{2-}$ we have a trigonal bipyramidal architecture [66] (Figure 14) and a similar structure has been reported for $Os_5(CO)_{16}$ [67]. The major interest of the five atom cluster is that it indicates that the growing of the 4 metal tetrahedral cluster to the 5 metal cluster can occur by formation of a polytetrahedron as in the case of very small particles. Even in more complex mixed clusters a similar trigonal bipyramidal arrangement has been found as

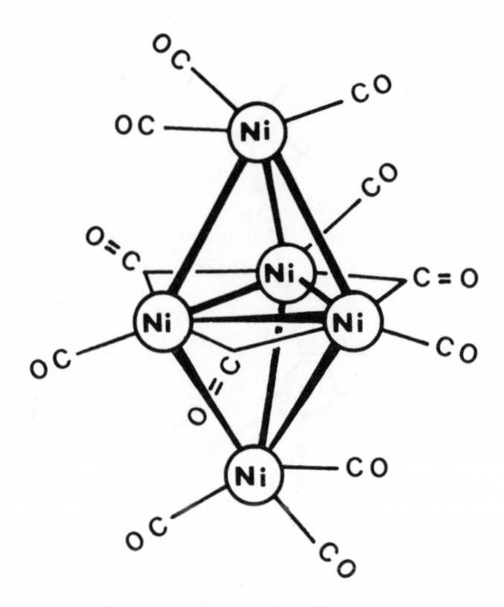

Figure 14. Structure of $[Ni_5(CO)_9(\mu_2\text{-}CO)_3]^{2-}$.

with the metallic clusters $[M_2Ni_3(CO)_{16}]^{2-}$, (M = Cr, Mo, W) [68]. Very recently, structural work on nude clusters has appeared in the literature (e.g. Bi_5^{3+}, Pb_5^{2-}, Sn_9^{4-}, etc.). In the case of Pb_5^{2-} a trigonal bipyramidal structure has been reported [69]. thus supporting that this architecture is the most stable one in the absence of any ligand, as should be a small crystallite. Work on the full characterisation of this important series of nude clusters is still lacking, although, owing to their relationship with boranes, much work has been done from a theoretical point of view [70]. For instance, the kind of bonding in the series Bi_3^{3+}, Bi_4^{4+}, Bi_5^{3+}, Bi_8^{2+} and Bi_9^{3+} is in agreement with a polytetrahedral arrangement of the metal cage [70, 71].

The situation is more complex when we consider complex clusters of higher nuclearity. Clusters with six metallic atoms usually do not show a polytetrahedral arrangement with the exception of $Os_6(CO)_{18}$ [72] which involves a capped trigonal bipyramid (Figure 15). This structure is maintained in some substitution products [73], however, introduction of electrons as in the species $[HOs_6(CO)_{18}]^-$, $[Os_6(CO)_{18}]^{2-}$ or $H_2Os_6(CO)_{18}$ [74] produces deformation of this latter cage towards a more or less distorted octahedron up to a pure octahedron as in $[Os_6(CO)_{18}]^{2-}$.

These structural data emphasize that in molecular clusters of high nuclearity the basic skeletal structures of the metal system is very sensitive to the electron structure of the cluster bonding system, therefore suggesting that many and different cages, not all related to only a good packing of the metal atoms, have rather similar energies.

The capped trigonal bipyramid geometry has been observed in another rather unusual kind of clusters, as the hydroxo or oxo compounds of Sn, Pb, Al, etc . . , which show weak metal–metal interactions. X-ray studies on the two forms of $[Pb_6O(OH)_6]$ $[ClO_4]$ H_2O have confirmed a structure consistent with an arrange-

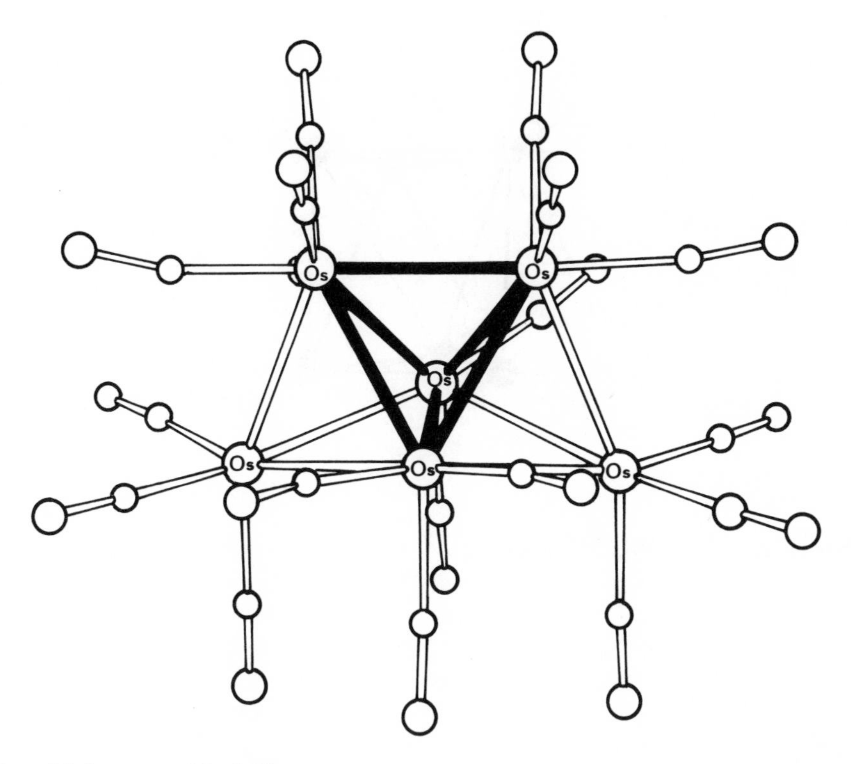

Figure 15. Structure of $Os_6(CO)_{18}$.

ment of the six lead atoms as three tetrahedra sharing faces [75]. However in other cases as $Sn_6O_4(OH)_4$ (or 3SnO, H_2O) an octahedral arrangement of the tin atoms has been observed [76]. In transition metal clusters we have a high tendency, as explained above, to assume more or less distorted octahedral structures.

Consequently whilst we have a pure octahedron in $Rh_6(CO)_{16}$ [7], we can observe either a small distortion as in $Au_6L_6^{2+}$ (L = PPh_3) [77] or a higher distortion as in $[Ni_3(CO)_3(\mu_2\text{-}CO)_3]_2^{2-}$ [78] (Figure 16). This latter kind of distortion, which however does not really produce a poly-tetrahedral metal cage, has been observed with more complex nuclearity, as in the polymeric anionic species of Chini *et al.* [79] $[Pt_3(CO)_3(\mu_2\text{-}CO)_3]_n^{2-}$ ($n = 2, 3, 4, 5$), that we shall consider it in detail later on.

Very interesting analogies with the growing of VSP should be expected with clusters with more than 6 metal atoms since it is roughly the value from which a metal could be linked in the center of a metal cage without any bonding of additional extra ligands.

In the cluster $[Rh_7(CO)_{16}]^{3-}$ [80] we have the bonding of the extra rhodium atom in the center of a triangular face of the original octahedron, as should be expected for the growing of small particles [13] (Figure 17).

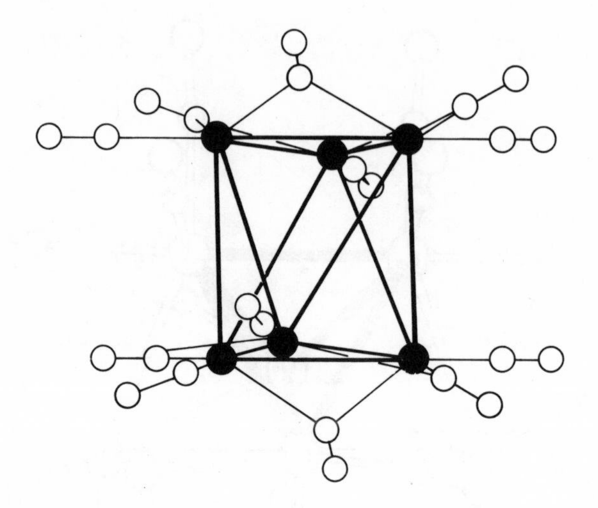

Figure 16. Structure of $[Ni_3(CO)_3(\mu_2\text{-}CO)_3]_2^{2-}$.

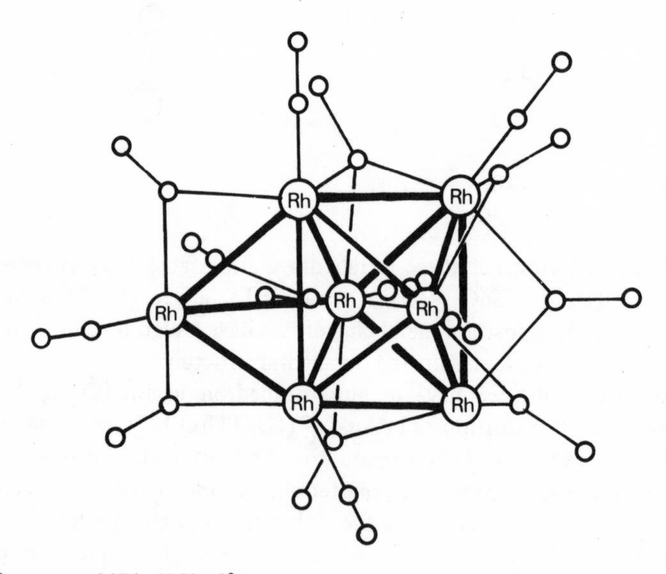

Figure 17. Structure of $[Rh_7(CO)_{16}]^{3-}$.

Also the structure of $Os_7(CO)_{21}$ is consistent with an octahedron of osmium atoms with one face capped by the remaining osmium atom [72].

The proposed structures for Bi_8^{2+} and Bi_9^{3+} are respectively a square antiprism and a tricapped trigonal prism [70, 71].

At higher nuclearity very interesting examples are given by gold compounds

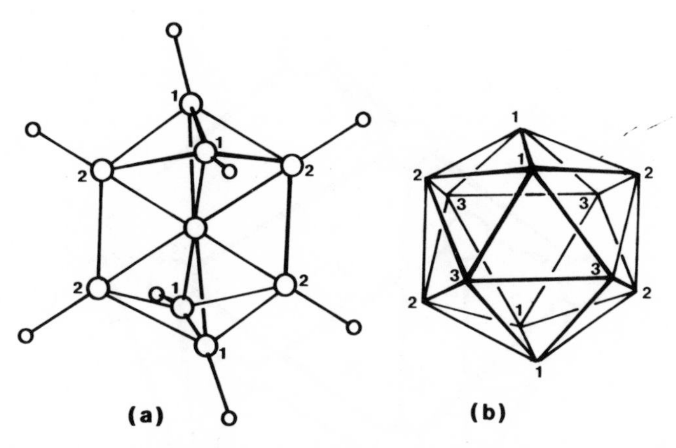

Figure 18. Structure of $[Au_9L_8]^{3+}$ unit (*a*) and the relation with a centred icosahedron from which corner 3 (equatorial rectangle) have been removed (*b*).

with aryl phosphines which have been recently reviewed by Malatesta [81].

Naldini, Cariati *et al.* were able to prepare the following compound: $[Au_9L_8]^{3+}$ (L = tri(*p*-totyl)phosphine) the structure of which is represented in Figure 18. In this structure, the central gold atom is bound to 8 peripherical gold atoms, each attached to a phosphine ligand. The center to periphery distances fall within two ranges: 2.689 and 2.729 Å and the gold–gold distances in the periphery are 2.752 and 2.863 Å. All these distances are shorter than the value found for the metal (2.884 Å) (see Table 2). The geometry of the cluster has been considered as that of a centred icosahedron from which an equatorial rectangle has been removed (see Figure 18).

They also prepared non ionic clusters with 11 gold atoms e.g.: $Au_{11}L_7X_3$ (L = $P(p\text{-}C_6H_4F)$; X = I). The X-ray structure (Figure 19) showed that a central gold atom is surrounded at an average distance of 2.68 Å by 10 gold atoms at bonding distances from each other (2.98 Å) and each bound to an L or an X ligand. The structure of the cluster has been considered as an incomplete centred icosahedron in which three atoms at the vertices of a triangular face have been stretched (see Figure 19). It is clear that cages based on icosahedron are fundamentally developed in these rather 'nude' clusters.

However the very recent characterisation of a Rh_{13} cluster cage (the anion $[Rh_{13}(CO)_{24}H_{5-n}]^{n-}$ (n = 2 or 3) [82] has shown a hexagonal close packing of the metal atoms (Figure 20). Such an arrangement is rather far away from an icosahedral cage, which should be the one preferred in the growth of small crystallites [13].

An icosahedron *leit motiv* is present again in a 15 metal atoms cluster [83] as $[Rh_{15}(CO)_{28}C_2]^-$ (see Figure 21). Despite the presence of two carbon atoms (which usually stabilise rather unusual metal cage geometries as in $Fe_5(CO)_{15}C$ [84] or in $Rh_8(CO)_{19}C$ [85]) this cluster possesses a centred tetracapped pentagonal prism with a Rh_{13} unit having a local 5 fold symmetry axis with a central 'metallic' rhodium.

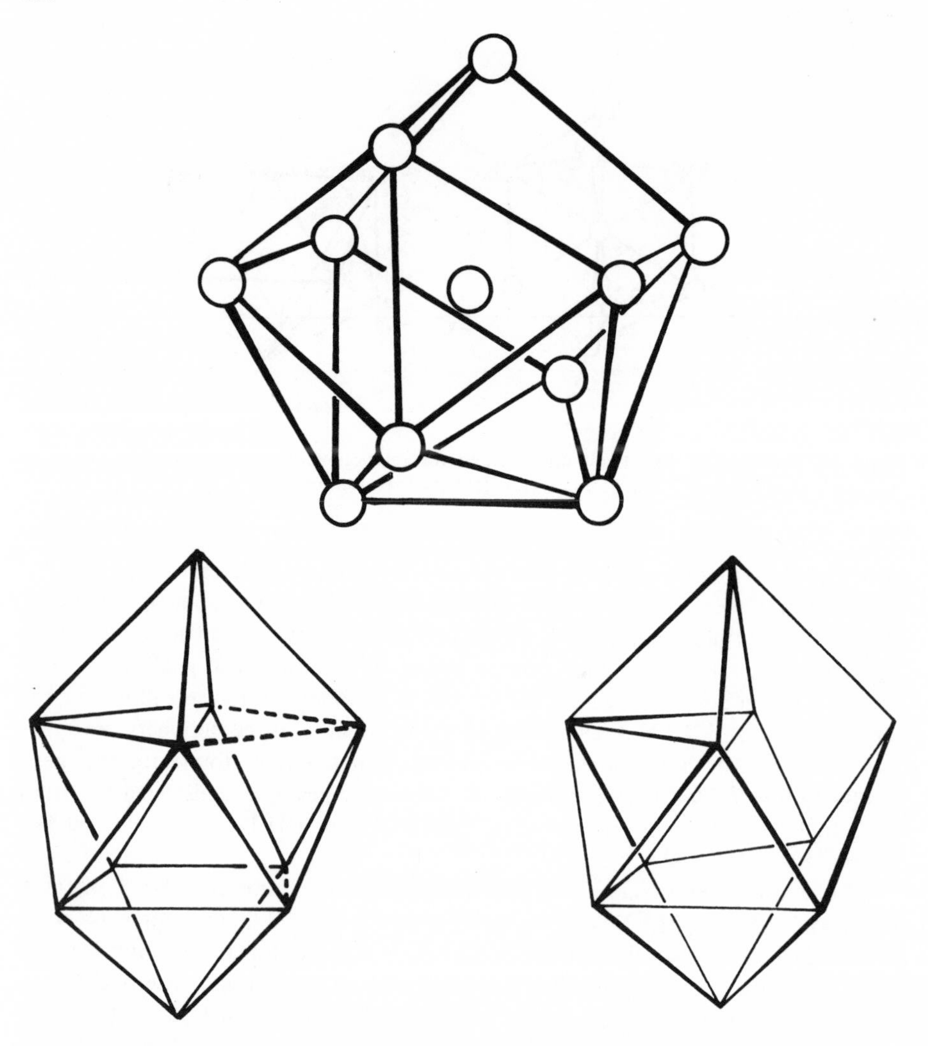

Figure 19. Metal frame in the structure of $A_{11}L_7X_3$ units (*a*) and the relation with an incomplete centred icosahedron (*b*) transformed by stretching 3 edges (*c*).

In this latter structure the two fused pentagonal pyramids have an eclipsed configuration instead of a staggered configuration corresponding to an icosahedron. This might be due to the presence of two carbon atoms which are known to be usually placed in the middle of a square pyramidal arrangement of metal atoms.

The growth of such a structure has been discussed [83]: the approach is very similar to that of growth of small particles.

In conclusion, although the geometry of the cage of metal clusters is dependent from different factors which are also of electronic origin (e.g. nature of the ligands, formal charge of the cluster unit, presence of carbide carbon atoms, etc . . .), it, appears that arrangements based on polytetrahedral icosahedron are very often the most stable, as is expected in the case of small particles.

5. ELECTRONIC STRUCTURES OF MOLECULAR CLUSTERS AND OF VERY SMALL PARTICLES

The previous Section arrived at the conculsion that we have some relations between geometry and electronic structure of a cluster.

A simple generalisation of the geometry and bonding of cluster compounds is related to the use of noble gas formalism; this simplified, but nevertheless used rule, has been applied to cluster compounds by Chini [2]. The number of electrons on each transition metal has been calculated by dividing the total number of electrons brought by the ligands by the number of metal atoms, and adding the electrons present in the valence shell of the metal, without considering eventual metal–metal bonds. This approach can give the geometry of the complex by deduction of the number of metal-metal bonds necessary to reach the 18 electron rule (Table 2).

It is obvious that with 17 electrons a dimer is formed, with 16 electrons a trimer, and with 15 electrons a tetramer. Since in this case one can envisage respectively one, three and six metal–metal bonds the noble gas configuration is obtained in each case with formation of linear, triangular and tetrahedral arrangement.

However this approach is not so easy to handle when the complexity of clusters increases; for instance, $Rh_6(CO)_{16}$, $Co_6(CO)_{16}$, $Ir_6(CO)_{16}$ share 86 electrons in the valence shell, which according to the noble gas rule formalism, would result in 11 metal–metal bonds. Since there is no simple geometry to accommodate 11 metal–metal bonds using 6 metal atoms, it is not surprising to observe an octahedral geometry with 12 metal–metal bonds. In this case the molecular clusters have an excess of two electrons which are located in bonding orbitals [2, 7]. If one increases the number of electrons in the hexanuclear frame, for example by introducing a carbon

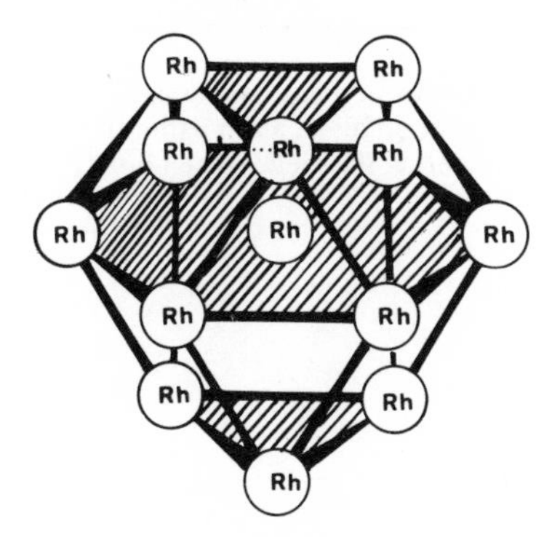

Figure 20. Metal frame in the structure of the anion $[Rh_{13}(CO)_{24}H_{5-n}]^{n-}$.

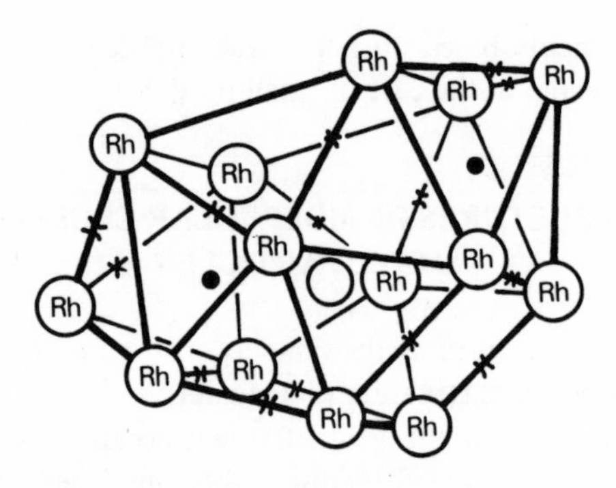

Figure 21. Metal frame of $[Rh_{15}(CO)_{28}C_2]^-$ (the carbide atoms are black, the central rhodium atom is white).

atom in the frame as in $[Rh_6(CO)_{15}C]^{2-}$ [86, 87], it is impossible to accommodate the 90 outer valence electrons in the stable orbitals of an octahedral cluster (54 of rhodium atoms, 30 of carbonyl groups, 4 of the central atom and 2 anionic). In the latter case the excess of electrons would be present in cluster non-bonding orbitals. Thus the geometry will change from an antiprismatic arrangement as for $Rh_6(CO)_{16}$ to a prismatic arrangement (trigonal prism) [86] (Figure 22*b*).

The hepta nuclear anionic cluster $[Rh_7(CO)_{16}]^{3-}$ [80] shows a small deviation from the noble gas rule, since it also possesses two electrons in excess. It has 98 electrons in its valence shell which would require 15 metal–metal bonds, but the observed geometry corresponds only to 14 metal–metal bonds. However the related octo-nuclear $Rh_8(CO)_{19}C$ [85] cluster follows the noble gas rule if one assumes electron pair bonds for metal–metal interactions (72 electrons for rhodium, 38 for CO groups, 4 for the central carbon, that is 18 electrons per rhodium atom).

Table 2

THE NOBLE GAS FORMALISM ACCORDING TO CHINI [2]

Number of transition metal atoms in the cluster	Number of electrons for each transition metal atom without considering the metal–metal bonds	Possible metal–metal bonds
2 $Fe_2(CO)_9$	17	
3 $Fe_3(CO)_{12}$	16	
4 $Co_4(CO)_{12}$	15	

Table 2 (Continued)

5 $Fe_5(CO)_{15}C$	$4 \times 15 + 1 \times 14$	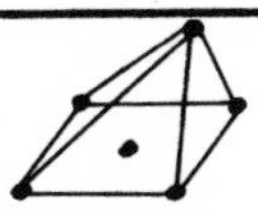
6 $Rh_6(CO)_{16}$	14, 33	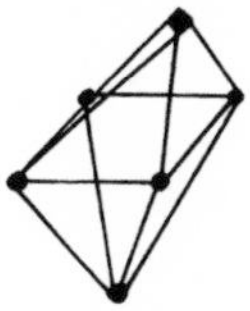

The geometry observed, although very unusual (Figure 2*a*), is in perfect agreement with the required number of metal–metal bonds, that is 15. On the contrary, with the 9 or 11 gold atoms clusters of Malatesta *et al.* [81], an electron calculation indicates that all the gold atoms achieve a configuration with 16 electrons, two electrons less than required by the noble gas rule. Moreover the 15 atom cluster $[Rh_{15}(CO)_{28}C_2]^-$ anion [83] with a 'metallic' central atom also does not seem to obey the noble gas rule. Obviously local counts, starting from the real geometry of the molecule, are not often meaningful. For example in the case of $Rh_8(CO)_{19}C$ one rhodium atom (in the Rh_8 frame) would attain 17 electrons with this approach. Consequently the complex should be paramagnetic. The observed diamagnetism shows that the bonds of these molecular clusters cannot be described as localized but should be described in terms of delocalized molecular orbitals.

A convenient correlation of the potential structures dictated by the number of electrons has been suggested by Wade [88] in a manner used to rationalize the higher boron hydride compounds. The basic analogy between the two fields arises from the utilisation of three hybrid orbitals to the cluster bonding system by each metal atom, and the employment of the remaining six bonding orbitals of the initial nine orbitals ($5d + 1 + 3p$) on each metal atom for bonding ligand orbitals. The orbitals available for ligand bondings are then filled and the remaining electron pairs are considered to be involved in cluster bonding.

Therefore the essence of the Wade approach is that the highest filled and the lowest empty molecular orbitals are primarily cluster orbitals, with the metal–ligand bonding orbitals at very low energy. The basic stereochemical arrangement of the cluster is established from the number of electron pairs involved in the cluster bonding, such that $n + 1$ pairs are associated with regular polyhedra with n vertices. Therefore five bonding pairs are associated with a tetrahedral skeleton, six with a trigonal bipyramid and seven with a regular octahedron.

When the system has the same number of skeletal electron pairs as the number of metal atoms in the cluster, the structure is related to that of the metal cluster of one metal atom less: one face of the polyhedra is capped by the remaining metal atom. In this way the observed structures of $Os_5(CO)_{16}$, $Os_6(CO)_{18}$ and $[Os_6(CO)_{18}]^{2-}$ can be easily explained [89]. Wade's theory is empirical but in recent years a molecu-

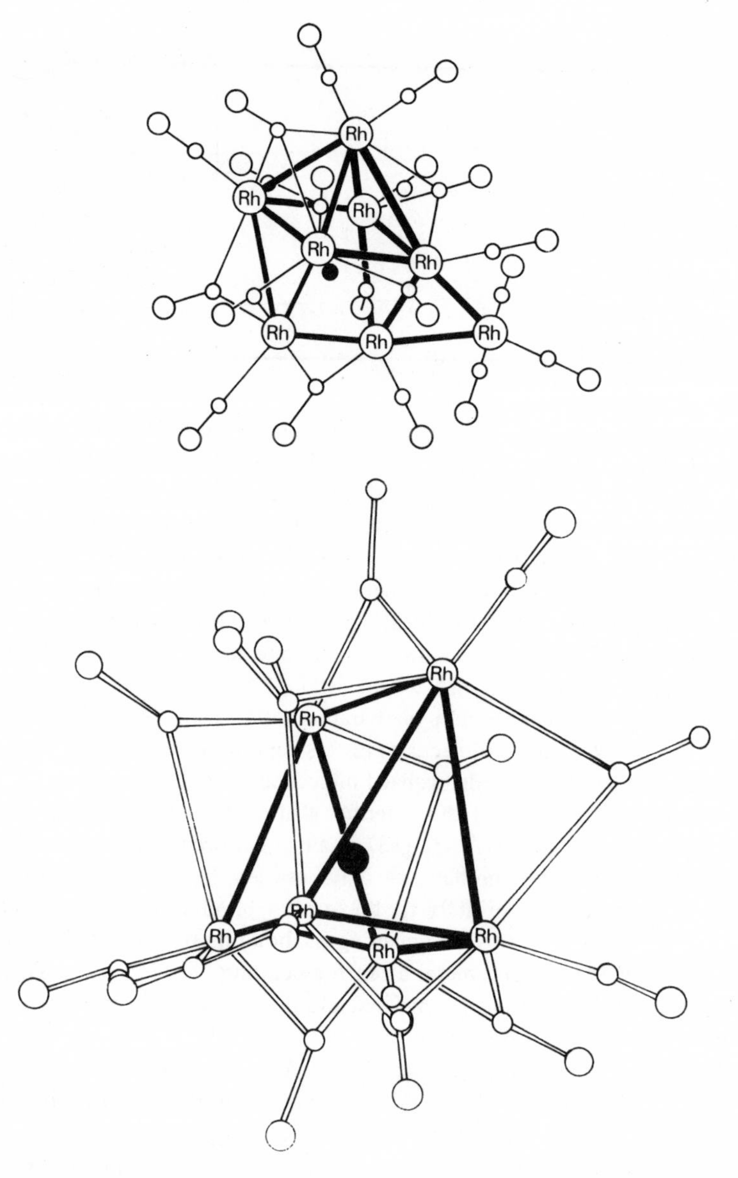

Figure 22. Structure of $Rh_8(CO)_{19}C$ (the carbide atom is black) (*a*); structure of $[Rh_6(CO)_{15}C]^{2-}$ (the carbide atom is black) (*b*).

lar orbital approach to molecular clusters has been developed. Mingos [90] has given some mathematical foundation to Wade's approach; this work has emphasized that the highest filled molecular orbitals are weakly antibonding with regard to the metal–metal framework. The molecular orbital approach, although often rather simple, produces a picture of molecular clusters in which a large electron delocalisation takes place.

In agreement with this description a very recent ESCA investigation [91] on gold clusters [81] has shown that the gold atoms look rather equivalent, although they are not such from the geometrical point of view. Obviously a large electronic delocalisation takes place in these clusters.

The great sensitivity of the basic skeletal structures of the metal cluster system to the electronic structure of the cluster bonding system implies a facility to these systems to accept and donate electrons [92] from the skeletal metal framework as in a pure metal system.

The facilities of molecular polymetal aggregates to act as electron 'sinks' can therefore support the view that, since very small particles are assumed to have in vacuum no extra ligands able to fill the orbitals of the metallic atoms of the surface, they must be rather electron deficient.

In order to reach a certain stability they must increase the bond order between the metal atoms when compared to the bond order in ordinary crystalline metals. It has effectively been found that the very small particles show high ionisation potential [38] and metal–metal distances smaller than in the bulk materal [13].

By increasing the size of the crystallites, the ionisation potential decreases whereas the metal–metal bond distance seems to increase (Figure 23).

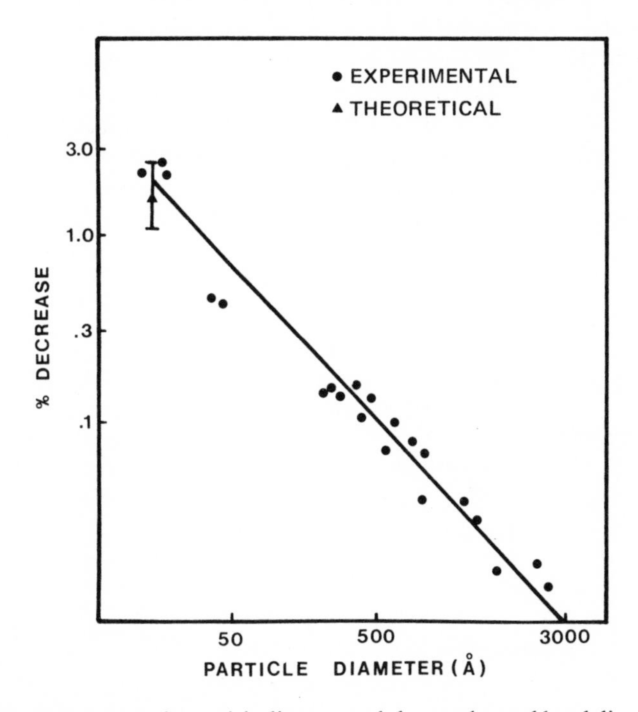

Figure 23. Relation between the particle diameter and the metal–metal bond distance.

6. METAL–METAL BOND CHARACTERISTICS IN VERY SMALL PARTICLES AND MOLECULAR CLUSTERS

As a consequence of the conclusions of the previous section an important aspect of the comparison between very small particles and molecular clusters concerns the metal–metal distances. In the case of very small particles, there is experimental evidence that the metal–metal distance is smaller than that typical of bulk material [93–98]. However, the lattice contraction is rather small since theoretical as well as experimental values seem to indicate only 3% contraction when the particle size decreases from 10^3 Å to 10 Å (Figure 23). Similar contraction (3–5%) have been observed by LEED at the surface of the 111 face of aluminium [5]; the first atomic layer being closer to the second layer than the normal distance in the bulk.

A larger variety of metal–metal distances have been observed in molecular clusters. First, it is necessary to reject the values obtained with molecular clusters with high oxidation state such as $(Re_2Cl_8)^{2-}$. The exceptional shortness of the Re–Re distance (2.24 Å) compared to 2.75 Å in the metal has been explained by the occurrence of a quadruple bond between the two rhenium atoms in a relatively high oxidation state [6]. The meaningful values must be taken out of zero-valent and low-valent cluster compounds. Some data are given in Table 3 and compared to metal–metal distance in pure metals. For the cluster compounds which do not contain an internal atom, the metal–metal distance depends strongly on the nature of the external ligands and total symmetry: for example the presence of a bridging carbonyl can decrease the Fe–Fe distance by about 0.13 Å [53]. The inverse effect seems to occur with bridging hydride which causes expansion of the metal–metal bond [99].

This fact obviously reflects the electron sink nature of the metal cage. In particular the studies of Dahl and co-workers [92] have established the amphoteric behaviour of a variety of trinuclear and tetranuclear metal clusters; extensive structural investigations on the variation of metal–metal bond distances (and indirectly also

Table 3

METAL–METAL DISTANCES IN BULK METALS AND IN MOLECULAR CLUSTERS WITH INCREASING NUMBER OF METAL ATOMS

Reference	Metal cluster	Metal–metal distance (Å)
153	Bulk gold (f.c.c.)	2.690
77–81	$[Au_6L_6]^{2+}$	3.012 (av. value)
81	$[Au_9L_8]^{3+}$	(Central atom to periphery): 2.689, 2.729 (In the periphery): 2.752, 2.868
81	$Au_{11}L_7X_3$	(Central atom to periphery): 2.68 (av. value) (In the periphery): 2.98 (av. value)

Table 3 (Continued)

Reference	Metal cluster	Metal–metal distance (Å)
153	Bulk rhodium (f.c.c.)	2.690
154	$Rh_2(CO)_3(\eta^5\text{-}Cp)_2$	2.68
155	$Rh_3(CO)_3(\eta^5\text{-}Cp)_4H$	2.70
156	$Rh_3(CO)_3(\eta^5\text{-}Cp)_3$	2.62
157	$Rh_4(CO)_{12}$	2.755 (av. value)
158	$Rh_6(CO)_{16}$	2.776 (av. value)
86	$[Rh_6(CO)_{15}C]^{2-}$	2.776 (Within the basal triangle)
		2.817 (Between the triangles)
80	$[Rh_7(CO)_{16}]^{3-}$	2.76 (av. value)
85	$Rh_8(CO)_{19}C$	2.81 (av. value)
160	$[Rh_{12}(CO)_{30}]^{2-}$	2.68 (Between the two octahedra)
		2.81 (Within the octahedra)
83	$[Rh_{15}(CO)_{28}C_2]^-$	2.90 (Central atom to periphery)
		2.738–3.332 (In the periphery)
153	Bulk iron (f.c.c.)	2.482
161	$Fe_2(CO)_9$	2.523
161	$Fe_2(CO)_4(\eta^5\text{-}C_5H_5)_2$	2.460
53	$Fe_3CO)_{12}$	2.550 (With bridging CO)
		2.685 (Without bridging CO)
162	$[Fe_4(CO)_{13}]^{2-}$	2.59 (With bridging CO) (av. value)
		2.50 (Without bridging CO)
162	$Fe_4(CO)_4(\eta^5\text{-}C_5H_5)_4$	2.52
163	$Fe_5(CO)_{15}C$	2.60 (In the square plane base)
		2.65 (From the apical Fe to the basal Fe)
164	$[Fe_6C(CO)_{16}]^{2-}$	2.70 (av. value)
153	Bulk cobalt (f.c.c.)	2.506
165	$Co_2(CO)_6(PBu)_3)_2$	2.66
166	$Co_2(CO)_8$	2.52
167	$Co_4(CO)_{12}$	2.49
107	$[Co_6(CO)_{15}]^{2-}$	2.49 (av. value)
	$[Co_6(CO)_{14}]^{4-}$	2.50 (av. value)
153	Bulk nickel (f.c.c.)	2.49
51	$Ni_2(CO)_2(\eta^5\text{-}C_5H_5)_2$	2.36
	$Ni_3(CO)_2(\eta^5\text{-}C_5H_5)_3$	2.389
52	$Ni_3(\eta^5\text{-}C_5H_5)_3N\text{-}tC_4H_9$	2.35
58	$Ni_4(\eta^5\text{-}C_5H_5)_4H_3$	2.46 (av. value)
168	$Ni_4(CO)_6(PR_3)_4$	2.508 ($R = CH_2\text{–}CH_2CN$)
66	$[Ni_5(CO)_9(\mu_2\text{-}CO)_3]^{2-}$	2.36 (Basal)
		2.32 (From apical Ni to basal Ni)
78	$[Ni_3(CO)_3(\mu_2\text{-}CO)_3]_2^{2-}$	2.38 (Within the triangle)
		2.77 (Between the triangles)

geometries) have been rationalized in terms of the electron acceptor or donor power of the cluster units.

With this assumption it can only be deduced from Table 3 that in these clusters the metal–metal distance is roughly of the same order of magnitude as that observed in bulk metal. In fact, for the clusters which have a low oxidation state, the values lie within ±0.10 Å of the metal value. If we consider now the Au_9 or Au_{11} clusters [81] for which a central atom is surrounded by 8 or 10 metal atoms, the distance between the central atom and the 'surface' atoms is significantly lower (average value 2.70 Å) than in the bulk metal (2.88 Å). However, this short distance metal–metal for a central atom does not seem to occur in the case of the anionic compound $[Rh_{15}(CO)_{28}C_2]^-$ [83]. The average distance between the central rhodium atom and the 12 nearest neighbours is equal to 2.90 Å compared with 2.69 Å for bulk rhodium. But in this case the situation is quite different since an excess of negative charge brought in the skeleton by the negative charge of the anion and the carbide atoms can explain the increase of this Rh–Rh distance; on the contrary, in the gold complexes a positive charge is localised on the cluster.

If we consider the strength of the metal–metal bond in metal clusters and metals or crystallites, requisite data for a complete comparison of metal–metal bond strength are not fully available [100].

In Table 4, we have reported the available data of ΔH for metals and some carbonyl clusters of increasing complexity [100–102].

Table 4

ENTHALPY CONTRIBUTION OF METAL –METAL BONDS IN METALS AND IN MOLECULAR CLUSTERS [100–102]

Metal	ΔH kcal/mole	Cluster	ΔH kcal/mole (metal–metal bond enthalpy contribution)
Mn	11–17	$Mn_2(CO)_{10}$	16
Re	32	$Re_2(CO)_{10}$	30
Fe	16	$[(\eta^3C_3H_5)Fe(CO)_3]_2$	13.5
Ru	27	$Ru_3(CO)_{12}$	28
Os	29	$Os_3(CO)_{12}$	31
Co	18	$Co_2(CO)_8$	11
Rh	23	$Rh_4(CO)_{12}$	26
		$Rh_6(CO)_{16}$	27.3
Ir	28	$Ir_4(CO)_{12}$	31

Trends in metal–metal bond energies in the two limiting cases are not too different.

In conclusion both metal–metal bond distances and metal–metal bond energies and their general trends support a close similarity between VSP and MC.

7. ELECTRONIC ASPECTS

7.1. Transfer of an electronic effect through metal–metal bonds of the surface of a crystallite or in the cage of molecular clusters

It is generally rather common in coordination chemistry to correlate qualitatively the stretching vibrations of various ligands to the electron density on the transition metal [103]. This results from a simple description of the metal–ligand bonding with ligands such as carbon monoxide or nitric oxide. In the case of CO, the bonding between CO and the transition metal implies the σ donation of σ lone pair of carbon and the π back-donation of metal $d\pi$ orbital to the π^* orbital of CO. This back donation reinforces considerably the metal–ligand bonding. All these effects can easily by detected by infrared spectroscopy through the shift of the ν(CO) vibration. For example with a complex such as $Ni(CO)_3L$, it is possible, by varying σ donor and π acceptor properties of L, to modify the electron density on the metal and consequently the extent of back-donation to the accepting ligand CO[103], (Table 5).

Table 5

ν(CO) FRÉQUENCIES OF $Ni(CO)_3L$ COMPLEXES ACCORDING TO TOLMAN [103]

Ligand L	ν(CO) cm^{-1}		Ligand	ν(CO) cm^{-1}	
	A_1 mode	E mode		A_1 mode	E mode
$P(t\text{-}Bu)_3$	2056	1971	$P(OEt)_3$	2076	1996
$P(nBu)_3$	2060	1976	$P(OMe)_3$	2079	2000
$PPhMe_2$	2065	1982	$P(OPh)_3$	2085	2012
PPh_2Me	2067	1987	PCl_3	2097	
PPh_3	2069	1990	PF_3	2110	

A highly basic ligand such as $P(n\text{-}C_4H_9)_3$ will result in a high electron density on Ni, leading to a low ν(CO) frequency. Conversely, a weakly basic ligand such as PF_3 will give place to a high ν(CO) frequency corresponding to a small electron density on Ni.

A similar kind of bonding occurs with the NO ligand [104] when it is coordinated under a linear form. In this case it is considered as NO^+ which is isoelectronic to CO.

With this approach, and considering the infrared spectra of simple clusters, a kind of long range interaction between ligands has been shown to occur via metal–metal bonds as for instance in the case of simple clusters [105] of formula $(CO)_5M$–$M'R_3$ (in which M =Mn, Re and M' = C, Si, Ge, Sn, Pb). In fact, it was shown that the inductive effects of R groups were operating on the electron density of the second metal through a localised σ metal–metal interaction.

By increasing the complexity of the cluster the long range electronic effect seems to occur in a completely delocalised way; for instance a kind of collective property has been shown to occur in many anionic cluster compounds (see Table 6).

Table 6

INFRARED SPECTRA IN THE ν(CO) REGION [106–109] OF THE PARENT CARBONYL CLUSTERS

$Co_6(CO)_{16}$	$[Co_6(CO)_{15}]^{2-}$	$[Co_6(CO)_{14}]^{4-}$
2103 w		
2061 s		
2057 sh	2042 m	
2026 w	1982 s	
2020 w	1959 sh	1970 w
2028 w	1778 s	1892 s
1806 w	1737 s	1730 s
1772 w	1685 m	

Direct evidence of similar long range electronic effects has been reported in chemisorption over various films of transition metals [110–113]. In this way, CO can be used as a molecular probe to indicate the electronic changes at the surface of very small particles. When the surface of a crystallite is covered with a small amount of CO, any further chemisorption of a Lewis base will result in an electronic effect very likely through metal atoms of surface of the crystallite (Figure 24).

This kind of effect has been shown to occur with L = amines [110, 111, 113], arenes [112], ethyl-isocyanide [110], water [111] and with M = Pt [110, 111], Fe, Co, Ni [110, 113], Rh and Pd [110]. The largest low frequency shifts are observed with the most basic ligands such as $N(CH_3)_3$ (Table 7). For obvious steric reasons, it is likely that the chemisorbed basic molecule is not coordinated to the same metal atom as CO, suggesting the occurrence of some collective or metal properties of the small particle. As a consequence the concept of 'long distance' or long range interaction has been introduced to explain the progressive shift of the ν(CO) vibration when introducing a progressive amount of the ligand L. Such collective effect on the surface can also explain the increase of the ν(CO) frequency when one increases the surface coverage of platinum by CO itself; at small coverage the number of free and mobile electrons available for back-donation on a π accepting ligand as CO will be high, whereas at increasing coverage they will decrease and the extent of back donation will be smaller [114, 115]. Although recent theoretical calculations seem to predict such an effect [116], another type of explanation based on coupling between carbonyl groups has been also advanced by Eischens [117] and cannot be completely rejected.

An inverse effect occurs with electron attracting ligands such as halogens or oxygen. Here the situation seems to be more complicated probably because surface oxidation occurs. Consequently, it is not established whether or not the metal in a higher oxidation state remains within the skeleton of the small particle. Infrared results seem to indicate that long distance effects no longer occur [111] suggesting than when surface platinum has reached an oxidation state different from zero, its collective properties seem to disappear and only localised surface complexes of oxidised platinum atoms are formed (see Figure 24).

Table 7

SHIFT OF THE ν(CO) FREQUENCY BY ADSORPTION OF LEWIS BASES ON PLATINUM CRYSTALLITES COVERED WITH CO[111]

Adsorbate	H_2O	NH_3	Py	$N(CH_3)_3$
1st I.P. (eV)	12.6	10.5	9.2	7.9
Frequency shift of the ν(CO) vibration (cm^{-1})	15	25	75	75

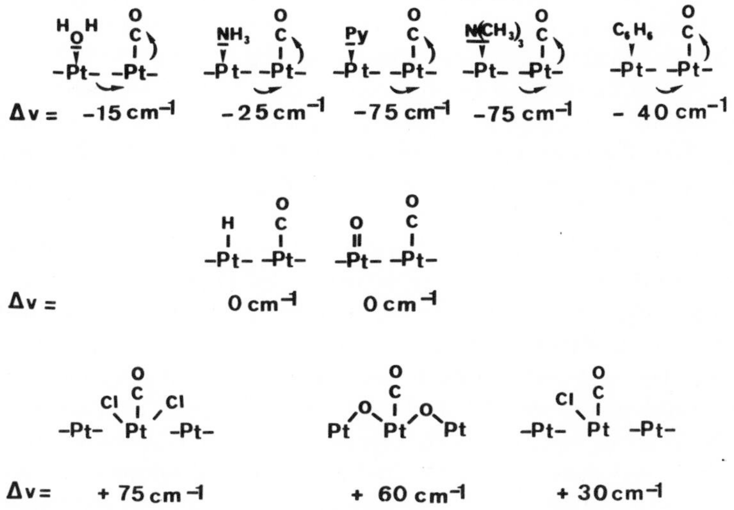

Figure 24. Mechanism of long range and short range effects in chemisorption evidenced by IR spectroscopy.

Finally, let us mention that electronic interaction through metal–metal bonds at the surface of small particles has been postulated to account for selectivity modifications obtained by introduction of selective 'poisons' or ligands, in heterogeneous metallic catalysts [118]. The effect is the highest with the smallest particles which would exhibit the highest electronic sensitivity; similar conclusions have been reached from theoretical calculations [119].

Therefore, in this field, the similarities between MC and VSP are very relevant.

7.2. Electronic interaction between very small particles and the surface of a carrier

We are concerned here with the possibility that electron transfer may occur between a supported metal and a carrier (which can be considered as a ligand), to such an extent that significant electronic modification of very small particles might occur. According to the conventional space-charge theory [120], electrons will migrate towards the metal or the support depending which has the higher work-function. In this way a potential gap equal to the work function difference is established. However a simple examination of this theory clearly indicates that no significant electron transfer will occur if the particle has a size of 20 Å or above [121]. The situation becomes much different in the case of very small particles of a few atoms.

It is known that the surface of some inorganic oxides have strong Lewis sites able to give paramagnetic charge transfer complexes with perylene (I.P. = 6.8 eV) or even benzene (I.P. = 9.2 eV) [122]. Various estimates have been made concerning the ionisation potential of very small aggregates. For instance for Ag_4 (tetrahedral) the ionisation potential lies in the region 4.7–6.0 eV; for Pd_4 Anderson estimates the value for the ionisation potential to be in the range 5.5–8 eV [121]. Therefore if some insulating oxides are able to produce a charge transfer with benzene, their electron affinity should be high enough to oxidise a Pd_4 aggregate to Pd_4^+.

Baetzold has theoretically investigated the charge transfer between very small particles of silver and a graphite support [123]. The average charge per silver atom transfered from the silver aggregate to graphite decreases sharply with the number of atoms; the charge transfered per atom becomes almost null above about 10 atoms in the aggregate.

Such an electronic transfer has been shown to occur in the case of palladium supported on various inorganic oxides. Figueras *et al.* [124] have used CO as a molecular probe to get some insight on the electron density of supported palladium. The results (Table 8) demonstrate the occurrence of an electron transfer from palladium to the most acidic solids. It must be pointed out that the most important electron transfer occurs with CaY zeolites, in this latter case the palladium particles, in order to fit in the supercages, have necessarily a diameter lower than 10 Å. Therefore, it seems that there are experimental and theoretical evidence that very small particles can give acid-base or electron transfer interactions with a support. However here again the situation depends strongly on the size of the very small particles, no effect being expected or even detectable above about 10 atoms.

As a matter of fact, direct and reversible protonation of metal clusters as $Ir_4(CO)_{12}$, $Os_3(CO)_{12}$, $Ru_3(CO)_{12}$ [125] has been observed, in this case a positively charged cage of metals is produced in a reversible way.

The interaction of carbonyl metal clusters with Lewis acids has been also observed. However in this latter case the interaction occurs mainly between some carbonyl groups (basic sites) and the Lewis acid (acid site). In any case a certain charge transfer occurs as evidenced by the shift of the mean value of carbonyl stretching frequencies [126].

7.3. Electronic interaction between two metals in alloys or in multimetallic clusters

Small particles of alloys are very important in heterogeneous catalysis for practical purposes as well as for fundamental implications [127, 128]. For instance, very

Table 8

EVIDENCE FOR AN ELECTRON TRANSFER BETWEEN VSP AND CARRIERS: INFLUENCE OF THE ACIDITY OF A SUPPORT ON THE ν(CO) FREQUENCY OF CO ADSORBED ON PALLADIUM (ACCORDING TO [124])

Carrier	ν(CO) frequency cm^{-1}
Na X zeolite	2035
MgO	2065
Al_2O_3	2075
$SiO_2-Al_2O_3K_{25}$	2050
$SiO_2-Al_2O_3K_{13}$	2100
HY zeolite	2105
CaY zeolite	2115

drastic effects can be observed upon alloying small particles of nickel with a very small amount of copper. Whereas pure nickel exhibit comparable activity for hydrogenation and hydrogenolysis, introduction of 1% copper decreases considerably the rate of hydrogenolysis, whereas specific activity for hydrogenation is almost unchanged [129]. Many effects have been claimed to occur: electronic and geometric. Before discussing them in the view of recent experimental evidence on their respective occurrence, it is necessary to consider the various models proposed for the electronic structure of alloys: 'the rigid band model' [130] assumes that both kinds of atoms form common bands, the position at the Fermi Level being determined by the alloy composition. For example, in the case of palladium/silver alloy, the additional *s* electron of Ag should be equally distributed among Ag and Pd atoms and thereby lead to a successive filling of empty *d* states of Pd until these are completely filled. This theory corresponds to a model whereby adsorption on alloys can be considered as a phenomenon related to the collective electronic structure of the solid with particular emphasis placed on the existence of *d* holes.

A very different model is called the 'minimum polarity model' [131]. According to this model each component of an alloy has an electronic structure similar to that in its pure phase. Consequently the addition of silver to palladium would dilute palladium without changing very much its individual properties. This model has been improved [132] by assuming short range interactions between palladium atoms and silver atoms in the vicinity of palladium.

Experimental evidence in favor of geometric as well as electronic interaction between two metals in small particles is given by means of a molecular probe as carbon monoxide. In the case of Pd/Ag alloys Sachtler [133] has clearly demonstrated that upon adding silver to palladium the bridged form of adsorbed CO completely disappears at the expense of the linear forms. Obviously the major effect is geometric and the palladium seems to keep its individual properties, the frequencies of the linear and bridged form being almost unchanged upon alloying with silver. However a close examination of the results indicates a weak but meaningful shift towards low frequencies of the ν(CO) vibrations (linear and bridged forms) with increasing silver content.

This observation would imply also an electronic interaction between the two metals, resulting in a higher electron density on palladium in the presence of silver.

Obviously the minimum polarity approach is good in the case of VSP. In fact similar effects have been shown to occur with particles (50–100 Å) of Ni–Cu alloys [134]. Using CO as a molecular probe it was shown that CO could be coordinated to nickel and to copper as well. When increasing copper concentration the bridged form of CO adsorbed on nickel was replaced by the linear form indicating a geometric effect of dilution. But simultaneously the ν(CO) frequencies of CO adsorbed on nickel and on copper were shifted to low frequencies, indicating a mutual electronic enrichment of both metals. Even in the alloy carbon monoxide was only weakly adsorbed on copper with a high stretching frequency (above 2100 cm^{-1}) indicating the maintenance of the poor coordinating properties of a surface of pure copper.

It is possible that a heteronuclear molecular cluster can offer a suitable explanation of these latter results; a greater variety of heteronuclear molecular clusters with different transition metals are known. However most of them possess three or four metal atoms and do not represent the best models for very small particles of alloys. Examples of heteronuclear clusters with more than four metal atoms are rare. An interesting example is given by the hexanuclear iridium–copper complex synthetized by Bruce *et al.* [135] of formula $Cu_4Ir_2(PPh_3)_2$ $(-C \equiv C{-}Ph)_8$. In this cluster the six metal atoms form a distorted octahedron in which two $Ph_3P{-}Ir$ moieties are mutually *trans*; four $-C \equiv C{-}Ph$ units are only σ bonded to each Ir atom and four $-C \equiv C{-}Ph$ units (one from each Ir) form π linkages to each of the four equatorial copper atoms (Figure 25).

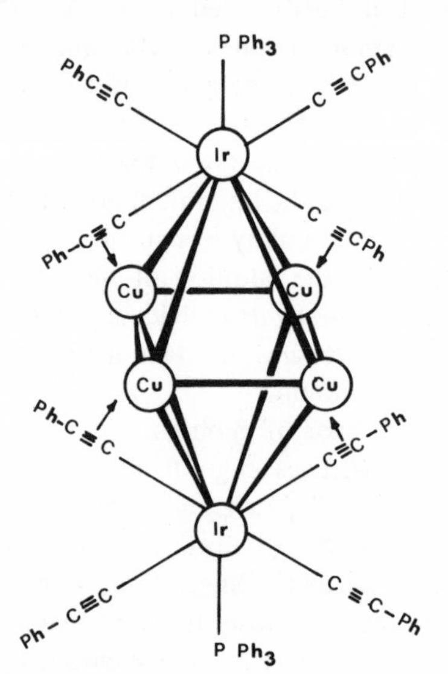

Figure 25. Structure of $Cu_4Ir_2(PPh_3)_2$ $(-C \equiv CPh)_8$.

Such a molecule is an interesting example because we may note that few ligands are interacting with copper atoms suggesting here that, even in mixed clusters between transition metals and group IB metals, each individual metal preserves its own properties. In fact it is known that copper in its low oxidation state has a tendency to form mainly dicoordinate and tricoordinate compounds. Analogously the crystals structures of gallium(III) and indium(III) complexes of formula $Mn_2(CO)_8(\mu\text{-}MMn(CO)_5)_2$ are the following [136]

```
                (CO)4
                 Mn
               /    \
(CO)5Mn—M <          > M—Mn(CO)5        M = Ga, In
               \    /
                 Mn
                (CO)4
```

The gallium and indium atoms maintain their typical tricoordination and do not interact with carbon monoxide which interacts only with the transition metal atoms (manganese). Such a behaviour means that non-transition and transition metal atoms maintain their own properties even when directly bound also in a complex cluster cage. In agreement with this point in a compound as $Sn[Co(CO)_4]_2[Mn(CO)_5]_2$ [137] the cobalt and manganese atoms maintain their typical coordination as the tin atom.

Of course from the values of CO stretching frequencies, small differences of the electronic densities on Co and Mn atoms (when compared with the parent compounds $Co_2(CO)_8$ and $Mn_2(CO)_{10}$) can be detected.

7.4. Influence of the size of very small particles and of molecular clusters on their electronic properties

Although this problem is related to the preceding part of this article, it deserves to be treated separately. Various results have been presented which indicate that very small particles seem to keep their metallic character at least to some extent. Let us mention the electronic interaction between two different chemisorbed species through metal–metal bonds on a surface [110–113], the variation of π back-donation on chemisorbed CO when the coverage varies [114], the electronic interaction between a very small particle and a carrier [124] and finally, although not very significant, the electronic interaction between two metals in alloys [133. 134].

The problem we are concerned with is related to the modification of these collective properties when the number of metal atoms decreases as in a small particle. In a macroscopic particle with metallic character the difference of energy between the levels is very small, which results in a band structure. The spacing between the levels δE depends on the size and the shape of the aggregate [5]:

$$\delta E = (E_M - E_0) \times (K_M L)^n$$

where $E_M - E_0$ is the band width, K_M is the wave vector at the Fermi level, L is the highest dimension of the particle and n is equal to -1 for linear shape -2 for films and -3 for spherical shape; (K_M is equal to a^{-1}, a being the interatomic distance).

When δE is higher than *ca.* KT, the distribution of states at the Fermi level cannot be considered any more as a quasi continuum and there will be a transition from the metallic to the molecular state of the aggregate. A rough calculation has been made by Friedel [5] assuming that the band width is equal to about 10 eV. At room temperature $\delta E \simeq KT$ for 10^3, 30 and 10 atoms for linear shape, film and spherical shape of the particles respectively. Using the same kind of rough calculation, Anderson [121] finds a deviation from the metallic state for particles having less than about 500 atoms.

In conclusion it seems that the evidence of collective electronic properties in metallic clusters appears with about 10 atoms. in agreement with this point, it is interesting to note here that molecular clusters present typical colour darkening when the number of metal atoms increases [2]. Usually the compounds are dark and even completely black when the number of metal atoms is higher than *ca.* 6–10. For instance $[Rh_{15}(CO)_{28}C_2]^-$ anion is a black crystalline solid and this might well be related to the appearance of a band structure as in a metallic state (on the contrary $Rh_4(CO)_{12}$ is orange-red). Although we have not at the moment any sure physico-chemical evidence of band electronic structure in high molecular weight clusters, the recent ESCA on gold clusters [91] are not against this hypothesis.

Indirect evidence can be obtained by chemisorption studies on very small particles. In fact chemisorption is related to the behaviour of metallic particles as a sink of electrons: when the size of the particles decreases the ratio surface atoms/bulk atoms increases as the inverse of the particle size. Therefore one should expect that the smaller the size, the smaller the number of free electrons, which means the higher the ionisation potential.

Whereas for bulk metal the ionisation potential would be equal to the work function and to the electron affinity, for very small particles one should expect a high ionisation potential. This result has been observed with various techniques: using NO^+ as a molecular probe [138], it was found that the ν(NO) vibration of NO adsorbed on platinum was particle size dependent, the smaller the size, the higher the ν(NO) vibration, corresponding to the smaller back-donation on the π accepting ligand, in agreement with a higher ionisation potential of the particles. The frequency shift is equal to 45 cm^{-1} when the particle size decreases from 35 Å to 15 Å. The shift is the most pronounced below 15 Å, and the value obtained in the case of a very small cluster of platinum in the supercages of Y zeolites is very high at 1828 cm^{-1} [139] compared to 1775 cm^{-1} for 35 Å particles (Figure 26).

Similar observations on the trends of ionisation potentials have been obtained by mass spectrometry on very small particles of selenium in the vapour phase. There is a continuous decrease of the I.P. when the number of atoms goes from 2 to 8 (about 1 eV) [38].

The above experimental observations have been confirmed by a large amount of theoretical calculations. In most cases semi-empirical molecular orbital procedure has been applied to the problem of very small particles. Recent calculations have treated

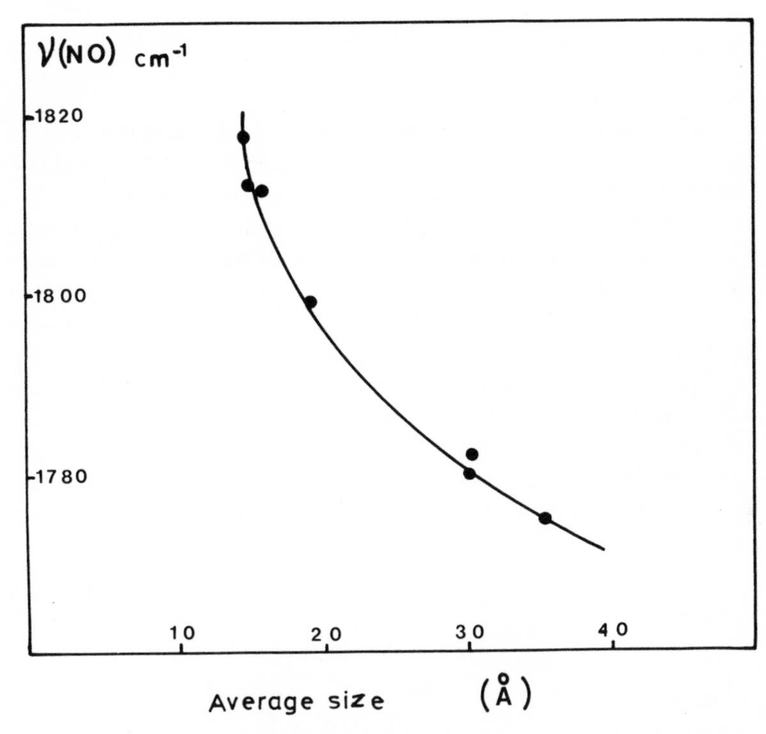

Figure 26. Dependence of ν_{NO} of chemisorbed NO from the size of the metal particle.

Li and Be aggregates [140] C particles [141], Fe chains [142], Ni [143] Ag [144] [145] and Na aggregates [146]. In all cases the electronic properties of the small particles are different from bulk properties. It was found with silver that the I.P. decreases from its single atom value towards the bulk work function as cluster size increases. In the case of Na particles [146] it was found that smaller particles exhibit the higher ionisation potential and excitation energy but lower bond energy.

The recent theoretical approach using SCF-Xα-SW method [44] applied by Fripiat *et al.* [43] to small particles of lithium has shown how the electronic structure and binding energy per atom vary with cluster size and geometry. A systematic trend of increasing cohesion with increasing cluster size and coordination number was observed. A Li_{13} icosahedron aggregate, which has the most stable geometry, has a binding energy per atom equal to about 60% of the bulk crystalline value. This theoretical result has to be related to other theoretical calculations [147] as well as experimental evidence [148] concerning the decrease of the melting temperature when the size of the particle decreases.

Influence of the size of molecular clusters on their electronic properties has not been, to our knowledge, the subject of any review. However, a very interesting example can be found in the anionic clusters prepared by Chini *et al.* [79] derived from platinum. This family of compounds has the formula:

$[Pt_3(\mu_2\text{-}CO)_3(CO)_3]_n^{2-}$

where n is equal to 2, 3, 4, 5.

The X-ray structure showed a monomeric unit of triangular geometry of three platinum atoms with a linear and a bridged carbonyl atom per platinum (Figure 27). The individual intra triangular Pt–Pt distances have an average value of 2.66 Å (compared with 2.775 Å for bulk platinum). The inter triangular Pt–Pt distances are equal to 3.04 Å ± 0.06. The interesting aspect of these polymeric molecules lies in the repetition of a fundamental structural unit ($Pt_3(CO)_6$) with a constancy of the anionic charge. The ν(CO) frequencies are very meaningful with this regard, since the ν(CO) frequencies for linear and bridged CO are increasing with n to reach a plateau for $n = 5$ (Figure 28). This produces an interesting observation on the relation between small particles and molecular clusters: the ν(CO) frequencies for $n = 5$ (that is for the $[Pt_{15}(CO)_{30}]^{2-}$ cluster) are equal to 2045 cm^{-1} and 1870 cm^{-1} which are exactly the values obtained in the case of CO adsorbed on metallic platinum (linear and bridged forms). This indicates the great analogy of behaviour between a particle

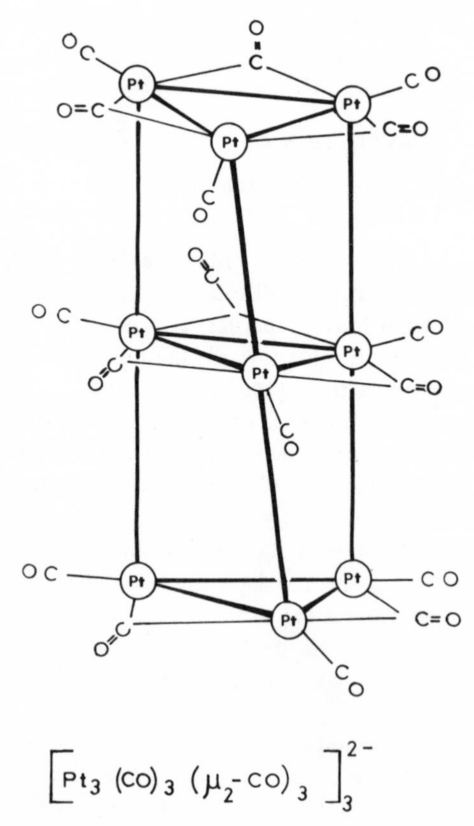

Figure 27. Structure of $[Pt_3(CO)_3(\mu_2\text{-}CO)_3]_3^{2-}$.

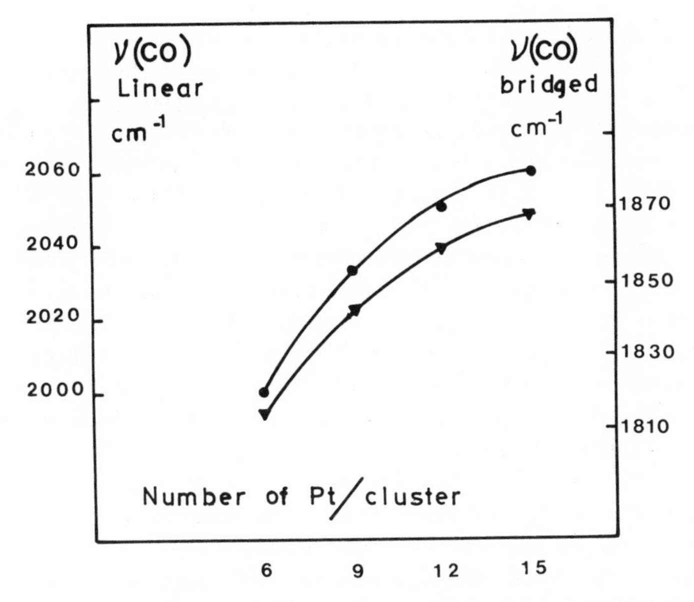

Figure 28. Dependence from the number of platinum atoms of ν_{CO} (terminal • and bridged ▼ CO) in $[Pt_3(CO)_3(\mu_2\text{-}CO)_3]_n^{2-}$ series ($n = 2, 3, 4, 5$).

of platinum which is fully covered with chemisorbed CO and a cluster compound of 15 platinum atoms with carbonyl ligands.

Moreover the progressive shift observed for ν(CO) frequencies, when increasing the number of platinum atoms, makes it necessary to invoke a molecular orbital model able to delocalize the negative charge over the entire platinum cluster system. At the same time, the donor properties of the platinum cage, change in the same way reported for platinum small particles chemisorbed by NO [138].

8. CONCLUSION

In this paper we have emphasized the structural and electronic similarities between VSP and MC, at least in those cases in which these similarities are rather evident. Of course, it would also be very important to find some more direct catalytic correlations, but at the moment, this aspect is actually rather poor.

Although the homogeneous catalytic activity of metal clusters has been recently the subject of a series of investigations [100], direct evidence that a real metal cluster cage is maintained as such under catalytic conditions is lacking in most cases. However, the recent work done with clusters as $Ni_4(CNR)_7$ [149], $Rh_6(CO)_{16}$ [150, 151], or $Ru_3(CO)_{12}$ [152] has shown that small clusters can display, under particular conditions, catalytic activity in reactions such as cyclisation of acetylene and butadiene [149], polymerization of allene [149], oxidation of ketones [150], oxidation with O_2 of CO to CO_2 [150], conversion of hydrogen and carbon monoxide to methanol and ethylene glycol [151] and hydrogenation of aromatics [152]. Mechanistically none of these catalytic reactions is even partially delineated. However in many cases there is direct spectroscopic evidence that cluster fragmentation does not occur.

Meanwhile very active and small crystallites have been obtained, for instance, by thermal decomposition and reduction with H_2 of molecular clusters as $[Pt_3(CO)_6]_n^{2-}$ ($n = 2$, 3, 4, 5) [11] or $(\eta^5\text{-}C_5H_5)_2Ni_2(CO)_2$ and $(\eta^5\text{-}C_5H_5)_3Ni_3(CO)_2$ [12]. It appears that the size of metallic particles obtained by these procedures does not exceed the size of the original molecular metal cluster. However it has not been proved in the case of platinum that the metal particles contain the same number of metal atoms as the molecular clusters from which they are formed. In the case of nickel particles [12] spectroscopic and chemisorption evidence suggest that the supported nickel atoms retain the original geometrical disposition of the parent molecular cluster. Meanwhile a certain reversibility to reform a carbonyl cluster by reaction with CO has been proved.

In conclusion, by both a homogeneous and heterogeneous catalytic approach, different groups are trying to develop the use of molecular metal clusters to produce new aspects of catalysis related to very small particles of metals.

On this line two recent review articles [41, 100] have emphasized that metal cluster complexes and particularly organometallic metal cluster complexes can be considered miniature versions of metal surfaces and particularly of metal surfaces of very small particles. This latter approach can be the origin of a series of more sophisticated models of surface species and surface reactivity, thus producing a new impulse to the knowledge of surface kinetics and consequently of the elementary steps of surface complex reactions.

9. REFERENCES

[1] J. A. Rabo, V. Schomaker and P. E. Pichert, *Proc. IIIrd International Conference on Catalysis - Amsterdam, 1964,* Vol. 2, p. 264, North-Holland, Amsterdam, (1965).

[2] *a*) P. Chini, *Inorganica Chim. Acta Rev.* 31, (1968).
b) F. W. Abel and F. G. A. Stone, *Quart. Rev.* **23**, 325, (1969).
c) P. Chini, *Pure and Applied Chem.*, **23**, 489, (1970).
d) R. B. King, *Progr. Inorg. Chem.*, **15**, 287, (1972).

[3] O. M. Poltorak, V. S. Boronin and A. N. Mitrofavona, *Proc. IVth International Conference on Catalysis, Moscow 1968,* Vol. 2, p. 267, Akademiai Kiabv, Budapest (1971).

[4] G. C. Bond, *Platinum Metal Rev.* **19**, 126, (1975), and references therein.

[5] Friedel, *Proc. Ecole d'ètè Thèorie des Mètaux et Catalyse, Lyon 1975,* Vol. 2, p. VIII, CNRS (ed.), (1975), and references therein.

[6] F. A. Cotton, *Quart. Rev.*, **20**, 389, (1966).

[7] B. Penfold, *Perspect. Struct. Chem.*, **2**, 71, (1968), and references therein.

[8] P. Gallezot, A. Alarcon-Diaz, J. A. Dalmon, A. J. Renouprez and B. Imelik, *J. Catalysis*, **39**, 334, (1975).

[9] *a*) C. R. Eady, B. F. G. Johnson, J. Lewis and T. Matheson, *J. Organometal. Chem.* **57**, C82, (1973).
b) M. I. Bruce, *Adv. Organomet. Chem.* **6**, 273, (1968), and references therein.
c) B. F. Johnson, J. Lewis, G. Williams, J. M. Wilson, *J. Chem. Soc. A*, 341, (1975).

[10] *a*) L. R. Anderson and D. E. Mainwaring, *J. Catalysis,* **35,** 162, (1974).
b) G. S. Smith, T. P. Chojnacki, S. R. Dasgupta, K. Iwatate and K. L. Waters, *Inorg. Chem,* **14,** 1419, (1975).

[11] M. Ichikawa, *J. Chem. Soc., Chem. Commun.,* 11, (1976).

[12] M. Ichikawa, *J. Chem. Soc., Chem. Commun.,* 26, (1976).

[13] J. J. Burton, *Catalysis Rev.* **9,** 209, (1974), and references therein.

[14] T. E. Whyte, *Catalysis Rev.,* **8,** 117, (1974), and references therein.

[15] A. D. O'Cinneide and J. D. A. Clarke, *Catalysis Rev.,* **7,** 214, (1973), and references therein.

[16] E. W. Muller, *J. Appl. Phys.,* **27,** 474, (1956).

[17] A. Y. Crewe, J. Wall and J. Langmore, *Science,* **168,** 1338, (1970).

[18] J. M. Basset, G. Dalmai-Imelik, M. Primet and R. Mutin, *J. Catalysis* **37,** 22, (1975).

[19] E. B. Prestridge and D. J. C. Yates, *Nature,* **234,** 345, (1971).

[20] P. de Montgolfier, B. Moraweck, G. A. Martin, A. J. Renouprez and G. Dalmai-Imelik, *Proc. IInd International Symposium on Fine Particles, Boston, 1973,* p. 43, W. E. Kuhn and J. Ehretsmann (Eds.), The Electrochemical Society, Princeton (1974), and references therein.

[21] A. J. Renouprez, C. Hoang-Van and P. A. Compagnon, *J. Catalysis,* **30,** 146, (1973).

[22] P. E. Selwodd, *Chemisorption and Magnetisation,* p. 17, Academic Press, (1975).

[23] R. Denton, B. Mulschelegel and D. J. Scalapino, *Phys. Rev.* **B7,** 3589, (1973).

[24] P. Bussiere, R. Dutartre, G. A. Martin and J. P. Mathieu, *Compt. Rend. Ac. Sci. (C),* **280,** 1133, (1975), and references therein.

[25] B. J. Wood and H. Wise, *Surface Sci.,* **52,** 151, (1975), and references therein.

[26] L. H. Scharpen, *J. Electron Spectroscopy and Related Phenomena,* **5,** 369, (1975), and references therein.

[27] J. E. Benson and M. Boudart, *J. Catalysis,* **4,** 704, (1965).

[28] D. E. Mears and R. C. Hansford, *J. Catalysis,* **9,** 125, (1967).

[29] J. Basset, A. Theolier, M. Primet and M. Prettre, *Proc. Vth International Conference on Catalysis, Palm Beach 1972,* Vol. 2, p. 915, Hightower (Ed.), North Holland (1973).

[30] G. R. Wilson and W. K. Hall, *J. Catalysis,* **17,** 190, (1970).

[31] R. A. Dalla Betta and M. Boudart, *Proc. Vth International Conference on Catalysis, Palm Beach 1972,* Vol. 2, p. 1319, Hightower (Ed.), North Holland (1973).

[32] M. Boudart and H. S. Awang, *J. Catalysis,* **39,** 44, (1975).

[33] M. Gillet, E. Gillet and A. Renou, *Thin Solid Films,* **29,** 217, (1975), and references therein.

[34] G. Dalmai-Imelik, C. Leclercq and A. Maubert-Muguet, *J. Solid State Chem.,* **16,** 129, (1976).

[35] C. G. Grandqvist and T. Ciaeson, *Abstracts Meeting on Random Packing Structure,* Orsay, (1974).

[36] C. L. Briant and J. J. Burton, *Nature Phys. Sci.,* **243,** 100, (1973), and references therein.

[37] J. R. Sambles, *Proc. Roy. Soc., Ser. A,* **324,** 379, (1971).

[38] A. Hoareau, J. M. Reymond, B. Cabaud and R. Uzan, *J. de Physique,* **36,** 737, (1975).

[39] P. Chini, G. Longoni and V. G. Albano, *Adv. Organomet. Chem.,* **14,** 285, (1976), and references therein.

[40] S. Owaka and D. F. Shriver, *Inorg. Chem.,* **15,** 915 (1976), and references therein.

[41] R. Ugo, *Catalysis Rev.,* **11,** 225, (1975).

[42] M. R. Hoare and P. Pal, *J. Cryst. Growth,* **17,** 77, (1972).

[43] J. G. Fripiat, K. T. Chon, M. Boudart, J. B. Diamond and K. H. Johnson, *J. Molec. Catalysis,* **1,** 59, (1975).

[44] J. C. Slater and K. H. Johnson, *Phys. Rev.,* **B5,** 844, (1972).

[45] Van Hardeveld and F. Hartog, *Surface Sci.,* **15,** 189, (1969).

[46] P. C. Fantucci and P. Balzarini, personal communication.

[47] J. G. Allpress and J. V. Sanders, *Surface Sci.,* 7, 1, (1967).

[48] S. Mader, *J. Vacuum Sci. Technology,* **8,** 247, (1971).

[49] J. Farges, unpublished data quoted from reference [50].

[50] J. Dixmier, *J. de Physique,* **35,** C4, (1974).

[51] *a*) E. O. Fisher and C. Palm, *Chem. Ber.,* **91,** 1725, (1958).

b) O. S. Mills and B. W. Shaw, *J. Organometal. Chem.*, **11**, 595 (1968).
[52] *a*) Sei Otsuka, A. Nakamura and T. Yoshida, *Annalen*, **719**, 54, (1968).
b) Sei Otsuka, A. Nakamura and T. Yoshida, *Inorg. Chem.*, **7**, 263, (1968).
[53] C. H. Wei and L. F. Dahl, *J. Amer. Chem. Soc.*, **91**, 1351, (1969), and references therein.
[54] E. R. Corey and L. F. Dahl, *J. Amer. Chem. Soc.*, **83**, 2203, (1961); *ibid. Inorg. Chem.* **1**, 521, (1962).
[55] J. R. Norton, J. P. Collman, G. Dolcetti and W. T. Robinson, *Inorg. Chem.* **11**, 382, (1972).
[56] C. W. Bradford, W. van Bronswijk, R. J. H. Clark and R. S. Nyholm, *J. Chem. Soc. A*, 2456, (1968).
[57] C. H. Wei and L. F. Dahl, *J. Amer. Chem. Soc.*, **89**, 4792, (1967).
[58] *a*) J. Muller, H. Darner, G. Huttner and H. Lorentz, *Angew. Chem. Int. Ed.*, **12**, 1005, (1973).
b) G. Huttner and H. Lorentz, *Chem. Ber.*, **107**, 996, (1974).
[59] *a*) P. Braunstein, J. Dehand, B. Muchenbach and J. F. Nennig, *Proc. XVI ICCC Dublin*, paper R38 (1974).
b) P. Braunstein, J. Dehand and J. F. Nennig, *J. Organomet. Chem.*, **92**, 117, (1975).
[60] L. F. Dahl and D. L. Smith, *J. Amer. Chem. Soc.*, **84**, 2450, (1962).
[61] R. Mason and K. M. Thomas, *J. Organometal. Chem.* **49**, C33, (1973).
[62] V. W. Day, R. O. Day, J. S. Kristoff, F. J. Hirsekorn and E. L. Muetterties, *J. Amer. Chem. Soc.*, **97**, 2571, (1975).
[63] R. Bau, D. Fontal, H. D. Kaesz and M. R. Churchill, *J. Amer. Chem. Soc.*, **89**, 6374, (1967).
[64] *a*) C. W. Bradford and R. S. Nyholm, *Chem. Commun.*, 867, (1968).
b) C. W. Bradford and W. Van Bronswijk, R. J. H. Clark and R. S. Nyholm *J. Chem. Soc. A*, 2889, (1970).
[65] H. Stoll and H. Preuss, *Phys. Status Solidi (b)*, **53**, 519, (1972).
[66] G. Longoni, P. Chini, L. D. Lower and L. F. Dahl, *J. Amer. Chem. Soc.*, **97**, 5034, (1975).
[67] C. R. Eady, B. F. G. Johnson, J. Lewis, B. E. Reichert and G. M. Sheldrick, *J. Chem. Soc. Chem. Commun.*, 271, (1976).
[68] J. K. Ruff, R. P. White and L. F. Dahl, *J. Amer. Chem. Soc.*, **93**, 2158, (1971), and references therein.
[69] J. D. Corbett and P. A. Edwards, *J. Chem. Soc. Chem. Commun.*, 984, (1975), and references therein.
[70] J. D. Corbett and R. E. Rundle, *Inorg. Chem.*, **3**, 1408, (1964).
[71] J. D. Corbett, *Inorg. Chem.*, **7**, 198, (1968), and references therein.
[72] R. Mason, K. M. Thomas and D. M. P. Mingos, *J. Amer. Chem. Soc.*, **95**, 3802, (1973).
[73] J. Lewis and C. Malatesta (personal communication).
[74] C. R. Eady, B. F. G. Johnson and J. Lewis, *J. Chem. Soc. Chem. Commun.*, 302, (1976).
[75] T. G. Spiro, D. H. Templeton and A. Zalkin, *Inorg. Chem.*, **8**, 856, (1969).
[76] W. Moser and R. A. Howie, *Nature*, **219**, 372, (1968).
[77] P. L. Bellon, M. Manassero, L. Naldini and M. Sansoni, *Chem. Commun.*, 1035, (1972).
[78] J. C. Calabrese, L. F. Dahl, P. Chini, G. Longoni and S. Martinengo, *J. Amer. Chem. Soc.*, **96**, 2616, (1974).
[79] J. C. Calabrese, L. F. Dahl, P. Chini, G. Longoni and S. Martinengo, *J. Amer. Chem. Soc.*, **96**, 2614, (1974).
[80] V. G. Albano, P. L. Bellon and G. Ciani, *Chem. Commun.*, 1024, (1969).
[81] L. Malatesta, *Gold Bulletin*, **8**, 48, (1975), and references therein.
[82] V. G. Albano, A. Ceriotti, P. Chini, G. Ciani, S. Martinengo and W. M. Anker, *J. Chem. Soc. Chem. Commun.*, 859, (1975).
[83] *a*) V. G. Albano, P. Chini, S. Martinengo, M. Sansoni and D. Strumolo, *J. Chem. Soc. Chem. Commun.*, 299, (1974).
b) *ibid J. Chem. Soc. Dalton*, 937 (1976).
[84] *a*) E. H. Braye, L. F. Dahl, W. Hubel and D. L. Wampler, *J. Amer. Chem. Soc.*, **84**,

4633, (1962).
b) J. G. Bullitt, F. A. Cotton and T. J. Mark, *Inorg. Chem.*, **11**, 671 (1972).
[85] V. G. Albano, M. Sansoni, P. Chini, S. Martinengo and D. Strumolo, *J. Chem. Soc. Dalton*, 305, (1975).
[86] V. G. Albano, M. Sansoni, P. Chini and S. Martinengo, *J. Chem. Soc. Dalton*, 651, (1973).
[87] V. G. Albano, P. Chini, S. Martinengo, P. S. McCaffrey, D. Strumolo and B. T. Heaton, *J. Amer. Chem. Soc.*, **96**, 8106, (1974).
[88] *a*) K. Wade, *Electron Deficient Compounds*, Nelson, London, (1971).
b) K. Wade, *Chemistry in Britain*, **11**, 177, (1975).
[89] J. Lewis and B. F. G. Johnson, *Pure and Applied Chem.*, **44**, 43, (1975).
[90] D. M. P. Mingos, *J. Chem. Soc. Dalton*, 133, (1974).
[91] C. Battistoni, F. Cariati, G. Mattogno, L. Naldini and A. Sgamelloti, *Proc. Meeting Ass. It. Chim. Inorg. – Iesolo (Italy)*, paper B16 (1975).
[92] *a*) M. A. Neumann, T. Toan and L. F. Dahl, *J. Amer. Chem. Soc.*, **94**, 3383, (1972).
b) T. Toan, W. P. Fehlhammer and L. F. Dahl, *J. Amer. Chem. Soc.*, **94**, 3389 (1972).
c) G. L. Sinon and L. F. Dahl, *J. Amer. Chem. Soc.*, **95**, 2164, (1973).
d) P. D. Frisch and L. F. Dahl, *J. Amer. Chem. Soc.*, **94**, 3409, (1972).
e) B. K. Teo, M. B. Hall, R. F. Fenske and L. F. Dahl, *J. Organometal. Chem.*, **70**, 413, (1974).
[93] C. W. B. Grigson, *Nature*, **212**, 749, (1966).
[94] F. W. C. Boswell, *Proc. Phys. Soc. A*, **64**, 465, (1971).
[95] F. G. Karioris, J. J. Woyci and R. R. Buckrey, *Adv. X-Ray Analysis*, **10**, 250, (1967).
[96] C. R. Berry, *Phys. Rev.*, **88**, 596, (9152).
[97] C. L. Briant and J. J. Burton, *Surface Sci.*, **51**, 345, (1975).
[98] D. Schroeer, R. F. Marzke, D. J. Erickson, S. W. Marshall and R. M. Wilenzick, *Phys. Rev.* **B2**, 4414, (1970).
[99] E. L. Muetterties, *Transition Metal Hydrides*, p. 24, M. Dekker, New York, (1971).
[100] E. L. Muetterties, *Bull. Soc. Chim. Belg.*, **84**, 959, (1975), and references therein.
[101] J. A. Connor, H. A. Skinner and Y. Virmani, *Faraday Symp. Chem. Soc.* **8**, 18, (1974).
[102] L. S. Brown, J. A. Connor and H. A. Skinner, *J. Chem. Soc. Faraday I*, **71**, 699, (1975), and references therein.
[103] C. A. Tolman, *J. Amer. Chem. Soc.*, **92**, 2953 (1970).
[104] N. G. Conelly, *Inorg. Chim. Acta Rev.*, 47, (1972).
[105] R. Ugo, S. Cenini and F. Bonati, *Inorg. Chim. Acta*, **1**, 451, (1967).
[106] P. Chini, *Inorg. Chem.*, **8**, 1206, (1969).
[107] *a*) V. G. Albano, P. Chini and V. Scatturin, *Chem. Commun.*, 163, (1968).
b) *ibid.*, 423 (1968).
[108] P. Chini, V. G. Albano and S. Martinengo, *J. Organometal. Chem.*, **16**, 471, (1969).
[109] V. G. Albano, P. L. Bellon, P. Chini and V. Scatturin, *J. Organometal. Chem.*, **16**, 461, (1969).
[110] R. Queau and R. Poilblanc, *J. Catalysis*, **27**, 200, (1972).
[111] M. Primet, J. M. Basset, M. V. Mathieu and M. Prettre, *J. Catalysis*, **29**, 213, (1973).
[112] A. Palazov, *J. Catalysis*, **30**, 13, (1973).
[113] R. W. Sheets and R. S. Hansen, *J. Phys. Chem.*, **76**, 7, (1972).
[114] J. M. Basset, *Catalysis: Heterogeneous and Homogeneous*, p. 69, Jannes and Delmon (Eds.), Elsevier, Amsterdam, (1975).
[115] R. Poilblanc and R. Queau, *Catalysis: Heterogeneous and Homogeneous*, p. 209, Jannes and Delmon (Eds.), Elesevier, Amsterdam, (1975).
[116] E. Miyazaki, *J. Catalysis*, **33**, 57, (1974).
[117] R. P. Eischens, *Accounts Chem. Res.*, **5**, 74, (1972), and references therein.
[118] I. De Aguirre and B. Duque, *Catalysis: Heterogeneous and Homogeneous*, p. 63, Jannes and Delmon (Eds.), Elsevier, Amsterdam, (1975).
[119] D. V. Sokolsky and Y. A. Dorfamn, *Catalysis: Heterogeneous and Homogeneous*, p. 33, Jannes and Delmon (Eds.), Elsevier, Amsterdam, (1975).
[120] F. Solymosi, *Catalysis Rev.*, **1**, 233, (1968).
[121] J. R. Anderson, *Structure of Metallic Catalysts*, p. 278, Academic Press, New York, (1975).

[122] J. M. Basset, C. Naccache, M. V. Mathieu and M. Prettre, *J. de Chim. Phys.*, **66**, 1522, (1969).
[123] R. C. Baetzold, *Surface Sci.*, **36**, 123, (1972).
[124] F. Figueras, R. Gomez and M. Primet, *Adv. Chem. Ser.*, **121**, 266, (1973).
[125] *a*) J. Knight and M. J. Mays, *J. Chem. Soc. A*, 711, (1970).
b) P. E. Cattermole, K. G. Orrell and A. G. Osborne, *J. Chem. Soc. Dalton*, 328, (1974).
[126] J. S. Kristoff and D. F. Shriver, *Inorg. Chem.*, **13**, 499, (1974), and references therein.
[127] E. M. Blues, *Hydrocarbon Processing*, **48**, 141, (1969).
[128] C. S. McCoy and P. Hunk, *Chem. Eng. Prog.*, **67**, 78, (1971).
[129] J. H. Sinfelt, J. L. Carter and D. J. C. Yates, *J. Catalysis*, **24**, 283, (1972); J. H. Sinfelt, *Platinum Met. Rev.* **20**, 114, (1976).
[130] N. F. Mott, *Proc. Phys. Soc. A*, **316**, 169,(1970).
[131] N. D. Lang and H. Ehrenreich, *Phys. Rev.*, **168**, 605, (1968).
[132] K. Christmann and G. Ertl, *Surface Sci.*, **33**, 254, (1972).
[133] Y. Soma-Noto and S. M. H. Sachtler, *J. Catalysis*, **32**, 315, (1974).
[134] J. A. Dalmon, M. Primet, G. A. Martin and B. Imelik, *Surface Sci.*, **50**, 95, (1975).
[135] O. M. Abu Salah, M. I. Bruce, M. R. Churchill and S. A. Bezman, *Chem. Commun.*, 859, (1972).
[136] H. Prent and H. J. Haupt, *Chem. Ber.*, **107**, 2860, (1974).
[137] J. P. Collmann, J. K. Hoyano and D. W. Murphy, *J. Amer. Chem. Soc.*, **95**, 3424, (1975).
[138] M. Primet, J. M. Basset, E. Garbowski and M. V. Mathieu, *J. Amer. Chem. Soc.*, **97**, 3655, (1975).
[139] P. Gallezot, J. Dakta, J. Massardier, M. Primet and B. Imelik, *Proc. VIth International Conference on Catalysis*, London, A11, (1976).
[140] H. Stoll and H. Preuss, *Phys. Status Solidi*, **53**, 519, (1972).
[141] R. Hoffmann, *Tetrahedron*, **22**, 521, (1966).
[142] H. Dunken, *Zeit. für Chem.*, **12**, 433, (1973).
[143] G. Blyholder, *J. Chem. Soc. Chem. Commun.*, 625, (1973).
[144] R. C. Baetzold, *J. Chem. Phys.*, **55**, 4363, (1971).
[145] R. C. Baetzold, *J. Catalysis*, **29**, 129 (1973).
[146] R. C. Baetzold and R. E. Mack, *Inorg. Chem.*, **14**, 686, (1975).
[147] J. P. Borel, *Compt. Rend. Ac. Sci. C*, **277**, 1275, (1973), and references therein.
[148] M. P. Blackman and J. R. Sambles, *Nature*, **226**, 938, (1970).
[149] *a*) V. W. Day, R. O. Day, J. S. Kristoff, F. J. Hirsekorn and E. L. Muetterties, *J. Amer. Chem. Soc.*, **97**, 2571, (1975).
b) M. G. Thomas, B. F. Beier and E. L. Muetterties, *J. Amer. Chem. Soc.*, **98**, 1296, (1976).
c) *Ibid.*, manuscript in preparation.
d) M. G. Thomas, E. L. Muetterties, R. O. Day and V. W. Day, *J. Amer. Chem. Soc.*, in submission.
[150] G. D. Hercer, J. S. Shu, T. B. Rauchfuss and D. M. Roundhill, *J. Amer. Chem. Soc.*, **97**, 1967, (1975), and references therein.
[151] *Belgian Pat. 815841* (to Union Carbide).
[152] L. Bianchi *et al.*, Unpublished results quoted from reference [41].
[153] *Internation Tables for X-ray Crystallography*, Vol. 3, p. 278, Kynoch Press, Birmingham, (1968).
[154] O. S. Mills and J. P. Nice, *J. Organometal. Chem.*, **10**, 337, (1967).
[155] O. S. Mills and E. F. Paulus, *Chem. Commun.*, 643, (1967).
[156] O. S. Mills and E. F. Paulus, *J. Organometal. Chem.*, **10**, 331, (1967).
[157] C. H. Wei, *Inorg. Chem.*, **8**, 2384, (1969).
[158] E. R. Corey, L. F. Dahl and W. Beck, *J. Amer. Chem. Soc.*, **85**, 1202, (1963).
[159] V. G. Albano, P. L. Bellon and M. Sansoni, *J. Chem. Soc. A*, 678, (1971).
[160] P. Chini and S. Martinengo, *Inorg. Chim. Acta*, **3**, 299, (1969), and references therein.
[161] F. A. Cotton and J. M. Troup, *J. Chem. Soc. Dalton*, 801, (1974).
[162] R. Doedens and L. F. Dahl, *J. Amer. Chem. Soc.*, **88**, 4847, (1966).
[163] J. G. Bullit, F. A. Cotton and T. J. Mark, *Inorg. Chem.*, **11**, 671, (1972).

[164] J. R. Churchill, J. Wormald and J. Kniglie, *J. Amer. Chem. Soc.*, **93**, 3073, (1971).
[165] J. A. Ibers, *J. Organometal. Chem.*, **14**, 423, (1968).
[166] G. Gardner-Sommer, H. P. Klug and L. E. Alexander, *Acta Cryst.*, **17**, 732 (1964).
[167] C. H. Wei and L. F. Dahl, *J. Amer. Chem. Soc.*, **88**, 1821, (1966).
[168] H. J. Bennet, F. A. Cotton and B. M. C. Winquist, *J. Amer. Chem. Soc.*, **89**, 5366, (1967).

Chapter 3

Asymmetric Hydrosilylation by Means of Homogeneous Catalysts with Chiral Ligands

IWAO OJIMA

Sagami Chemical Research Center, Sagamihara, Japan

KEIJI YAMAMOTO

Department of Chemical Engineering, Tokyo Institute of Technology, Tokyo, Japan

and

MAKOTO KUMADA

Department of Synthetic Chemistry, Kyoto University, Kyoto, Japan

1. INTRODUCTION

Asymmetric synthesis has acquired much interest for a long time, and many types of stoichiometric reactions have been explored [1]. However, the most effective use of chiral reagents is their use as catalysts. Stereoselective reactions catalyzed by chiral transition metal complexes may be of great promise. In general, an attractive approach to 'homogeneously catalyzed asymmetric reactions' has been made possible mainly by the development of preparative routes to optically active phosphines as ligands. Yet, one of the crucial problems is how to develop a chiral ligand which will enable the catalyst for a particular reaction to be as efficient in stereoselectivity as possible. At the present time the choice of the chiral ligand for this purpose is quite empirical. In this sense, some of the most exciting developments in asymmetric synthesis have been made on the catalytic asymmetric hydrogenation of prochiral olefins using soluble rhodium(I) complexes with chiral phosphine ligands. Knowles *et al.* [2] have found that the catalytic hydrogenation of α-acylamidocinnamic acids affords the corresponding amino acid derivatives with 80–95 percent enantiomeric excess (% e.e.) using rhodium complexes with (*R*)-*o*-anisylcyclohexylmethylphosphine or (*R*)-*bis*(*o*-anisylphenylphosphino)ethane. Similar results have been obtained by Kagan and Dang [3] by the use of a rhodium catalyst with (−)-2,3-*O*-isopropylidene-2,3-dihydroxy-1,4-bis(diphenylphosphino)butane (DIOP) as ligand.

What has turned out to be an attractive approach to the catalytic asymmetric synthesis is not restricted within hydrogenation. In fact, it has applied to date to most of the catalytic reactions in which an unsaturated substrate is activated by metal complexes: they include hydroformylation [4], dimerization or oligomerization [5], cyclopropanation [6], and hydrosilylation [7], of olefins.

This review deals with recent advances in catalytic asymmetric hydrosilylation of olefins, carbonyl and imino compounds in the presence of transition metal complexes of chiral phosphine ligands with particular emphasis on the asymmetric reduction of prochiral carbonyl compounds, which has been extensively studied in the last few years by several research groups and proved to provide an effective reduction method for organic syntheses.

First, for reasons of clarity, the currently-accepted mechanism of transition-metal complex catalyzed-hydrosilylation reactions will be described briefly. Furthermore, consideration of selective, if not asymmetric, reduction of certain carbonyl compounds by way of rhodium(I)-catalyzed hydrosilylation (Section 4) is included in this review because the catalytic process and stereochemical course of this reaction correlate closely with those of their asymmetric reduction under similar conditions that will be described in the succeeding section.

2. HYDROSILYLATION USING HOMOGENEOUS CATALYSTS

The hydrosilylation of carbon–carbon multiple bonds has been studied extensively over many years, since the reaction is one of the most important laboratory, and also industrial, methods for preparing certain organosilicon compounds.

Although addition of silicon hydrides to olefins and acetylenes takes place in the presence of a variety of catalysts, chloroplatinic acid ('Speier's catalyst') is by far the most commonly used catalyst for the reaction. This subject including the reaction mechanism has been well reviewed elsewhere [8].

A proposed mechanism [9] for the hydrosilylation of olefins catalyzed by platinum(II) complexes (chloroplatinic acid is thought to be reduced to a platinum(II) species in the early stages of the catalytic reaction) is similar to that for the rhodium(I) complex-catalyzed hydrogenation of olefins, which was advanced mostly by Wilkinson and his co-workers [10]. Besides the Speier's catalyst, it has been shown that tertiary phosphine complexes of nickel [11], palladium [12], platinum [13], and rhodium [14] are also effective as catalysts, and homogeneous catalysis by these Group VIII transition metal complexes is our present concern. In addition, as we will see later, hydrosilanes with chlorine, alkyl or aryl substituents on silicon show their characteristic reactivities in the metal complex-catalyzed hydrosilylation. Therefore, it seems appropriate to summarize here briefly recent advances in elucidation of the catalysis by metal complexes, including activation of silicon–hydrogen bonds.

Oxidative addition of a hydrosilane with its hydrogen–silicon bond to a metal complex, which is in a low oxidation state and usually coordinatively unsaturated, constitutes one of the essential steps in the hydrosilylation since in this step the activation of hydrosilane by the 'catalyst' occurs. Thus, many transition-metal ions and complexes, especially Group VIII metals containing π-acid ligands such as carbon monoxide, tertiary phosphines or olefins exhibit the catalytic action [15]. A sequence of elemental (stoichiometric) reactions in a typical d^8 metal complex-catalyzed hydrosilylation may be described as follows [16]:

(i) activation of an olefin substrate (olefin-complex formation)

$$-\overset{|}{\underset{|}{M}}- + R'CH{=}CH_2 \underset{}{\overset{K}{\rightleftharpoons}} \begin{matrix} R'CH \\ \| \\ CH_2 \end{matrix}\!\!-\!\overset{|}{\underset{|}{M}}- \tag{1}$$

(ii) activation of a hydrosilane (oxidative addition)

$$\begin{matrix} R'CH \\ \| \\ CH_2 \end{matrix}\!\!-\!\overset{|}{\underset{|}{M}}- + HSiR_3 \underset{k_{-1}}{\overset{k_1}{\rightleftharpoons}} \begin{matrix} R'CH \\ \| \\ CH_2 \end{matrix}\!\!-\!\overset{SiR_3}{\underset{|}{M}}\!\!\begin{matrix} H \\ - \end{matrix} \tag{2}$$

(iii) *cis* addition of M–H to C=C bond (*cis* ligand insertion)

$$\begin{matrix} R'CH \\ \| \\ CH_2 \end{matrix}\!\!-\!\overset{SiR_3}{\underset{|}{M}}\!\!\begin{matrix} H \\ - \end{matrix} \underset{k_{-2}}{\overset{k_2}{\rightleftharpoons}} R'CH_2CH_2-\overset{SiR_3}{\underset{|}{M}}\!\!< \tag{3}$$

(iv) product liberation (reductive elimination)

$$R'CH_2CH_2-\overset{SiR_3}{\underset{|}{M}}\!\!< + R'CH{=}CH_2 \xrightarrow{k_3} R'CH_2CH_2SiR_3 + \begin{matrix} R'CH \\ \| \\ CH_2 \end{matrix}\!\!-\!\overset{|}{\underset{|}{M}}- \tag{4}$$

Detailed kinetic studies [17] have been done of oxidative addition of some hydrosilanes to an iridium(I) complex, which is isoelectronic with the "Wilkinson's complex" [chlorotris(triphenylphosphine)rhodium(I)] but ineffective as catalyst. Hydrosilanes $Me_n(EtO)_{3-n}SiH$ were found to undergo a readily reversible oxidative addition to the complex.

$$H(CO)(Ph_3P)_2Ir + Me_n(EtO)_{3-n}SiH \underset{k_{-1}}{\overset{k_1}{\rightleftharpoons}} H_2[Me_n(EtO)_{3-n}Si](CO)(Ph_3P)_2Ir \tag{5}$$

$$(Ph_3P)_3RhCl + Me_nCl_{3-n}SiH \rightleftharpoons HCl(Me_nCl_{3-n}Si)(Ph_3P)_2Rh + Ph_3P \tag{6}$$

The order of stabilities of the adducts was the same as that observed previously for additions of the hydrosilanes to d^8 complexes [14, 32*c*, 18], i.e., negative substituents such as alkoxy groups or chlorine atoms on silicon stabilize the adducts. Furthermore, rate measurements have indicated that the structure of hydrosilanes does not affect markedly the rate (k_1) of the forward reaction (oxidative addition), but affects strongly the rate (k_{-1}) of the reverse reaction (reductive elimination). This latter fact, in addition to a possible dependence of the stability of metal-silicon bonds on metal species (*e.g.*, rhodium *vs.* platinum) will be reflected in the catalysis by particular metal complexes, which is clearly shown in the following sections.

Another approach has been to study the stereochemistry of an optically active hydrosilane, α-naphthylmethylphenylsilane (abbreviated as R_3Si^*H), in the oxidative addition and subsequent reductive elimination [19]. Thus, a complex (−)-*cis*-$PtH(Si^*R_3)(PPh_3)_2$(I) is formed from (+)-R_3Si^*H and ethylenebis(triphenylphosphine)platinum(0), with little loss of optical activity, and most likely with retention of configuration at silicon. The same is true for platinum(II) species.

$$Pt(PPh_3)_2C_2H_4 + (+)\text{-}R_3Si^*H \longrightarrow (-)\text{-}cis\text{-}PtH(Si^*R_3)(PPh_3)_2 + C_2H_4 \quad (7)$$

$$[\alpha]_D^{25} + 32.9° \qquad\qquad (I)\ [\alpha]_D^{25} - 18.5°$$

The loss of ethylene may be synchronous with the oxidative addition of the hydrosilane to give the platinum(II) species, in which the metal atom inserts into the silicon–hydrogen bond in a three center process.

It is suggested that the cleavage of Si–Pt bonds in complex I involving the formation of R_3Si^*H may also occur with complete retention of configuration at silicon.

These facts well reinforce the previous finding about overall retention of configuration at silicon during the hydrosilylation catalyzed by a platinum(II) complex [20].

$$(+)\text{-}R_3Si^*H + n\text{-}C_6H_{13}CH{=}CH_2 \xrightarrow{Pt(II)} (-)\text{-}R_3Si^*(n\text{-}C_8H_{17}) \quad (8)$$

Although numerous investigations, from both preparative and mechanistic points of view, had been done on the olefin hydrosilylations, it was not until 1972 that the rhodium- or platinum-catalyzed hydrosilylation of simple ketones was found to occur. The reaction may be regarded as an equivalent means to catalytic reduction because the silyl group can readily be removed from the resulting silicon–oxygen bond in the product.

$$R^1R^2C{=}O + HSiR_3 \xrightarrow[\text{or } Pt(II)]{Rh(I)} R^1R^2CHOSiR_3 \xrightarrow{H_3O^+} R^1R^2CHOH \quad (9)$$

Thus, organosilicon hydrides can serve as reducing agents like metal hydrides so far as the catalytic hydrosilylation is effectively achieved. To date, rhodium(I)-phosphine complexes have been found to be by far the most useful catalysts for these purposes, especially for the asymmetric reduction of carbonyl and imino compounds.

3. CATALYTIC ASYMMETRIC HYDROSILYLATION OF PROCHIRAL OLEFINS

Asymmetric hydrosilylation of prochiral olefins, *e.g.* 1,1-disubstituted olefins, catalyzed by chiral phosphine complexes of platinum(II) [21], nickel(II) [22], and palladium(II) [23] has been reported. When a platinum complex of the type $[L^*PtCl_2]_2$, where L^* stands for (*R*)-benzylmethylphenylphosphine (BMPP), was used as catalyst, methyldichlorosilane, but not trichlorosilane, reacted satisfactorily with α-methylstyrene, 2,3-dimethyl-1-butene, and 2-methyl-1-butene to give the corresponding optically active addition products, $RMeC^*HCH_2SiMeCl_2$, R = Ph, *i*-Pr, and Et, respectively, equation (10). Terminal additions of methyldichlorosilane to the 1,1-disubstituted olefins also took place with little difficulty in the case of a nickel(II) complex, $L_2^*NiCl_2$, as catalyst, while similar palladium(II) complex, $L_2^*PdCl_2$, did not work at all for the hydrosilylation of these olefins. With the nickel catalyst considerable amounts of those addition products of methylchlorosilane were always obtained, which certainly came from a chlorine–hydrogen interchange of methyldichlorosilane under the reaction conditions [11], equation (11).

$$RMeC{=}CH_2 + MeCl_2SiH \xrightarrow[40^\circ]{Pt^*(II)} RMeC^*HCH_2SiMeCl_2 \qquad (10)$$

$$RMeC{=}CH_2 + MeCl_2SiH \xrightarrow[90^\circ]{Ni^*(II)} RMeC^*HCH_2SiMeCl_2 + RMeC^*HCH_2Si^{(*)}MeClH \qquad (11)$$

$$R = Ph,\ i\text{-}Pr,\ \text{and Et}$$

Styrene cannot be used for the asymmetric hydrosilylation catalyzed by these platinum(II) and nickel(II) complexes, because the terminal addition product (β-adduct) always predominates. In contrast, a palladium(II)-phosphine complex-catalyzed hydrosilylation of styrene with trichlorosilane gave an α-adduct exclusively, equation (12). Results of the asymmetric hydrosilylation of 1,1-disubstituted olefins and of styrene catalyzed by these chiral complexes are summarized in Table 1.

$$PhCH{=}CH_2 + HSiCl_3 \xrightarrow[25^\circ]{Pd^*(II)} PhMeC^*HSiCl_3 \qquad (12)$$

Although these asymmetric reactions are synthetically of little use, the observed definite relationship between olefin structure and predominant configuration of the addition product may allow one to speculate some stereochemical outcome of the present hydrosilylation.

On the basis of the fact that (*R*)-BMPP coordinated to the metal center can induce asymmetric addition of methyldichlorosilane across the carbon–carbon double bond of 2-substituted propenes to afford an enantiomeric excess of (*R*)-2-substituted propylmethyldichlorosilanes, the following processes should be involved in these reactions: (*a*) insertion of the metal center into the silicon–hydrogen bond (oxidative addition of the hydrosilane); (*b*) addition of the resulting hydridometal moiety to the coordinated olefin preferentially from its *re* face (in a *cis* manner) to convert the olefin into an alkyl–metal species; and (*c*) transfer of the silyl group from the metal center to the alkyl carbon to form the product. Since process (*b*) most likely involves diastereomeric transition states or intermediates, the overall asymmetric bias onto the *R* configuration at the chiral carbon would have already been determined prior to process (*c*). A schematic view of such a process is given in Scheme 1.

Table 1

ASYMMETRIC HYDROSILYLATION OF PROCHIRAL OLEFINS CATALYZED BY CHIRAL PHOSPHINE COMPLEXES: $[L^*PtCl_2]_2$ (**1**), *cis*-$(C_2H_4)L^*PtCl_2$ (**2**), $L_2^*NiCl_2$ (**3**), AND $L_2^*PdCl_2$ (**4**) [20–22]

Olefin	Silane	Catalyst[a]	Product	Yield (%)	Optical yield[b,c] (%)
$PhMeC{=}CH_2$	$MeCl_2SiH$	**1a**	$PhMeCHCH_2SiMeCl_2$	56	5.2 (*R*)
		1b		64	1.1 (*R*)
		2		43	6.1 (*R*)
		3		31	20.9 (*R*)
i-$PrMeC{=}CH_2$	$MeCl_2SiH$	**1a**	*i*-$PrMeCHCH_2SiMeCl_2$	83	1.4 (*R*)
		2		76	1.2 (*R*)
		3		21	6.2 (*R*)
$EtMeC{=}CH_2$	$MeCl_2SiH$	**1a**	$EtMeCHCH_2SiMeCl_2$	68	1.2 (*R*)
		2		69	1.1 (*R*)
		3		17	2.5 (*R*)
$PhCH{=}CH_2$	Cl_3SiH	**4a**	$PhMeCHSiCl_3$	87	5.1 (*S*)
		4b		87	3.3 (*R*)
		4c		70	0.3 (*R*)

[a] **1a**, L* = (*R*)-(+)-$(PhCH_2)MePhP$ (BMPP); **1b**, L* = (*R*)-(—)-MePh(*n*-Pr)P (MPPP); **2**, L* = (*R*)-BMPP; **3**, L* = (*R*)-BMPP; **4a**, L* = menthyldiphenylphosphine (MDPP); **4b**, L* = neomethyldiphenylphosphine (NMDPP); 4c, L* = (*R*)-BMPP.
[b] Based on the maximum rotation of authentic samples and calibrated for the optical purity of the chiral phosphine used.
[c] Preferred configuration is indicated in parentheses.

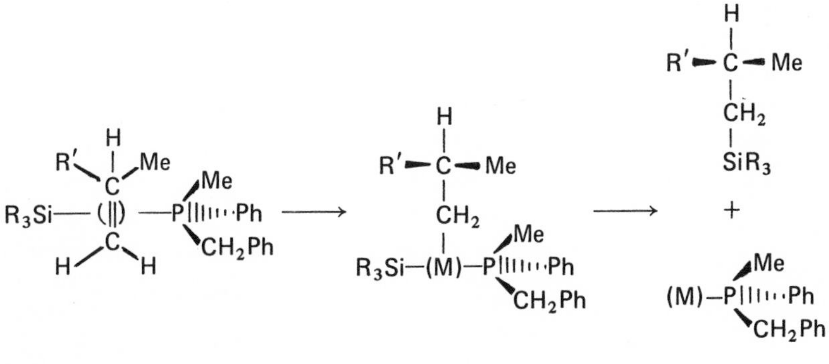

$$R_3Si—(M)(\eta^2\text{-}CH_2{=}CR'Me)—P(Me)(Ph)(CH_2Ph) \longrightarrow R'C(H)(Me)—CH_2—(M)(SiR_3)—P(Me)(Ph)(CH_2Ph) \longrightarrow R'C(H)(Me)CH_2SiR_3 + (M)—P(Me)(Ph)(CH_2Ph)$$

R = Ph, *i*-Pr, and Et; M = Pt, Ni, or Rh

Scheme 1

In the case of palladium(II) complex catalysts (**4** in Table 1), the reaction seems to proceed in somewhat different fashion. Formation of α-phenylethyltrichlorosilane as a sole product in the palladium(II) complex-catalyzed hydrosilylation of styrene may well be explained in terms of incorporation of the silyl group exclusively into the benzylic position of a π-α-methylbenzyl-metal intermediate, which in turn must exhibit diastereomeric interactions with the chiral phosphine ligands.

Asymmetric hydrosilylation of cyclic conjugated dienes catalyzed by palladium(II) complexes either with menthyldiphenylphosphine (MDPP) (**4a**) or with neomenthyldiphenylphosphine (NMDPP) (**4b**) deserves to be mentioned from a mechanistic point of view. It is of particular interest that the reaction gave 2-cycloalkenyltrichlorosilanes with an enantiomeric excess of the (*S*)-(−)-isomer, regardless of the phosphine epimers used in the catalyst (Table 2). This is not the case with styrene as shown in Table 1. In the light of current views of the mechanism of metal-catalyzed hydrosilylation, mentioned above, the product may be formed (see Scheme 2) *via* addition of a hydridopalladium moiety to the cyclic diene to convert it into a π-alkenyl–metal, followed by intramolecular transfer of the silyl group from the metal center to the π-enyl carbon. Since the cyclic π-enyl–palladium bonding has a local plane of symmetry and cannot induce asymmetry, the last process involves those diastereomeric transition states or intermediates (including the catalyst complex), which cannot necessarily reflect the difference in epimeric structures of the phosphine ligands in the stereochemical course of the reaction.

Table 2

ASYMMETRIC HYDROSILYLATION OF CYCLIC DIENES WITH $HSiCl_3$ CATALYZED BY CHIRAL PHOSPHINE-PALLADIUM(II) COMPLEXES[a] [22]

Olefin	Catalyst[b]	Yield (%)	$[\alpha]_D^{20}$[c]	Configuration
Cyclopentadiene	**4a**	69	−11.84	*S*
	4b	81	−7.54	*S*
1,3-Cyclohexadiene	**4a**	64	−11.08	*S*
	4b	56	−3.66	*S*

[a] Reactions at 120° for 58 hr in the presence of 0.2 mol% catalyst.
[b] See Table 1, footnote *a*.
[c] Value for the methylated sample.

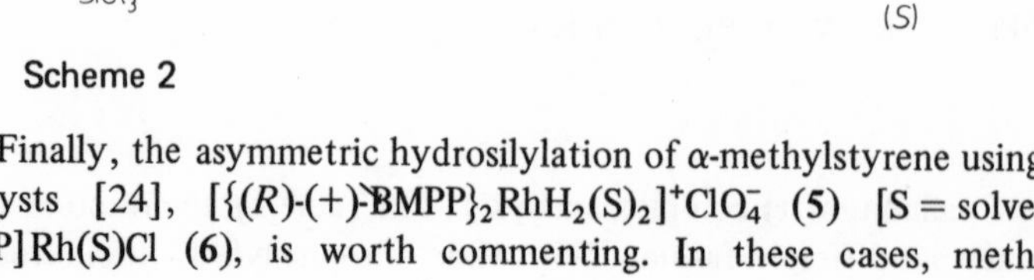

Scheme 2

Finally, the asymmetric hydrosilylation of α-methylstyrene using chiral rhodium catalysts [24], $[\{(R)\text{-}(+)\text{-BMPP}\}_2RhH_2(S)_2]^+ClO_4^-$ (**5**) [S = solvent] and [(−)-DIOP]Rh(S)Cl (**6**), is worth commenting. In these cases, methyldichlorosilane scarcely reacted even under forced conditions, despite a facile addition of this hydrosilane to olefins in the presence of platinum(II) and nickel(II) complexes. The low catalytic activity of rhodium towards chlorohydrosilanes may be ascribed to the formation of very stable silyl–rhodium bonds [14]. However, trialkylsilanes such as trimethylsilane and phenyldimethylsilane were found to be moderately reactive (Table 3). As is seen from Table 3, changes in structure of hydrosilanes used do not significantly affect the extent of asymmetric induction. The results again coincide with the proposed mechanism shown in Scheme 1.

Table 3

ASYMMETRIC HYDROSILYLATION OF α-METHYLSTYRENE CATALYZED BY CHIRAL RHODIUM COMPLEXES[a] [24]

Silane	Catalyst[b]	Yield (%)	Optical yield (%)
$HSiMe_3$	5	63	7.0[c] (*R*)
$HSiMe_2Ph$	5	25	5.2[c] (*R*)
$HSiMe_3$	6	63	10.4 (*S*)
$HSiMe_2Ph$	6	19	6.7 (*S*)

[a] Reactions at 120° for 40 hr in the presence of 0.05 mol% catalyst.
[b] 5, $[\{(R)\text{-BMPP}\}_2RhH_2(S)_2]^+ClO_4^-$ (S = solvent); 6, [(−)-DIOP] RhCl(S).
[c] Calibrated for the optical purity of BMPP (70%).

4. SELECTIVE REDUCTION OF CARBONYL COMPOUNDS *via* HYDROSILYLATION CATALYZED BY A RHODIUM(I) COMPLEX

In contrast to a number of studies on the homogeneous hydrogenation of carbon–carbon multiple bonds [25], there had been few papers about hydrogenation of simple ketones before Schrock and Osborn [26] reported in 1970 a catalytic activity of cationic rhodium complexes with relatively basic phosphines as ligands. In fact, the Wilkinson's rhodium(I) complex usually lacks activity towards hydrogenation of carbonyl groups, and rather catalyzes decarbonylation of aldehydes. The catalytic cycle of the hydrogenation of ketones proposed by Schrock and Osborn is depicted in Scheme 3.

Taking advantage of these findings, a catalytic asymmetric hydrogenation of ketones has been effected [27]. It should be noted in this connection that asymmetric reduction of benzil to benzoin has been found to occur with *bis*(dimethylglyoximato)cobalt(II)–quinine system as a catalyst [28].

A similar trend can be seen in the homogeneous hydrosilylation of carbon-hetero atom multiple bonds, which has not received sufficient attention. Although the catalytic hydrosilylation of carbonyl compounds using zinc chloride [29], nickel metal [30] or chloroplatinic acid [31] as catalyst has so far been studied, the reaction conditions were rather drastic and side reactions were often observed.

Recently, it has been disclosed [32, 33] that chlorotris(triphenylphosphine)-

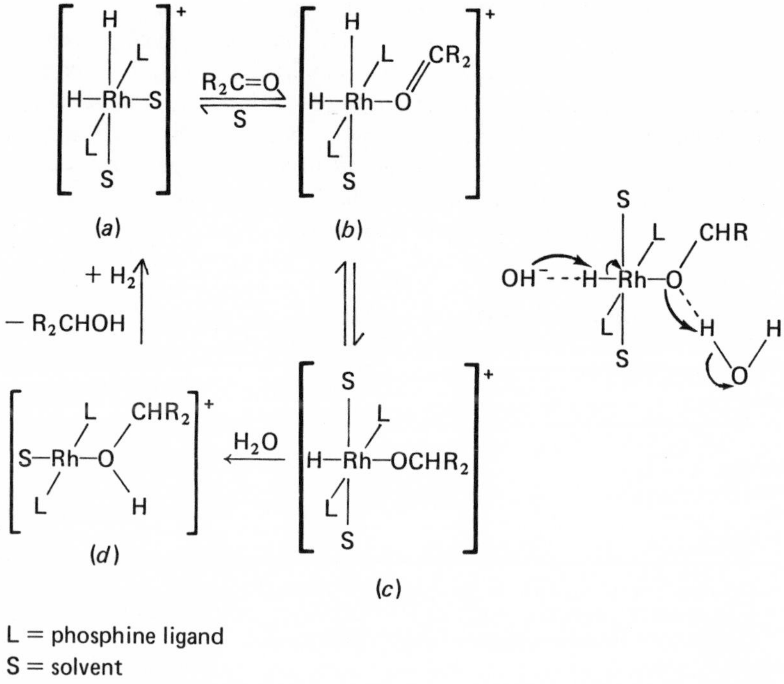

L = phosphine ligand
S = solvent

Scheme 3

rhodium(I) (7) can serve as quite an effective catalyst for the hydrosilylation of a variety of carbonyl compounds. Dichlorotris(triphenylphosphine)ruthenium(II) also has proved to be useful in the addition of dihydrosilanes to ketones [33] although it is less effective than the rhodium catalyst [34]. Synthetic usefulness of this reaction has soon been found in the selective reduction of terpene carbonyl compounds and Schiff bases.

4.1. Selective reduction of α,β-unsaturated carbonyl compounds

It is most striking that the hydrosilylation of α,β-unsaturated carbonyl compounds using *mono*-hydrosilanes was found to proceed in a manner of 1,4-addition, while *di*-hydrosilanes very specifically underwent 1,2-addition to carbonyl functionalities [35]. Since the resulting silyl enol ethers and allylic silyl ethers can readily be converted by hydrolysis to saturated carbonyl compounds and α,β-unsaturated alcohols, respectively, these reactions may furnish a unique method for selective reduction of carbonyl compounds, equation (13). The results are summarized in Table 4.

α,β-Unsaturated carbonyl compounds containing an isolated double bond in the same molecule also underwent selective reduction without any isomerization nor reduction at the isolated double bond. It was also demonstrated that a dihydrosilane–

Table 4

SELECTIVE REDUCTION OF α,β-UNSATURATED KETONES AND ALDEHYDES USING HYDROSILANE–RHODUM(I) COMPLEX COMBINATIONS [35*a*]

α,β-Unsaturated carbonyl compounds	Silane	Conditions[a]	Ratio of[b] 1,4/1,2	Yield (%)
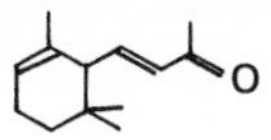	Et_3SiH Ph_2SiH_2	50°, 2 hr r.t., 30 min	100/0 0/100	96 97
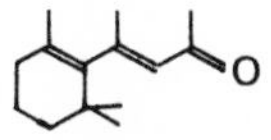	$EtMe_2SiH$ Ph_2SiH_2	r.t., 3 hr r.t., 30 min	92/8 0/100	94 98
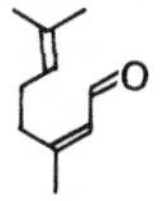	Et_3SiH Ph_2SiH_2	r.t., 1 hr 0°, 30 min	100/0 0/100	97 97
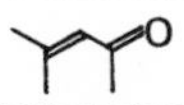	Et_3SiH $PhSiH_3$	50°, 1 hr r.t., 1 hr	99/1 1/99	95 95
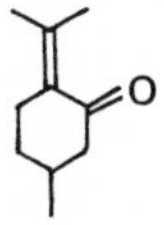	$EtMe_2SiH$ Ph_2SiH_2	r.t., 2 hr r.t., 20 min	70/30 0/100	97 98
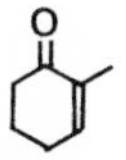	$EtMe_2SiH$ Ph_2SiH_2	45°, 4 hr 0°, 30 min	98/2 1/99	90 97
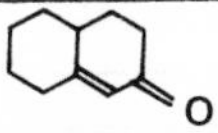	Et_3SiH Et_2SiH_2	80°, 25 hr 0°, 30 min	94/6 3/97	95 98

[a] 0.1 mol% $(Ph_3P)_3RhCl$ (7) was used.
[b] Products ratio was determined by NMR or glpc.

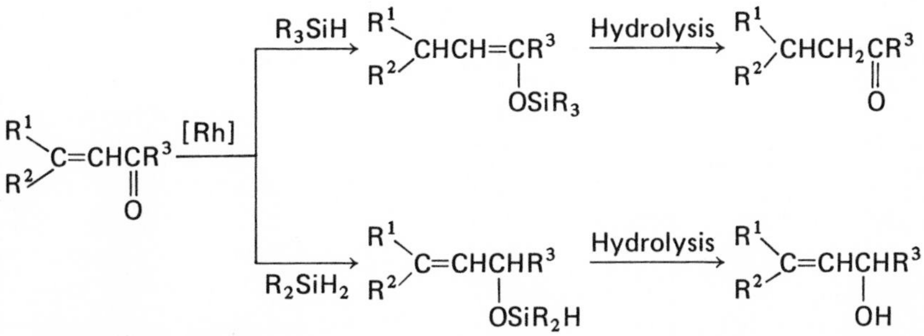

$$R^1R^2C{=}CHC({=}O)R^3 \xrightarrow{[Rh]} \begin{cases} \xrightarrow{R_3SiH} R^1R^2CHCH{=}C(OSiR_3)R^3 \xrightarrow{\text{Hydrolysis}} R^1R^2CHCH_2C({=}O)R^3 \\ \xrightarrow{R_2SiH_2} R^1R^2C{=}CHCH(OSiR_2H)R^3 \xrightarrow{\text{Hydrolysis}} R^1R^2C{=}CHCH(OH)R^3 \end{cases} \quad (13)$$

rhodium(I) complex combination in the reduction of pulegone and piperitone displays an exceedingly higher selectivity than common metal hydrides such as lithium aluminum hydride and sodium borohydride. Two typical cases for comparison of the selectivities obtained by this method with those attained by usual metal hydrides are shown in Table 5.

Table 5

REDUCTION OF PULEGONE AND PIPERITONE WITH METAL HYDRIDES[a] [35*b*]

Reducing agent	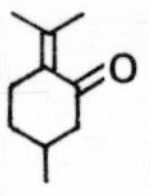 O	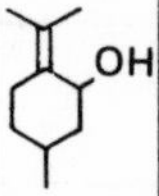 OH	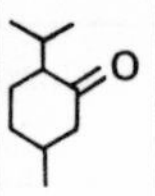 O	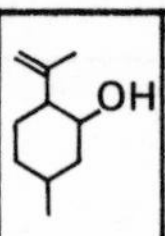 OH	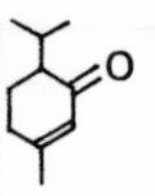 O	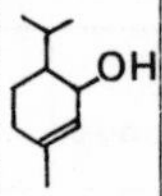 OH	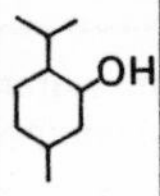 OH	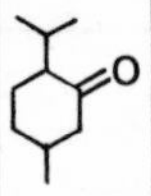 O
Et_2SiH_2-$(Ph_3P)_3RhCl$		100				100		
Ph_2SiH_2-$(Ph_3P)_3RhCl$		100				100		
$LiAlH_4$[b]	49	51				100		
$LiAl(O{-}t\text{-}Bu)_3H$[b]	39	43	18		29	24		47
$LiBH_4$[b]	7	93			23	48.5	28.5	
$NaBH_4$[b]	18	36		46	60	22.5	18.5	

[a] Performed using 1.0–1.3 equivalent of metal hydride at room temperature for 24 hr.
[b] See J. W. Wheeler and R. H. Chung, *J. Org. Chem.*, **34**, 1149 (1969).

4.2. Stereoselective reduction of terpene ketones

Stereoselective reduction of terpene ketones *via* hydrosilylation [36] exhibited significant difference in stereochemistry from other reduction by metal hydrides. Results of the reduction of camphor and menthone *via* hydrosilylation are listed in Table 6. Included also in the Table are reported selectivities obtained by using conventional reducing agents. It is seen in the Table that the bulkiness of silanes exerts remarkable influence on the stereochemical course of the reduction, *i.e.*, a bulky hydrosilane favors the production of the more stable alcohols. This trend is quite

Table 6

STEREOSELECTIVITIES IN THE REDUCTION OF TERPENE KETONES [36]

Reducing agent	Camphor *iso*-Borneol/Borneol	Menthone Neomenthol/Menthol
$PhSiH_3$	90/10	90/10
Et_2SiH_2	91/9	83/17
α-$NpPhSiH_2$[a]	88/12	73/27
$PhMeSiH_2$	75/25	87/13
Ph_2SiH_2	73/27	85/15
$EtMe_2SiH$	——	50/50
Et_3SiH	30/70	36/64
n-Pr_3SiH	30/70	——
$PhMe_2SiH$	——	0/100
$LiAlH_4$	91/9	29/71
$Al(O{-}i\text{-}Pr)_3$	——	30/70
$NaBH_4$	——	51/49
B_2H_6	52/48	——
Disiamylborane	65/35	——
Dicyclohexylborane	93/7	——
Diisopinocampheylborane	100/0	——

[a] See reference [45].

unusual since it has been shown by Brown and Varma [37] that, in the reduction of monocyclic and bicyclic ketones, a bulkier hydroborane produces predominantly the less stable of the two possible alcohols in accordance with the concept of 'steric approach control' [38]. The observed trend suggests that the transition state for the catalytic hydrosilylation cannot be accommodated to a simple four-centered process, which involves attack of the hydrido-rhodium moiety on the carbonyl carbon from the less hindered site of the ketone to form an alkoxy–rhodium complex (*cf.* Scheme 3). An alternative explanation can be advanced by taking account of the intermediacy of an α-siloxyalkyl–rhodium complex rather on the basis of 'product development control' as shown in Scheme 4.

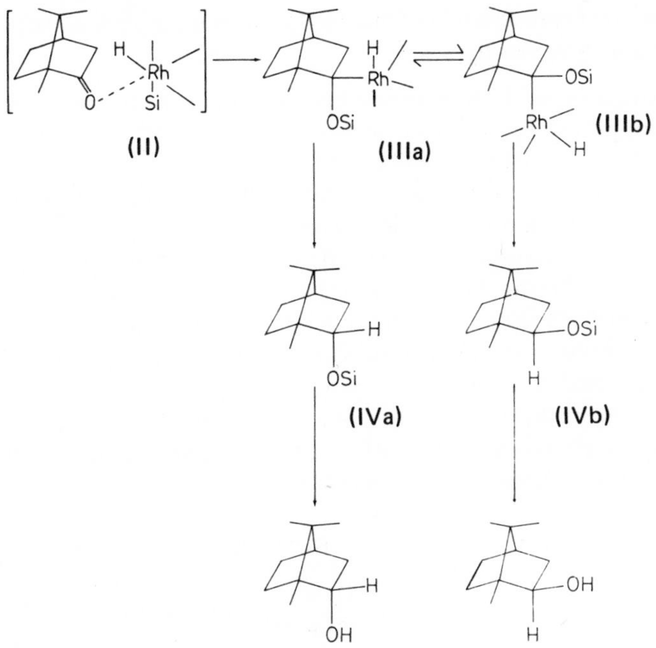

(II) (IIIa) (IIIb)

(IVa) (IVb)

Scheme 4

According to the proposed mechanism, addition of the silyl–rhodium moiety to the coordinated carbonyl group converts it into the α-siloxyalkyl–rhodium complex (III), which most likely is an equilibrium mixture of complexes III*a* and III*b*. Then, transfer of the hydride ligand in III from the metal center to the alkyl carbon affords the products, IV*a* and IV*b*, respectively. The formation of the α-siloxyalkyl–rhodium intermediate is quite probable in view of the well-documented 'soft-hard conception', and must be characteristic of the ketone hydrosilylation.

Since the steric course of the reaction is governed only by the size of the silyl group, it can be said that the bulkier the substituents on silicon, the more pronounced may be the formation of complex III*a* which is a precursor of the more stable alcohol. The unique selectivity of the reaction can be well understood by inspecting the Dreiding model of complex III.

5. ASYMMETRIC REDUCTION OF KETONES *via* CATALYTIC HYDROSILYLATION

Asymmetric reduction of ketones by hydrosilylation in the presence of a chiral catalyst followed by hydrolysis has been studied by several research groups independently. In this Section, results so far obtained are properly compiled and plausible mechanisms of the asymmetric hydrosilylation of prochiral ketones are discussed.

5.1. Asymmetric hydrosilylation of prochiral ketones using platinum(II) complexes with chiral phosphine ligands

The first paper on the asymmetric hydrosilylation of ketones appeared in 1972, which described the use of chiral phosphine–platinum(II) complexes of the type $[L^*PtCl_2]_2$ (L^* = (*R*)-BMPP or (*R*)-MPPP) [39].

Catalytic activities of several platinum, palladium, nickel as well as rhodium complexes in hydrosilylation of acetophenone were briefly examined by using methyldichlorosilane, and the platinum(II) complex, $[(PhMe_2P)PtCl_2]_2$, was found to be the most effective for this particular hydrosilane. Chlorotris(triphenylphosphine)rhodium(I) (**7**) has no catalytic activity at all in sharp contrast to the case in which the hydrosilylation is carried out by using alkylhydrosilanes, R_nSiH_{4-n} ($n = 1$, 2 or 3), as described in the previous Section.

Addition of methyldichlorosilane to a series of alkyl phenyl ketones catalyzed by these chiral phosphine–platinum(II) complexes gives partially optically active silyl ethers of 1-phenylalkanols, equation (14).

$$PhCOR + HSiMeCl_2 \xrightarrow{[L^*PtCl_2]_2} PhRC^*HOSiMeCl_2 \xrightarrow[(2)\ H_3O^+]{(1)\ MeLi} PhRC^*HOH \qquad (14)$$

The results examined are summarized in Table 7. It is noteworthy that the platinum(II) complex (**1a**) catalyzes the asymmetric addition of methyldichlorosilane to the ketones leading predominantly to (*S*)-1-phenylalkanols, whereas complex (**1b**) to the (*R*)-enantiomers except for pivalophenone. The results clearly indicate that, with the two phosphine complexes employed, it is the only chiral nature at the phosphorus atom that is transmitted to the diastereomeric transition states giving rise to an opposite configuration of the product. This is not the case for the asymmetric hydrosilylation of, for example, α-methylstyrene with methyldichlorosilane as shown in Table 1. Furthermore, the latter reaction catalyzed by complex **1a** gives rise to the (*R*)-adduct predominantly, whereas (*S*)-1-phenylalkanol is the preferred enantiomer in the asymmetric addition of methyldichlorosilane to acetophenone, equation (15) and (16).

Table 7

ASYMMETRIC HYDROSILYLATION OF ALKYL PHENYL KETONES RCOPh WITH $MeCl_2SiH$ CATALYZED BY $[L^*PtCl_2]_2$ (1) AT ROOM TEMPERATURE[a] [39]

R	Yield (%)	1-Phenylalkanol $[\alpha]_D^{20}$, deg.	Optical yield (%)[b] (configuration)
L* = (*R*)-BMPP (79% optical purity)			
Me	81	− 2.61 (neat)	7.6 (*S*)
Et	81	− 2.21 (neat)	10.0 (*S*)
n-Pr	77	− 3.0 (benzene)	8.4 (*S*)
i-Pr	55[c]	− 1.7 (ether)	4.5 (*S*)
i-Bu	65	− 4.74 (*n*-heptane)	18.6 (*S*)
t-Bu	33	− 3.8 (benzene)	18.6 (*S*)
L* = (*R*)-MPPP (93% optical purity)			
Me	71	+ 2.24 (neat)	5.5 (*R*)
Et	83	+ 1.94 (neat)	7.4 (*R*)
n-Pr	74	+ 2.0 (benzene)	4.7 (*R*)
i-Pr	57[c]	+ 1.1 (ether)	2.5 (*R*)
i-Bu	54	+ 0.03 (*n*-heptane)	0.1 (*R*)
t-Bu	24	− 1.5 (benzene)	6.2 (*S*)

[a] Catalyst, 0.07 mol%.
[b] Calibrated for the optical purity of chiral phosphines. For maximum rotations, see R. McLeod, F. J. Welch and H. S. Mosher, *J. Amer. Chem. Soc.*, 82, 876 (1960).
[c] Contaminated with *ca.* 10% of $Ph(MeCl_2SiO)C{=}CMe_2$.

$$\text{Ph(Me)C=CH}_2 + \text{HSiMeCl}_2 \xrightarrow[\text{(front-side)}]{(re\text{-face})} \text{Ph(Me)C(H)CH}_2\text{SiMeCl}_2 \quad (5.2\%\ \text{e.e.},\ R) \tag{15}$$

$$\text{Ph(Me)C=O} + \text{HSiMeCl}_2 \xrightarrow[\text{(back-side)}]{(re\text{-face})} \text{Ph(Me)C(H)OSiMeCl}_2 \quad (7.6\%\ \text{e.e.},\ S) \tag{16}$$

These facts may well imply that the stereoselectivity for addition of a hydrosilane to the enantiotopic faces of a ketone is different from that of an olefin which is undoubtedly π-coordinated to the chiral catalyst; the difference is again explained in terms of the intermediacy of an α-siloxyalkyl–platinum complex similar to the α-siloxyalkyl–rhodium complex described in the hydrosilylation of terpene ketones (see Scheme 4).

5.2. Asymmetric reduction of prochiral ketones *via* hydrosilylation catalyzed by rhodium(I) complexes with chiral phosphine ligands

In Section 4, it is described that chlorotris(triphenylphosphine)rhodium(I) (7) is quite an effective catalyst for the hydrosilylation of carbonyl compounds. For this reason, extensive studies on asymmetric hydrosilylation of prochiral ketones to date have been based on employing rhodium(I) complexes with chiral phosphine ligands. The catalysts all prepared *in situ* are rhodium(I) complexes of the type, $(BMPP)_2Rh(S)Cl$ (**8**) [40] and (DIOP)Rh(S)Cl (**6**) [41], and a cationic rhodium(III) complex, $[(BMPP)_2RhH_2(S)_2]^+ClO_4^-$ (**5**) [42], where S represents a solvent molecule. An interesting polymer-supported rhodium complex (V) [41], and several chiral ferrocenylphosphines [43], recently developed as chiral ligands, have also been employed for asymmetric hydrosilylation of ketones. Included in this section also are selective asymmetric hydrosilylation of α,β-unsaturated carbonyl compounds and of certain keto esters.

polymer–(O,O-acetal)–CH(H)(CH$_2$PPh$_2$)–CH(H)(CH$_2$PPh$_2$) $\xrightarrow[\text{(PhH)}]{1/2[(C_2H_4)_2RhCl]_2}$ polymer–(O,O-acetal)–[CH$_2$PPh$_2$]$_2$Rh(Cl)(S) (17)

(V)

5.2.1. CHIRAL RHODIUM(I) COMPLEXES AS CATALYSTS

Results of the asymmetric reduction of prochiral ketones using a chiral rhodium(I) complex of the type $L_2^*Rh(S)Cl$ are summarized in Table 8.

Up to now, although we are not in a position to correlate reaction conditions, especially reaction temperatures, with optimal optical yields of the reduction of particular alkyl phenyl ketones, some features observed in the asymmetric hydrosilylation may be pointed out. As regards the catalysis by the two complexes (**6**) and (**8**), both configuration and optical yield of the resulting alcohols depend markedly on the structure of hydrosilanes employed. For example, asymmetric reduction of alkyl phenyl ketones using $[(S)\text{-BMPP}]_2Rh(S)Cl$ as catalyst and diethylsilane as silane component gave rise to (*S*)-1-phenylalkanols predominantly, while with phenyldimethylsilane to (*R*)-enantiomers uniformly in much higher optical yields.

Somewhat different but still distinct effects of hydrosilanes on the stereoselectivity were observed when a rhodium(I) complex with DIOP was employed. As is seen from Table 8, monohydrosilanes such as phenyldimethylsilane reacted with ketones under rather forced conditions to afford, after hydrolysis, the corresponding alcohols in low optical yield. However, both chemical and optical yields were remark-

Table 8

ASYMMETRIC REDUCTION OF PROCHIRAL KETONES *via* HYDROSILYLATION CATALYZED BY $L_2^*RhCl(S)$ (6) OR (8)

Ketone	Silane	Ligand (L*)	Alkanol		Reference
			Optical yield (%)	Configuration	
PhCOMe	Et_2SiH_2	(*R*)-BMPP[a]	16	*R*	40
	Ph_2SiH_2	(*R*)-BMPP[a]	47	*R*	*b*
	$PhMe_2SiH$	(*S*)-BMPP[c]	44	*R*¶	40
	$PhMeSiH_2$	(+)-DIOP	13(12)[d]	*S*	41
	Ph_2SiH_2	(+)-DIOP	28(29)[d]	*S*	41
	α-$NpPhSiH_2$	(+)-DIOP	58(58)[d]	*S*	41
	$(EtO)_3SiH$	(+)-DIOP	10	*S*	41
	Et_3SiH	(+)-DIOP	3.8	*S*	41
	Et_2SiH_2	(—)-DIOP	26	*R*	44
	Ph_2SiH_2	(—)-DIOP	30	*R*	44
	α-$NpPhSiH_2$	(—)-DIOP	55	*R*	45
	$PhMe_2SiH$	(—)-DIOP	2.8	*R*	44
PhCOEt	Ph_2SiH_2	(*R*)-BMPP[a]	42	*R*	40
	Et_2SiH_2	(*S*)-BMPP[c]	17	*S*	40
	$PhMe_2SiH$	(*S*)-BMPP[c]	50	*R*¶	40
	α-$NpPhSiH_2$	(+)-DIOP	56	*S*	45
	Et_2SiH_2	(—)-DIOP	38	*R*	44
	Ph_2SiH_2	(—)-DIOP	28	*R*	44
	Me_3SiH	(—)-DIOP	4.4	*R*	44
	$PhMe_2SiH$	(—)-DIOP	4.8	*R*	44
PhCO(*i*-Pr)	Et_2SiH_2	(*S*)-BMPP[c]	23	*S*	40
	$PhMe_2SiH$	(*S*)-BMPP[c]	56	*R*¶	40
	$PhMeSiH_2$	(+)-DIOP	20(6.5)[d]	*S*	41
	Ph_2SiH_2	(+)-DIOP	35(28)[d]	*S*	41
	α-$NpPhSiH_2$	(+)-DIOP	24	*S*	41
	Et_2SiH_2	(—)-DIOP	9.8	*R*	44
	Ph_2SiH_2	(—)-DIOP	27	*R*	44
PhCO(*n*-Bu)	Ph_2SiH_2	(+)-DIOP	15	*S*	57
PhCO(*i*-Bu)	Et_2SiH_2	(—)-DIOP	37	*R*	44
	Ph_2SiH_2	(—)-DIOP	19	*R*	44
PhCO(*t*-Bu)	$EtMe_2SiH$	(*R*)-BMPP[a]	56	*R*	40
	$PhMe_2SiH$	(*R*)-BMPP[a]	54	*S*¶	40
	Et_2SiH_2	(—)-DIOP	25	*S*¶	44
	Ph_2SiH_2	(—)-DIOP	41	*R*	44
	Me_3SiH	(—)-DIOP	14	*R*	44
	$PhMe_2SiH$	(—)-DIOP	16	*R*	44

Table 8 (Continued)

Ketone	Silane	Ligand (L*)	Alkanol		Reference
			Optical yield (%)	Configuration	
$PhCOCH_2NMe_2$	Et_2SiH_2	(—)-DIOP	51	S^e	44
	Ph_2SiH_2	(—)-DIOP	28	S^e	44
$PhCOCH_2CH_2NMe_2$	Et_2SiH_2	(—)-DIOP	52	*R*	44
	Ph_2SiH_2	(—)-DIOP	14	*R*	44
(1-indanone structure)	Et_2SiH_2	(—)-DIOP	17	*R*	44
	Ph_2SiH_2	(—)-DIOP	8	*R*	44
MeCOEt	α-$NpPhSiH_2$	(—)-DIOP	42	*R*	45
MeCO(*n*-Bu)	Et_2SiH_2	(*S*)-BMPP[c]	30	*S*	40
MeCO(*t*-Bu)	Et_2SiH_2	(*S*)-BMPP[c]	39	*S*	40
$MeCOCH_2Ph$	$(EtO)_3SiH$	(+)-DIOP	5.3	*S*	41

[a] Optical purity 77%, calibrated for optical yields.
[b] T. Onoda and S. Tomita, *Japan Kokai*, **74**, 110, 631 [*Chem. Abstr.* 27539h (1975)].
[c] Optical purity 62%.
[d] Polymer-supported rhodium catalyst in parenthesis.
[e] *cf.* Data for PhCO(*i*-Bu).
¶ Indicates a reverse of preferred configuration from general trend.

ably improved by the use of appropriate dihydrosilanes. Thus, the catalyzed hydrosilylation may well proceed at room temperature to give silyl ethers of 1-phenylalkanols in high chemical yield, and the optical yields were also much higher than with monohydrosilanes, *e.g.* the best result (58% e.e.) was obtained when α-naphthylphenylsilane was reacted with acetophenone. Diethylsilane and diphenylsilane were found to be the best silanes for propiophenone and isobutyrophenone, respectively. Furthermore, with the DIOP–rhodium(I) system, the reversal of preferred configuration of 1-phenylalkanols was scarcely observed except for the reaction of diethylsilane with pivalophenone.

All the facts described here clearly indicate that the steric requirement for a match of the chiral ligand, a hydrosilane and a ketone is of definite importance in order to bring about an effective asymmetric induction. However, it is not understandable why the two catalysts, $(BMPP)_2Rh(S)Cl$ (**8**) and (DIOP)Rh(S)Cl (**6**), behave differently in activating monohydrosilanes.

It is noteworthy that the immobilized chiral rhodium complex V displays a high catalytic activity as well as stereoselectivity similar to that of the soluble chiral catalyst (**6**), the results being indicated in parentheses in Table 8.

Finally, when dialkylsilanes were used it was also possible to reduce dialkyl ketones to the corresponding optically active secondary alcohols without any side reactions [39].

As for the prediction of preferred configuration of products in the asymmetric hydrosilylation, an empirical rule has been proposed [40*b*] for the BMPP–rhodium(I) complex system in consideration of a relationship between the configuration of the chiral phosphine and that of the products, silyl ethers of 1-phenylalkanols.

Since the intermediacy of α-siloxyalkyl–rhodium complex III (see Scheme 4) is strongly suggested in the hydrosilylation of terpene ketones catalyzed by chlorotris-(triphenylphosphine)rhodium(I) (7), such an intermediate most likely plays a key role in the asymmetric induction of the present system. The fundamental premise is to assume a square-pyramidal structure of the α-siloxyalkyl–rhodium intermediate VI, as shown in equation (18), on the basis of the established structures of dihydrido [46] and of silylhydrido [14, 32*c*, 47] complexes derived from the Wilkinson's rhodium(I) complex.

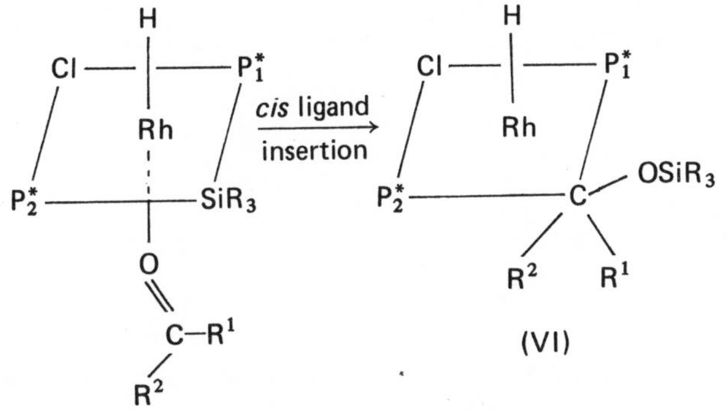

$$\mathrm{H(Cl)(P_1^*)(P_2^*)(SiR_3)Rh\cdots O{=}C(R^1)R^2} \xrightarrow[\text{insertion}]{\textit{cis}\ \text{ligand}} \mathrm{H(Cl)(P_1^*)(P_2^*)Rh{-}C(OSiR_3)(R^1)(R^2)}\ \text{(VI)} \qquad (18)$$

Under proper chiral influences, which are given by the *bis*[(*S*)-BMPP]–rhodium moiety having these phosphines *trans* to each other, the emerging chirality at the carbonyl carbon atom which is now bound to the rhodium center in VI should depend upon the relative bulkiness of substituents (R^1 and R^2) of ketones and the siloxy group. For example, when the siloxy group is bulkier than either of the substituents ($R^1 > R^2$) of the ketone, *i.e.*, $\equiv$SiO $>$ L(R^1) $>$ S(R^2), inspection of the Dreiding model of VI leads one to conclude that the siloxy group should occupy a quasi-apical position which is sterically the least hindered site with respect to the coordination sphere, and at the same time orient itself between the two smallest groups (methyl and benzyl) of the chiral phosphines (P_1^* and P_2^*). It follows that a substitutent, L, will lie between methyl and phenyl groups of P_1^* and the substituent, S, between benzyl and phenyl groups of P_2^*. The conformation which satisfies these requirements is depicted as C_1 in Scheme 5. In a similar manner, when the order of bulkiness is L $>\equiv$ SiO $>$ S, the most stable conformation will be C_2, and when it is L $>$ S $>\equiv$ SiO, the most stable conformation, C_3. As is immediately seen from Scheme 5, the alcohol derived from C_1 has the same configuration as that derived from C_3, whereas the alcohol derived from C_2 has the inverted configuration provided that the priority sequence [48] of the substituents on the carbon atom is valid in each case.

According to the proposed rule, the relationship between the configuration of the chiral phosphine and that of the resulting alcohol should fall into six different

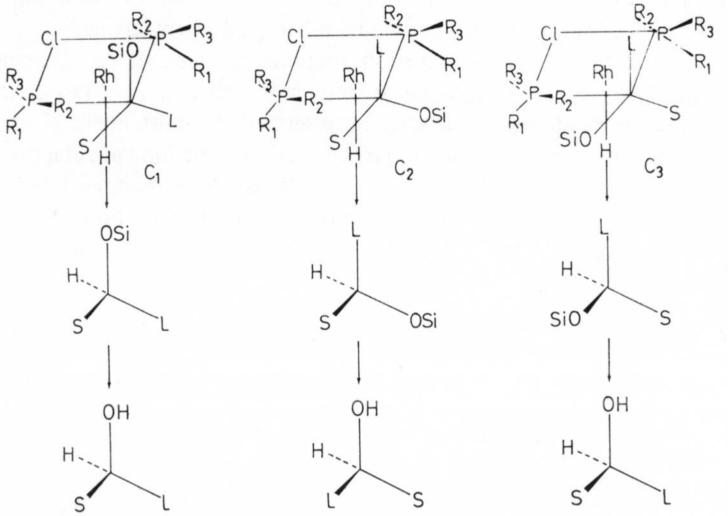

$R_1 = Ph$, $R_2 = PhCH_2$, and $R_3 = Me$

Scheme 5

cases as shown in Table 9 on account of both bulkiness and the priority sequence.

The preferred configurations (*R* and *S*) of the resulting alcohols are consistently predicted on the basis of the rule mentioned above, when the bulkiness of the siloxy groups is estimated just empirically as follows: *t*-Bu > $PhMe_2SiO$ > Ph > $EtMe_2SiO$ ~ Ph_2HSiO ~ Et_2HSiO > *i*-Pr > Et > Me for alkyl phenyl ketones. Thus, correspondence between experimental results and the prospects shown in Table 10 indicates no

Table 9

RELATIONSHIP BETWEEN THE BULKINESS SEQUENCE AND CONFORMATIONAL REQUIREMENT AT CARBONYL CARBON [40*b*]

Case	Bulkiness	Configurational relationship[a]	Conformation of complex VI
	Priority sequence L > S		
A	≡ SiO > L > S	Opposite	C_1
B	L > ≡ SiO > S	Same	C_2
C	L > S > ≡ SiO	Opposite	C_3
	Priority sequence L < S		
A′	≡ SiO > L > S	Same	C_1
B′	L > ≡ SiO > S	Opposite	C_2
C′	L > S > ≡ SiO	Same	C_3

[a] Relationship between the configuration of the chiral phosphine and that of the resulting alcohols.

Table 10

PREDICTION ON THE CONFIGURATIONAL RELATIONSHIP

Substituents			Classification[a]	Configurational relationship[b]
L	S	SiO		
Ph	Me	Et_2HSiO	B	Same
Ph	Et	Et_2HSiO	B	Same
t-Bu	Ph	$EtMe_2SiO$	C′	Same
t-Bu	Ph	$PhMe_2SiO$	B′	Opposite
Ph	Me	$PhMe_2SiO$	A	Opposite
Ph	Et	$PhMe_2SiO$	A	Opposite
Ph	*i*-Pr	$PhMe_2SiO$	A	Opposite
Ph	Et	Et_2HSiO	B	Same
Ph	*i*-Pr	Et_2HSiO	B	Same

[a] See Table 9. [b] See Table 8.

exception. It is of particular interest that the prediction is based only upon steric bulk considerations.

The proposed rule on the BMPP–rhodium(I) complex system, however, should be applied with great care to other systems.

More recently, an alternative mechanism has been proposed [45] for asymmetric hydrosilylation of prochiral ketones using (+)-DIOP–rhodium(I) complex (**6**) and α-naphthylpenylsilane, the latter undergoing concomitant conversion into an optically active, bifunctional alkoxysilane, which will be discussed separately (see Section 7.1). According to this proposed mechanism, diastereomeric silylhydrido–rhodium(III) complexes having trigonal bipyramidal structure are assumed as intermediates, which distinguish enantiotopic faces of a prochiral ketone in terms of 'steric approach control'.

A similar argument about prediction of predominant configuration of products in asymmetric hydrogenation and hydrosilylation has been reported [49]. Inspection of CPK-molecular models for (+)-DIOP–rhodium(I) system makes the author suggest VII as a model suffering the least steric hindrance in asymmetric addition of a dihydrosilane to a ketone.

In spite of an approximation and rather arbitrary choice about conformations of DIOP, model VII can be used successfully to predict the configuration of the major enantiomer produced.

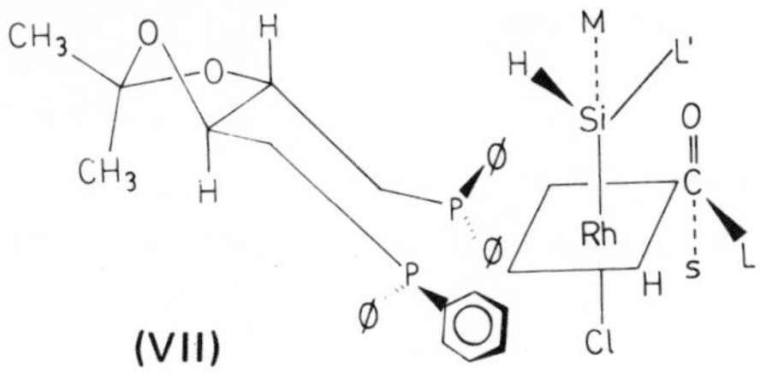

(VII)

5.2.2. CATIONIC CHIRAL RHODIUM(III) COMPLEX AS CATALYST

$[\{(R)\text{-BMPP}\}_2RhH_2(S)_2]^+ClO_4^-$ (**5**) was found to catalyze efficiently the asymmetric addition of trialkylsilanes to alkyl phenyl ketones [42] to give, after hydrolysis, predominantly (*S*)-1-phenylalkanols as shown in Table 11.

The extent of asymmetric hydrosilylation depends strongly upon the structure of hydrosilanes employed in a similar manner to the cases of other chiral rhodium complex-catalyzed reactions: with dimethylphenylsilane optical yields are generally more than several times as high as with trimethylsilane. Most remarkable is the fact that the addition of dimethylphenylsilane to pivalophenone gave the silyl ether of (*S*)-2,2-dimethyl-1-phenylpropanol, while that of trimethylsilane led to the (*R*)-enantiomer.

In order to explain the marked effect of silane structure on the stereochemical outcome, it seems reasonable, in the absence of contrary evidence, to assume that diastereomeric α-siloxyalkyl–rhodium intermediates are formed in the rate determining step, where a predominant configuration and the extent of enantiomeric excess of the product would have already been determined. It is, therefore, evident that the steric demands of not only the chiral phosphine ligand but also the substituents on the silicon bound to the rhodium catalyst exhibit a remarkable effect on the selection of enantiotopic faces of a prochiral ketone. An extreme example is given by the reversal of preferred configuration in the reactions of pivalophenone with dimethylphenylsilane (VIII) on the one hand and with trimethylsilane (IX) on the other.

Table 11

ASYMMETRIC HYDROSILYLATION OF RCOPh WITH R'_3SiH CATALYZED BY $[\{(R)\text{-BMPP}\}_2RhH_2(S)_2]^+ClO_4^-$ (5) (S = SOLVENT)[a] [42]

R	Yield (%)	1-Phenylalkanol $[\alpha]_D^{20}$, deg.	Optical yield (%) (configuration)
		$R'_3SiH = PhMe_2SiH$	
Me	97	−9.61	31.6 (*S*)
Et	94	−8.47	43.1 (*S*)
i-Pr	62	−18.51[b]	56.3 (*S*)
t-Bu	84	−11.20[c]	61.8 (*S*)
$PhCH_2$	44	*nil.*	———
		$R'_3SiH = Me_3SiH$	
Me	100	−1.55	5.1 (*S*)
Et	92	−1.26	6.4 (*S*)
i-Pr	98	−1.23[b]	3.7 (*S*)
t-Bu	81	+5.10[c]	28.1 (*R*)
$PhCH_2$	70	*nil.*	———

[a] Conditions = 50°, 40 hr in benzene solution with 0.05 mol% catalyst 5, (*R*)-BMPP optical purity 70%.
[b] Specific rotation in ether.
[c] In benzene.

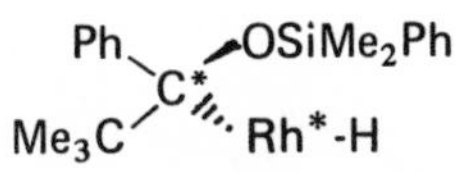

Ph OSiMe$_2$Ph
C*
Me$_3$C Rh*-H

(VIII)

Ph Rh*-H
C*
Me$_3$C OSiMe$_3$

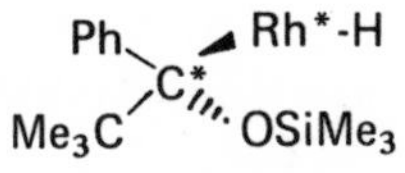

(IX)

It is still conceivably possible to argue that the hydrosilylation of ketones proceeds, in an analogous way to that of olefins, *via* alkoxyrhodium intermediates which arise from insertion of the ketone carbonyl into the hydridorhodium moiety. In fact, the intervention of an alkoxyrhodium has been proposed by Schrock and Osborn [26] for the hydrogenation of ketones catalyzed by cationic rhodium complexes as depicted in Scheme 3. However, such a mechanism involving alkoxyrhodium intermediates would not give rise to the observed changes in optical yield on changing the silane structure. In addition, the fact that the asymmetric hydrogenation of acetophenone catalyzed by the same chiral cationic rhodium complex (**5**) has been found to give (*R*)-1-phenylethanol [27] may reinforce the argument on the differrence in key steps between hydrogenation and hydrosilylation of ketones.

At this point, a question arises: Does the true catalyst species operating in the hydrosilylation from the cationic rhodium complex precursor, $[\{(R)\text{-BMPP}\}_2RhH_2(S)_2]^+ClO_4^-$ (**5**), happen to be the same as that from the non-cationic rhodium complex precursor, $[(R)\text{-BMPP}]_2Rh(S)Cl$ (**8**) which can lose the chloride anion by the action of hydrosilane to give a cationic species? In order to obtain some information, it seems appropriate to compare the results of Tables 8 and 11. The empirical order of bulkiness aforementioned may also apply to Table 11, if a relative bulkiness of trimethylsiloxy group be comparable to that of phenyl in the coordination sphere of the cationic rhodium catalyst.

Consequently, it can be said on the basis of the comparisons that the actual catalyst from the non-cationic rhodium complex (**8**) may well be closely related to that from the cationic rhodium complex (**5**), though the effect of the counter anion, ClO_4^-, should be taken into account.

5.2.3. CHIRAL FERROCENYLPHOSPHINES

In previous Sections, several kinds of chiral phosphines were described that have been demonstrated to be fairly to exceedingly effective as ligands of rhodium catalysts for asymmetric hydrogenation and/or hydrosilylation of certain unsaturated compounds.

Recently, a novel type of chiral phosphines has been reported [43*a*] which are very unique in that they have planar chirality. Representative examples are (*S*)-α-[(*R*)-2-diphenylphosphinoferrocenyl]ethyldimethylamine (PPFA), (*R*)-α-[(*S*)-2-dimethylphosphinoferrocenyl]ethyldimethylamine (MPFA), and (*S*)-α-[(*R*)-1′,2-*bis*(diphenylphosphino)ferrocenyl]ethyldimethylamine (BPPFA), which were prepared by lithiation, followed by treatment with an appropriate diorganochlorophosphine, of chiral α-dimethylaminoethylferrocene reported a few years before [43*b*].

Selected data on the asymmetric hydrosilylation of ketones catalyzed by

PPh$_2$ CHMeNMe$_2$ (S)–(R)–PPFA

CHMeNMe$_2$ Fe PMe$_2$ (R)–(S)–MPFA

PPh$_2$ PPh$_2$ Fe CHMeNMe$_2$ (S)–(R)–BPPFA

rhodium complexes containing these chiral ferrocenylphosphines as ligands are given in Table 12.

The marked effect of hydrosilanes on the stereoselectivity, which is very characteristic of the asymmetric hydrosilylation of ketones as described in the previous Sections, is seen here again. Fairly good optical yields comparable to those obtained in other chiral rhodium complex-catalyzed reactions were attained. For example, the reaction of acetophenone with diphenylsilane catalyzed by (*R*)-(*S*)-MPFA–rhodium complex gave higher optical yield than when (*R*)-BMPP or DIOP was used as ligands.

It should be noted that, in addition to the expected steric effects, some attractive interaction (*cf.* [2]) between the amino group in the BPPFA ligand and an appropriate prochiral substrate might contribute to the asymmetric induction. This is

Table 12

ASYMMETRIC HYDROSILYLATION OF KETONES CATALYZED BY FERROCENYLPHOSPHINE–RHODIUM COMPLEXES [24, 43]

Ketone	Silane	Ligand[a]	Optical yield (%) (configuration)
PhCOMe	Ph_2SiH_2	MPFA	49 (*R*)
	Ph_2SiH_2	BPPFA	28 (*R*)
	α-NpPhSiH_2	MPFA	52 (*R*)
PhCOEt	Ph_2SiH_2	MPFA	38 (*R*)
	Ph_2SiH_2	BPPFA	24 (*R*)
	Et_2SiH_2	BPPFA	25 (*R*)
	$PhMe_2SiH$	PPFA	10(9)[b] (*R*)
PhCO(*t*-Bu)	Ph_2SiH_2	MPFA	25 (*S*)
	Ph_2SiH_2	BPPFA	3.7 (*S*)
	Et_2SiH_2	BPPFA	18 (*S*)
	$PhMe_2SiH$	PPFA	10 (*R*)
MeCO(*t*-Bu)	Ph_2SiH_2	MPFA	41 (*R*)

[a] MPFA: (*R*)-α-[(*S*)-2-dimethylphosphinoferrocenyl]ethyldimethylamine.
BPPFA: (*S*)-α-[(*R*)-1',2-*bis*(diphenylphosphino)ferrocenyl]ethyldimethylamine.
PPFA: (*S*)-α-[(*R*)-2-diphenylphosphinoferrocenyl]ethyldimethylamine.
[b] PPFA/Rh = 1 in parenthesis.

inferred from the fact that the hydrogenation of α-acetamidocinnamic acid catalyzed by (*S*)-(*R*)-BPPFA–rhodium complex afforded (*S*)-*N*-acetylphenylalanine in extremely high optical yield (89–93%), whereas only much lower optical yield (38%) was observed in the hydrogenation of methyl α-acetamidocinnamate [43*c*], suggesting involvement of ammonium carboxylate formation in the transition state or intermediate.

5.3. Selective asymmetric hydrosilylation of α,β-unsaturated carbonyl compounds

In 1958, Russian chemists [31] reported that chloroplatinic acid-catalyzed hydrosilylation of α,β-unsaturated carbonyl compounds takes place in a 1,4-fashion. Recently, it has been disclosed [35] that highly selective 1,2- as well as 1,4-addition of hydrosilanes to α,β-unsaturated terpene ketones can be achieved by using chlorotris(triphenylphosphine)rhodium(I) (7), the selectivity depending markedly on the nature of the hydrosilane employed as described in Section 4.1. This achievement has resulted in studies on two kinds of selective asymmetric hydrosilylation of α,β-unsaturated carbonyl compounds by making use of either selective 1,4-addition or 1,2-addition: the 1,4-addition induces asymmetry on a β-carbon to afford optically active saturated carbonyl compounds, while the 1,2-addition gives optically active allylic alcohols.

Both complexes [(−)-DIOP]Rh(S)Cl (**6**) and $[\{(R)\text{-BMPP}\}_2RhH_2(S)_2]^+ClO_4^-$ (**5**) catalyze the asymmetric hydrosilylation of α,β-unsaturated ketones with monohydrosilanes to provide optically active ketones *via* 1,4-addition [50].

For instance, addition of dimethylphenylsilane to (*E*)-4-phenylpent-3-en-2-one in the presence of the cationic rhodium complex (**5**) dissolved in benzene gave only a 1,4-adduct, 2-phenyldimethylsiloxy-4-phenylpent-2-ene which was converted by hydrolysis to 4-phenylpentan-2-one, equation (19).

$$\text{Ph(Me)C=CH(COR}^1) + HSiMe_2R^2 \xrightarrow{[Rh^*]} PhMeC^*HCH{=}C(R^1)(OSiMe_2R^2) \xrightarrow{H_2O} PhMeC^*HCH_2COR^1 \quad (19)$$

The results obtained for the asymmetric hydrosilylation of α,β-unsaturated ketones are summarized in Table 13. It should be noted that in all cases the addition gives rise to (*R*)-ketones preferencially, that is, the addition takes place in a sense of selecting a *si–si* face of carbon–carbon double bonds of α,β-unsaturated ketones in an *E* form whether R^1 is methyl or phenyl group.

Attempted asymmetric hydrosilylation of β-methylcinnamaldehyde with trimethylsilane or phenyldimethylsilane resulted in formation of a mixture of optically inactive 1,4- and 1,2-adducts.

Of particular significance is that the present reaction may provide a facile route to the preparation of optically active silyl enol ethers, which are well suited for generation of metal enolate species.

Another asymmetric reduction of α,β-unsaturated ketones may be achieved if a

Table 13

ASYMMETRIC HYDROSILYLATION OF α,β-UNSATURATED KETONES $PhMeC{=}CHCOR^1$ WITH $HSiMe_2R^2$ CATALYZED BY CHIRAL PHOSPHINE–RHODIUM COMPLEXES[a] [50]

R^1	R^2	Catalyst[b]	Yield (%)	Optical yield (%) (configuration)
Me	Ph	**5**	76	16 (*R*)
Me	Ph	**6**	92	6.4 (*R*)
Me	Me	**5**	90	1.4 (*R*)
Ph	Ph	**5**	72	10 (*R*)
Ph	Ph	**6**	83	3.3 (*R*)
Ph	Me	**5**	94	9.5 (*R*)
Ph	Me	**6**	87	10 (*R*)

[a] Catalyst, 0.1 mol%.
[b] **5**: (*R*)-BMPP–Rh^+; **6**: (–)-DIOP–Rh.

chiral rhodium(I) complex catalyst and a dihydrosilane are employed for the hydrosilylation. A selective asymmetric reduction of β-ionone, 2-methylcyclohexenone, and mesityl oxide *via* 1,2-hydrosilylation has been reported [51] which employed (*R*)-BMPP–rhodium(I) complex (**8**) or (+)-DIOP–rhodium complex (**6**) as chiral catalyst, equation (20). Results are summarized in Table 14.

Table 14

SELECTIVE ASYMMETRIC REDUCTION OF α,β-UNSATURATED KETONES USING CATALYTIC HYDROSILYLATION[a] [51]

Ketone	Silane	Catalyst[b]	Product	Optical yield (%)
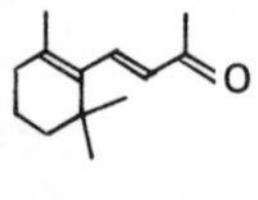	Ph_2SiH_2	8	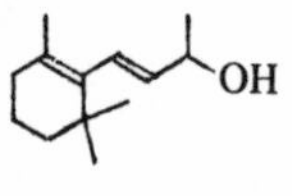	21
	α-$NpPhSiH_2$	8		34
	Ph_2SiH_2	6		1.3
	α-$NpPhSiH_2$	6		1.3
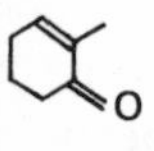	Ph_2SiH_2	8	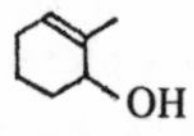	41
	α-$NpPhSiH_2$	8		43
	Ph_2SiH_2	6		15
	α-$NpPhSiH_2$	6		52
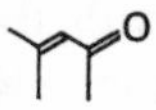	Ph_2SiH_2	8	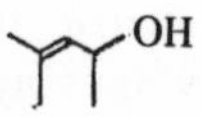	34
	α-$NpPhSiH_2$	8		35
	Ph_2SiH_2	6		18
	α-$NpPhSiH_2$	6		38

[a] Catalyst (0.3 mol%) in benzene solution.
[b] **6**: (+)-DIOP–Rh; **8**: (*R*)-BMPP–Rh.

$$\begin{array}{c}\diagdown\ \ |\\ C{=}C{-}\underset{\underset{O}{\|}}{C}{-}R + H_2SiR^1_2 \xrightarrow{[Rh^*]} C{=}C{-}\underset{\underset{OSiHR^1_2}{|}}{C^*H}{-}R\end{array} \qquad (20)$$

$[Rh^*] = [(R)\text{-BMPP}]_2Rh(S)Cl$ (**8**) or $[(+)\text{-DIOP}]Rh(S)Cl$ (**6**)

The optical yields attained in these reactions are well comparable to those realized in the asymmetric hydrosilylation of simple ketones as described in Section 5.2.1. Since these allylic alcohols cannot be obtained by catalytic hydrogenation of α,β-unsaturated ketones, the present asymmetric reduction may be of particular use for certain of organic syntheses.

5.4. Asymmetric reduction of keto esters *via* hydrosilylation catalyzed by chiral rhodium(I) complexes

Most of the work on asymmetric reduction of α-keto esters has centered upon heterogeneous catalytic hydrogenation and metal hydride reduction of chiral esters of α-keto acids. Relatively little is known about the asymmetric reduction of α-keto esters by chiral reducing agents, and the only reports are on the reduction of benzoylformic acid and its esters by the use of chiral magnesium alkoxides [52] and lithium aluminum hydride–chiral alcohol complexes [53]. Recently, a catalytic asymmetric hydrogenation of benzoylformates using *bis*(dimethylglyoxymato)cobalt(II)–quinine complex to give mandelates in 11.5–19.5% optical yield has been reported [54]. In this section is described a highly asymmetric reduction of certain keto esters as an application of the asymmetric catalytic hydrosilylation.

5.4.1. ASYMMETRIC REDUCTION OF α-KETO ESTERS *VIA* HYDROSILYLATION

The asymmetric reduction of α-keto esters, typically *n*-propyl pyruvate and ethyl benzoylformate, has been achieved [55] under conditions of chiral rhodium complex-catalyzed hydrosilylation as shown in equation (21).

$$R^1COCOOR^2 + R_2SiH_2 \xrightarrow{[Rh^*]} R^1\underset{\underset{OSiHR_2}{|}}{C^*}HCOOR^2 \qquad (21)$$

R^1 = Me, Ph; R^2 = Et, *n*-Pr, etc.

$[Rh^*] = (\text{BMPP})_2Rh(S)Cl$ (**8**) or $(\text{DIOP})Rh(S)Cl$ (**6**)

Results are summarized in Table 15. Generally, optical yields of lactates are remarkable and much higher than those obtained by other methods.

Further investigation revealed that the ester group as well as the hydrosilane used has a considerable effect on the extent of asymmetric induction. The optical yield attained in the case of *n*-propyl pyruvate using α-naphthylphenylsilane (85.4%) is the highest of this type of asymmetric reduction so far reported.

The preferred configuration can be predicted on the basis of a stereochemical

Table 15

CATALYTIC ASYMMETRIC REDUCTION *via* HYDROSILYLATION OF ALKYL PYRUVATES AND ETHYL BENZOYLFORMATE [55]

α-Keto ester	Silane	Ligand[a] in **6** or **8**	Optical yield (%) (configuration)
$MeCOCO_2Et$	Ph_2SiH_2	(—)-DIOP	72.5 (*R*)
	α-$NpPhSiH_2$	(—)-DIOP	76.1 (*R*)
$MeCOCO_2$(*n*-Pr)	Et_2SiH_2	(*R*)-BMPP	30.3 (*R*)
	$PhMeSiH_2$	(*R*)-BMPP	50.0 (*R*)
	Ph_2SiH_2	(*R*)-BMPP	60.3 (*R*)
	Ph_2SiH_2	(+)-DIOP	76.5 (*S*)
	α-$NpPhSiH_2$	(—)-DIOP	85.4 (*R*)
$MeCOCO_2$(*n*-Bu)	Ph_2SiH_2	(+)-DIOP	74.2 (*S*)
	α-$NpPhSiH_2$	(+)-DIOP	83.1 (*S*)
$MeCOCO_2$(*i*-Bu)	Ph_2SiH_2	(—)-DIOP	63.1 (*R*)
	α-$NpPhSiH_2$	(—)-DIOP	72.1 (*R*)
$PhCOCO_2Et$	Et_2SiH_2	(*R*)-BMPP	6.4 (*R*)
	Ph_2SiH_2	(*R*)-BMPP	10.3 (*S*)
	Ph_2SiH_2	(+)-DIOP	9.7 (*S*)
	α-$NpPhSiH_2$	(+)-DIOP	39.2 (*S*)
$PhCOCOOC_6H_{11}$	Ph_2SiH_2	(+)-DIOP	42.5(*S*)
	α-$NpPhSiH_2$	(+)-DIOP	47.2(*S*)

[a] **6**: (DIOP)RhCl(S) (S = Solvent); **8**: $(BMPP)_2RhCl(S)$.

model similar to that described in the previous Section (5.2.1). The most stable conformations of the intermediate α-siloxyalkyl–rhodium hydride complexes, according to inspection of Dreiding models for these two catalyst systems (**8**) and (**6**), are to be depicted as X and XI, respectively. It should be mentioned that the stereochemical requirements in the coordination sphere are very different from each other due principally to *trans* configuration in X and *cis* in XI.

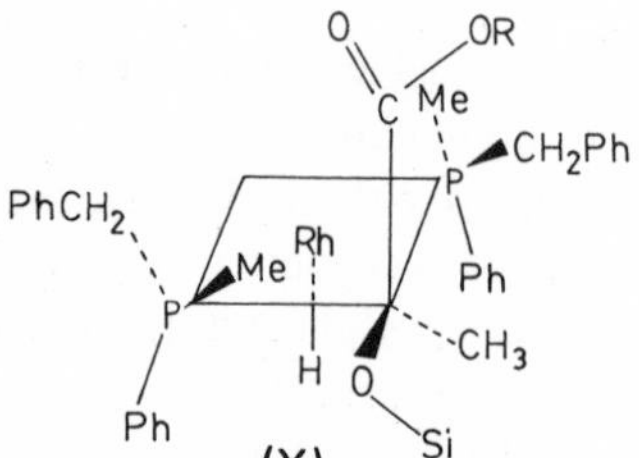

(X)

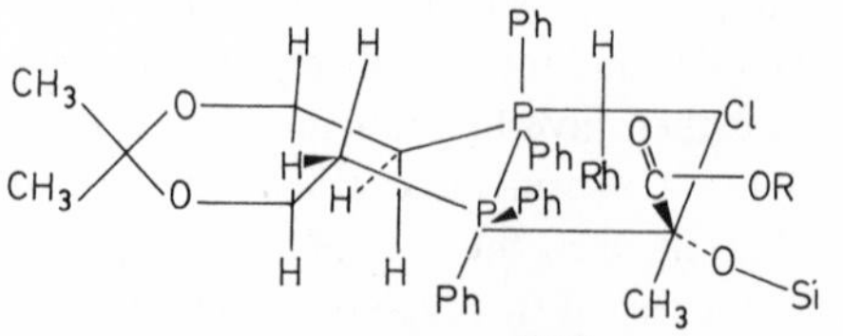

(XI)

5.4.2. 'DOUBLE ASYMMETRIC REDUCTION' OF (−)-MENTHYL PYRUVATE AND (−)-MENTHYL BENZOYLFORMATE

Conceptually, there are several distinct ways in which the asymmetric reduction of an α-keto ester to the corresponding optically active α-hydroxy ester can be achieved: (*a*) reduction of a chiral ester with an achiral reducing agent; (*b*) reduction of an achiral ester with a chiral reducing agent; and (*c*) a combination of a chiral ester and a chiral reducing agent ('double asymmetric reduction').

Some years ago, these possibilities were examined [53] with (−)-menthyl benzoylformate and ethyl benzoylformate. A simple asymmetric reduction involving either (−)-menthyl benzoylformate with an achiral agent, $LiAlH_4$(LAH)-cyclohexanol, process (*a*), or ethyl benzoylformate with a chiral reducing agent, LAH-(+)-camphor, process (*b*), gave (*R*)-mandelic acid after hydrolysis, in relatively low optical yields (10 and 4% e.e., respectively). On the other hand, the 'double asymmetric reduction', process (*c*), resulted in 49% asymmetric synthesis. This result is more than would be anticipated on the basis of a simple additive effect.

Very recently, catalytic 'double asymmetric reduction' has been accomplished [55] in the presence of rhodium complexes.

Thus, (−)-menthyl pyruvate was hydrosilylated using rhodium complexes with (*R*)-BMPP, (+)-DIOP or (−)-DIOP as chiral phosphine ligand, giving a lactic acid derivative, equation (22). The results of this 'double asymmetric induction' are listed in Table 16. The optical yields attained by the system are rather low as compared with those obtained in case of the asymmetric reduction of *n*-propyl pyruvate (*cf.* Table 15). Thus, the effect of (−)-menthyl group is by no means remarkable in these cases.

$$CH_3COCOO\text{-(menthyl)} + R_2SiH_2 \xrightarrow{[Rh^*]} CH_3C^*H(OSiHR_2)COO\text{-(menthyl)} \longrightarrow CH_3C^*H(OH)COO(n\text{-}Pr) \quad (22)$$

Contrary to these results, (−)-menthyl benzoylformate exhibited a remarkable effect on the following processes: (*a*) a simple asymmetric reduction of (−)-menthyl benzoylformate *via* the hydrosilylation catalyzed by a rhodium complex with achiral phosphine ligands, $(Ph_3P)_3RhCl$ (**7**); (*b*) double asymmetric reduction of (−)-menthyl benzoylformate using [(−)-DIOP]Rh(S)Cl (**6**); and (*c*) the same reaction using [(+)-DIOP]Rh(S)Cl (**6**).

From the results also shown in Table 16, the stereochemical control by the (−)-menthyl group itself was found to produce predominantly (−)-menthyl (*S*)-mandelate (21% e.e.), an antipode of that obtained from the lithium aluminum hydride reduction. In process (*b*), it was shown that a counteracting asymmetric induction by the chiral catalyst of (−)-DIOP exceeded the effect of (−)-menthyl group to give the (*R*)-mandelate with rather low stereoselectivity (37% e.e.). Production of the (*R*)-mandelate was scarcely favored in case of the asymmetric reduction of ethyl benzoyl-

Table 16
'DOUBLE ASYMMETRIC REDUCTION' OF (—)-MENTHYL PYRUVATE AND BENZOYLFORMATE [55]

Menthyl ester	Silane	Ligand[a] in 6 or 8	Optical yield (%) (configuration)
Pyruvate	Et_2SiH_2	(*R*)-BMPP	16.4 (*R*)
	Et_2SiH_2	(+)-DIOP	42.3 (*S*)
	Ph_2SiH_2	(+)-DIOP	62.4 (*S*)
	Ph_2SiH_2	(—)-DIOP	65.8 (*R*)
	α-NpPhSiH_2	(+)-DIOP	85.6(*S*)
	α-NpPhSiH_2	(—)-DIOP	82.8(*R*)
Benzolyformate	Ph_2SiH_2	(Ph_3P)[b]	21 (*S*)
	Ph_2SiH_2	(—)-DIOP	37 (*R*)
	Ph_2SiH_2	(+)-DIOP	60 (*S*)
	α-NpPhSiH_2	(+)-DIOP	77 (*S*)

[a] See Table 15. [b] $(Ph_3P)_3RhCl$ (7).

formate with the use of [(—)-DIOP]Rh(S)Cl (**6**) despite the absence of a negative effect of the (—)-menthyl group. These results typically indicate that the bulkiness of the ester group is an essential factor for determining the effectiveness and the direction of the asymmetric induction. In process (*c*), it was demonstrated that an efficient double asymmetric induction was realized (60–77% e.e.) by a synergetic effect of (+)-DIOP as a chiral ligand and the (—)-menthyl group, both of which favored the production of the (*S*)-mandelate. It should be noted that the effect of the (—)-menthyl group in the hydrosilylation is opposite on determining the preferred configuration to that in LAH-cyclohexanol reduction, the latter case being well understood by the Prelog's generalization [56] in which the two carbonyl groups of the α-keto ester are in the anti-coplanar conformation.

According to the Prelog's stereochemical considerations, the two carbonyl groups of (—)-menthyl benzoylformate should be in the *syn*-coplanar conformation for the hydrosilylation. It has previously been shown that the reaction proceeds *via* the α-siloxyalkyl–rhodium complex. Consequently, the results may be best explained by postulating a chelating effect of the silyl group which arises from coordination of the remaining carbonyl to the silyl moiety as shown in Scheme 6.

syn-coplanar $\xrightarrow[Ph_2SiH_2]{(Ph_3P)_3RhCl}$ $\xrightarrow{H^+}$ (*S*)-mandelate

Scheme 6

5.4.3. ASYMMETRIC REDUCTION OF β- AND γ-KETO ESTERS

The marked increase in optical yield in the reaction of the pyruvates when compared with simple prochiral ketones can probably be ascribed to a ligand effect of the ester moiety in the transition states. Further support for this view comes from the results for asymmetric hydrosilylation of β- and γ-keto esters, equations (23) and (24), and also of hexan-2-one, summarized in Table 17 [55*c*, 57]. A remarkable optical yield is observed only in the case of levulinates.

$$CH_3COCH_2COOCH_3 + R^1R^2SiH_2 \xrightarrow{[Rh^*]} CH_3\underset{\displaystyle OSiHR^1R^2}{\underset{|}{C^*}}HCH_2COOCH_3 \qquad (23)$$

$$CH_3COCH_2CH_2COOR + R^1R^2SiH_2 \xrightarrow{[Rh^*]} CH_3\underset{\displaystyle OSiHR^1R^2}{\underset{|}{C^*}}HCH_2CH_2COOR \xrightarrow{TsOH-MeOH} \text{γ-valerolactone } (CH_3-C^*H-CH_2-CH_2-C(=O)-O, \text{ ring}) \qquad (24)$$

The hydrosilylation of levulinates followed by acid solvolysis afforded γ-valerolactone. Thus this process may provide a convenient route to the optically active (up to 84% e.e.) γ-valerolactone.

6. ASYMMETRIC REDUCTION OF CARBON–NITROGEN DOUBLE BOND *via* HYDROSILYLATION

Reduction of Schiff bases to secondary amines has been accomplished for some years both by catalytic hydrogenation and by chemical reduction. Especially, the heterogeneous catalytic hydrogenation of the Schiff bases of chiral α-keto esters has been extensively studied, while $LiAlH_3(OR^*)$, R_2^*BH, and $Li[R_2^*(n\text{-}Bu)BH]$ have been used for the asymmetric reduction of imines [1*a*].

6.1. Reduction of Schiff bases *via* catalytic hydrosilylation

It has recently been reported [58] that Schiff bases undergo rhodium(I) complex-catalyzed hydrosilylation under mild conditions, equation (25).

$$R^1R^2C{=}NR^3 + {\equiv}Si{-}H \xrightarrow{\text{catalyst}} R^1R^2CH{-}N(R^3)Si{\equiv} \xrightarrow{MeOH} R^1R^2CHNHR^3 \qquad (25)$$

Table 17

ASYMMETRIC REDUCTION OF β- AND γ-KETO ESTERS *via* HYDROSILYLATION USING (+)-DIOP–Rh (6) [55c, 57]

Keto ester	Silane	Yield (%)	Optical yield (%) (configuration)
$MeCOCH_2CO_2Me$	Ph_2SiH_2 α-$NpPhSiH_2$	84 89	14.4 (*S*) 21.1 (*S*)
$MeCOCH_2CH_2CO_2Me$	Ph_2SiH_2 α-$NpPhSiH_2$	99 99	39.6 (*S*) 76.2 (*S*)
$MeCOCH_2CH_2CO_2Et$	Ph_2SiH_2 α-$NpPhSiH_2$	100 99	38.1 (*S*) 80.5 (*S*)
$MeCOCH_2CH_2CO_2(n\text{-}Pr)$	Ph_2SiH_2 α-$NpPhSiH_2$	96 94	34.6 (*S*) 82.7 (*S*)
$MeCOCH_2CH_2CO_2(i\text{-}Bu)$	Ph_2SiH_2 α-$NpPhSiH_2$	96 96	39.2 (*S*) 84.4 (*S*)
$MeCOCH_2CH_2CO_2C_6H_{11}$	Ph_2SiH_2 α-$NpPhSiH_2$	94 98	43.1 (*S*) 83.5 (*S*)
$MeCOCH_2CH_2CH_2Me$	Ph_2SiH_2 α-$NpPhSiH_2$	90 85	14.7 (*S*) 25.9 (*S*)

The resulting *N*-silylamines are readily converted by methanolysis into the corresponding amines. Some selected data are summarized in Table 18. There is one precedent of Lewis acid-catalyzed addition of triethylsilane to benzylideneaniline [59]. However, the reaction in the presence of $(Ph_3P)_3RhCl$ (7) or $PdCl_2$ as catalyst has been found to proceed under much milder conditions in exceedingly high yield. Dihydrosilanes react more smoothly than monohydrosilanes. Based on the reaction with dihydrosilanes such as Et_2SiH_2, $PhMeSiH_2$ and Ph_2SiH_2, a fairly extensive series of complexes may be arranged with respect to their catalytic activity, the order being: $(Ph_3P)_3RhCl \gg (Ph_3P)_2Rh(CO)Cl > Py_2RhCl(DMF)BH_4 >$ [(1,5-hexadiene)-$RhCl]_2 >$ [(1,5-cyclooctadiene)$RhCl]_2 > PdCl_2 > (Ph_3P)_2PdCl_2$. Chloroplatinic acid cannot be practically used because of many side reactions. In a practical sense, it can be said that dihydrosilane–$(Ph_3P)_3RhCl$ combination is the most effective for the reaction.

6.2. Asymmetric reduction of Schiff bases

An effective asymmetric reduction of Schiff bases to optically active secondary amines *via* catalytic hydrosilylation using (+)-DIOP–rhodium complex (**6**) has been

Table 18

REDUCTION OF SCHIFF BASES *via* HYDROSILYLATION[a] [58]

Schiff base	Silane	Conditions	Yield (%)[b]
PhCH=NMe	Et_2SiH_2 Ph_2SiH_2 Et_3SiH Et_3SiH	30°, 0.5 hr 0°, 3 hr 100°, 20 hr 55°, 24 hr[c]	95 92 65 90
PhCH=N(*n*-Bu)	Et_2SiH_2 $PhMeSiH_2$	55°, 1.5 hr 55°, 0.5 hr	96 93
PhCH=NPh	Et_2SiH_2 Et_3SiH	50°, 2 hr 100°, 15 hr	96 91
PhMeC=NPh	Et_2SiH_2	55°, 72 hr	85
$PhCH{=}NCH_2Ph$	Et_2SiH_2	25°, 5 hr	94
(H)=NCH_2Ph	Et_2SiH_2	50°, 20 min	96
$Ph_2C{=}NH$	Et_2SiH_2 $PhMe_2SiH$	50°, 10 min 80°, 3 hr	98 98

[a] Catalyst, 0.5 mol% $(Ph_3P)_3RhCl$ (7).
[b] Determined by glpc.
[c] $PdCl_2$ catalyzed reaction.

Table 19

ASYMMETRIC REDUCTION OF OPEN CHAIN SCHIFF BASES *via* HYDROSILYLATION USING (+)-DIOP–Rh (6)[a] [60]

Schiff bases	Silane	Temperature (°C)	Yield (%)	Optical yield (%) (configuration)
Ph(Me)$C{=}N{-}CH_2Ph$	Ph_2SiH_2 Ph_2SiH_2 $(MeSiHO)_n$	24 2 24	98 97 —	50 (*S*) 65 (*S*) 3.4 (*S*)
Ph(Me)C=N–Ph	Ph_2SiH_2 Ph_2SiH_2	24 5	— 90	40 (*S*) 47 (*S*)
$PhCH_2$(Me)$C{=}N{-}CH_2Ph$	Ph_2SiH_2 $(MeSiHO)_n$	24 24	40 —	12 (*S*) 14 (*S*)

[a] Catalyst (2 mol%) in benzene solution.

Table 20

ASYMMETRIC REDUCTION OF CYCLIC SCHIFF BASES, (R[1], R[2], R[3]-substituted 3,4-dihydroisoquinoline) *via* HYDROSILYLATION WITH Ph_2SiH_2[a] [60]

R[1]	R[2]	R[3]	Yield (%)	Optical yield (%) (configuration)
H	H	$PhCH_2$	78	22.5
MeO	MeO	Me	93	5.7 (*R*)
MeO	MeO	3,4-(MeO)$_2C_6H_3CH_2$	98	38.7 (*R*)

[a] See Table 19, footnote *a*.

$$\text{R}^1,\text{R}^2\text{-3,4-dihydroisoquinoline (1-R}^3\text{)} \xrightarrow{[\text{Rh}^*]} \xrightarrow{\text{H}^+} \text{R}^1,\text{R}^2\text{-1-R}^3\text{-1,2,3,4-tetrahydroisoquinoline (NH)} \quad (26)$$

reported [60]. Results are summarized in Tables 19 and 20. From Table 19, it is seen that diphenylsilane generally affords better results than a polymethylsiloxane. The optical yield depends considerably upon the reaction temperature, *e.g.*, α-phenylethylidene-*N*-benzylamine was reduced to α-phenylethylbenzylamine in 50% optical yield at ambient temperature, whereas a higher optical yield (65%) was attained at 2 °C using diphenylsilane.

In Table 20 are listed results of asymmetric reduction of certain cyclic Schiff bases by the use of diphenylsilane-(+)-DIOP–rhodium complex combination, equation (26). The optical yields are significantly affected by the bulkiness of substituents R.

7. ASYMMETRIC SYNTHESIS OF BIFUNCTIONAL ORGANOSILICON COMPOUNDS *via* HYDROSILYLATION

The fundamental work of Sommer and his coworkers on the stereochemistry of organosilicon compounds originated from the preparation of optically pure methyl-α-naphthylphenylsilane (XII) and closely related compounds, which were obtained *via* fractional crystallization of diastereomeric menthoxymethyl-α-naphthylphenylsilane [61]. However, few examples of the preparation of optically active organosilanes either by kinetic resolution or by asymmetric synthesis have been recorded so far in the literature. The reports on kinetic resolution are that of bifunctional organosilanes upon mentholysis of chloro-α-naphthylphenylsilane [62] and partial reduction of various racemic methoxysilanes by a chiral reducing complex of lithium aluminum hydride [63]. A single paper [64] deals with asymmetric synthesis at a silicon center by the reaction of *bis*(*N*-methylacetamido)methylphenylsilane with optically active amino acids to form unequal amounts of diastereomeric pairs of 2-siloxazolidone-5, which are claimed to undergo a second order asymmetric transformation.

The first asymmetric synthesis at a prochiral silicon center in the sense of catalytic asymmetric reactions has recently been effected in the catalytic hydrosilylation of ketones with dihydrosilanes [45, 65].

In the preceding Sections it was described that chiral phosphine–rhodium complexes are effective in causing stereoselective addition of a hydrosilane to a variety of prochiral carbonyl compounds to give silyl ethers of the corresponding alkanols with fairly high enantiomeric bias at the carbon atom. The present section describes an application of the catalytic asymmetric hydrosilylation of ketones to the preparation of some new asymmetric bifunctional organosilanes.

7.1. Asymmetric synthesis at silicon using hydrosilylation catalyzed by chiral rhodium complexes

Addition of certain dihydrosilanes, $H_2SiR^1R^2$, to such symmetric ketones as diethyl ketone and benzophenone in the presence of chiral rhodium complexes gave silyl ethers in an optically active form associated with the silicon atom, equation (27).

$$R_2CO + H_2SiR^1R^2 \xrightarrow{[Rh^*]} R_2CHOSi^*HR^1R^2 \tag{27}$$

The silyl ethers were converted into known methyl-α-naphthylphenylsilane (XII) in order to determine respective optical yields, see equations (28) and (29).

$$\underset{\text{(XIII)}}{\text{H–Si(}\alpha\text{-Np)(Me)–H}} \xrightarrow[\text{Et}_2\text{C=O}]{[\text{Rh}^*]} \underset{(S)\text{-(XV)}}{\text{Et}_2\text{CHO–Si(}\alpha\text{-Np)(Me)–H}} \xrightarrow[\text{(retention)}]{\text{PhMgCl}} \underset{(R)\text{-(XII)}}{\text{Ph–Si(}\alpha\text{-Np)(Me)–H}} \tag{28}$$

$$\underset{\text{(XIV)}}{\text{H–Si(}\alpha\text{-Np)(Ph)–H}} \xrightarrow[\text{Et}_2\text{C=O}]{[\text{Rh}^*]} \underset{(S)\text{-(XVI)}}{\text{Et}_2\text{CHO–Si(}\alpha\text{-Np)(Ph)–H}} \xrightarrow[\text{(retention)}]{\text{MeMgBr}} \underset{(S)\text{-(XII)}}{\text{Me–Si(}\alpha\text{-Np)(Ph)–H}} \tag{29}$$

The results obtained for the reaction catalyzed by the cationic rhodium complex (**5**) are summarized in Table 21. It is particularly noteworthy that the (*R*)-enantiomer of methyl-α-naphthylphenylsilane (XII) was produced in excess through the hydro-

Table 21

ASYMMETRIC SYNTHESIS AT SILICON *via* HYDROSILYLATION OF SYMMETRIC KETONES

Silane	Ketone	Ligand[a] in 5 or 6	Optical yield (%)[b] (configuration)	Reference
α-NpMeSiH_2	Me_2CO	(*R*)-BMPP	7.4 (*R*)	65
	Et_2CO	(*R*)-BMPP	8.6 (*R*)	65
	Et_2CO	(+)-DIOP	12 (*S*)	45
	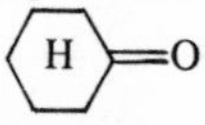 H =O	(*R*)-BMPP	7.2 (*R*)	65
	Ph_2CO	(*R*)-BMPP	19.7 (*R*)	65
	Ph_2CO	(—)-DIOP	18.8 (*R*)	66
	Ph_2CO	(+)-DIOP	7 (*S*)	45
α-NpEtSiH_2	Et_2CO	(+)-DIOP	21 (*S*)	45
	Ph_2CO	(+)-DIOP	33 (*S*)	45
α-NpPhSiH_2	Me_2CO	(+)-DIOP	30 (*R*)	45
	Et_2CO	(*R*)-BMPP	7.2 (*S*)	65
	Et_2CO	(+)-DIOP	46 (*R*)	45
	Et_2CO	(—)-DIOP	53 (*S*)	66
	$(n\text{-Pr})_2CO$	(+)-DIOP	39 (*R*)	45
	$(i\text{-Bu})_2CO$	(+)-DIOP	36 (*R*)	45
	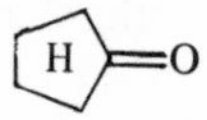 H =O	(+)-DIOP	35 (*R*)	45
	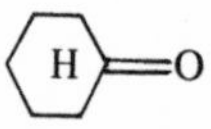 H =O	(*R*)-BMPP	3.1 (*S*)	65
	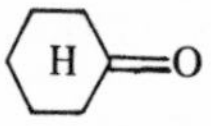 H =O	(+)-DIOP	32 (*R*)	45
	Ph_2CO	(*R*)-BMPP	27.7 (*S*)	65
	Ph_2CO	(+)-DIOP	31 (*R*)	45
	Ph_2CO	(—)-DIOP	46 (*S*)	66

[a] $[(BMPP)_2RhH_2(S)_2]^+ClO_4^-$ (**5**); (DIOP)RhCl(S) (**6**) (S = solvent).

[b] Based on α-NpPhMe$\overset{*}{\text{Si}}$H or α-NpPhEt$\overset{*}{\text{Si}}$H derived from reaction products.

silylation of ketones using methyl-α-naphthylsilane (XIII) while the (*S*)-enantiomer was obtained in excess when α-naphthylphenylsilane (XIV) was employed. If, for example, phenylation of 3-pentyloxy-α-naphthylmethylsilane (XV) and methylation of 3-pentyloxy-α-naphthylphenylsilane (XVI) proceed largely with retention of configuration at the silicon center, the above equations (28) and (29) would be valid to rationalize the present results [65]. This clearly suggests that, in both cases of forming XV and XVI, one enantiotopic hydrogen [*pro S*] attached to the silicon atom of the dihydrosilanes (XIII and XIV) participates preferentially in the hydrosilylation of symmetric ketones catalyzed by the chiral complex (**5**) over the other (*pro R*).

A considerable variation of optical yields on changing ketone structure should also be mentioned (Table 21); benzophenone gave appreciably higher optical yield than other ketones in the reaction with both XIII and XIV. It seems reasonable that the difference in bulkiness of the ketone, which is coordinated to the chiral phosphine–rhodium complex like solvent, must influence the stereoselectivity to some extent because of modifying the effective bulk of the rhodium complex. Then, benzophenone is of advantage under the given conditions to attain higher asymmetric induction over the less bulky ketones.

The addition reaction of XIII and XIV to a few aldehydes in the presence of the cationic rhodium complex (**5**) gave the silyl ethers of primary alkanols in moderate yields without deactivation of the catalyst by decarbonylation [67]. No appreciable asymmetric induction, however, was observed.

Comparable results have been obtained independently [45] using the DIOP–rhodium(I) complex (**6**). The addition of an asymmetric organosilane, (*R*)-α-NpPhMeSiH (XII), to acetone in the presence of $(Ph_3P)_3RhCl$ (**7**) was found to proceed with retention of configuration at the silicon center, equation (30).

$$\underset{(R)\text{-(XII)}}{\alpha\text{-Np}-\overset{\text{Ph}}{\underset{\text{H}}{\text{Si}}}-\text{Me}} \xrightarrow[(Ph_3P)_3RhCl]{Me_2C{=}O} \alpha\text{-Np}-\overset{\text{Ph}}{\underset{\text{O}(i\text{-Pr})}{\text{Si}}}-\text{Me} \xrightarrow[Et_2O]{LiAlH_4} \underset{(R)\text{-(XII)}}{\alpha\text{-Np}-\overset{\text{Ph}}{\underset{\text{H}}{\text{Si}}}-\text{Me}} \qquad (30)$$

(96% retention) (97% retention)

The asymmetric synthesis from prochiral organosilanes with several symmetric ketones using the DIOP–rhodium(I) complex (**6**) are also summarized in Table 21.

Similar reactions with aldehydes as substrates have also been performed [45]. Compared with the cationic rhodium complex (**5**), (+)-DIOP–rhodium(I) (**6**) causes appreciable asymmetric induction; however, optical yields are much lower than those obtained with ketones.

Such techniques have been extended to an asymmetric reaction at both silicon and carbon centers using either prochiral ketones or terpene ketones and silanes [68]. The reaction product was treated with a Grignard reagent to provide optically active organosilane XII and an alcohol. The results are given in Table 22. The optical purities of organosilanes obtained were greatly improved in some cases. The alcohols were obtained generally in better optical purity than organosilanes. However, there is no relation between the optical purity of the asymmetric silane and the alcohol. According to the proposed mechanisms, oxidative addition of a silane $R^1R^2SiH_2$ to the complex (**6**) will lead to the formation of two diastereomeric intermediates (XVII *a*, *b*), which are most likely in equilibrium in different quantities as shown in Scheme 7. These intermediates react with the prochiral ketone with different rate constants, k_a^r, k_a^s for XVII*a* and k_b^r, k_b^s for XVII*b*.

The optical purity at the silicon center (P_{Si}) depends on the relative rates of the reaction of the intermediates XVII*a* and XVII*b*: $P_{Si} = (k_a^r + k_a^s)[\text{XVII}a]/(k_b^r + k_b^s)$

Table 22

ASYMMETRIC SYNTHESIS AT SILICON *via* HYDROSILYLATION OF PROCHIRAL KETONES WITH α-$NpPhSiH_2$ [45, 68]

Ketone	Ligand[a] in 6	Optical purity (%)	
		Silane[b]	Alkanol
PhCOMe	(—)-DIOP	32 (*S*)	55 (*R*)
PhCOEt	(+)-DIOP	30 (*R*)	56 (*S*)
MeCOEt	(—)-DIOP	40 (*S*)	42 (*R*)
MeCOEt	(+)-DIOP	39 (*R*)	——
MeCO(*t*-Bu)	(+)-DIOP	35 (*R*)	——
(—)-Menthone	Ph_3P	67 (*R*)	*c*
(—)-Menthone	(+)-DIOP	82 (*R*)	——
(—)-Menthone	(—)-DIOP	46 (*R*)	——
(—)-Menthone	MDPP	4 (*S*)	——
(—)-Menthone	NMDPP	16 (*S*)	——
(+)-Camphor	Ph_3P	62 (*S*)	*c*
(+)-Camphor	(+)-DIOP	20 (*S*)	——
(+)-Camphor	(—)-DIOP	31 (*S*)	——

[a] Catalyst of the type $L_2^*Rh(S)Cl$; L* = monodentate phosphine, S = solvent.
[b] α-NpPhMeSiH or α-NpPhEtSiH obtained by alkylation.
[c] For stereoselectivity of the resulting alkanols, see Table 6.

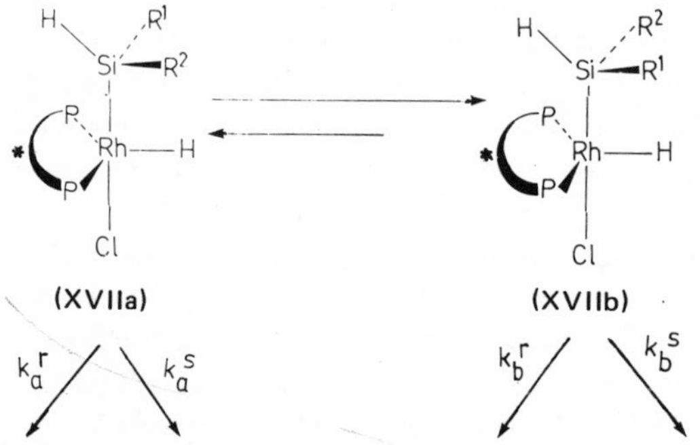

(XVIIa) (XVIIb)

Scheme 7

[XVII*b*]. Similarly, the optical purity at the carbon atom center (P_C) can be given as follows: $P_C = (k_a^r[\mathrm{XVII}a] + k_b^r[\mathrm{XVII}b])/(k_a^s[\mathrm{XVII}a] + k_b^s[\mathrm{XVII}b])$. This may well explain why there is no relation between the optical purity of the asymmetric silane and alcohol.

Table 23

ASYMMETRIC ALCOHOLYSIS OF DIHYDROSILANES [72]

Silane	Alcohol	Ligand[a] in 7 or 6	Optical yield (%)[b] (configuration)
α-NpPhSiH_2	(−)-Menthol	Ph_3P	48 (*R*)
	(+)-Cinchonine	Ph_3P	13 (*S*)
	(−)-Ephedrine	Ph_3P	47 (*R*)
	(+)-Ephedrine	Ph_3P	54 (*S*)
α-NpEtSiH_2	(−)-Menthol	Ph_3P	21 (*S*)
α-Np(PhCH_2)SiH_2	(−)-Menthol	Ph_3P	21 (*R*)
Ph(*i*-Bu)SiH_2	(−)-Menthol	Ph_3P	16 (*S*)
SiH_2 (cyclic)	(−)-Menthol	Ph_3P	40 (*S*)
α-NpPhSiH_2	MeOH	(+)-DIOP	3 (*R*)
	i-PrOH	(+)-DIOP	8 (*R*)
	t-BuOH	(−)-DIOP	7 (*S*)
	c-$C_6H_{11}OH$	(+)-DIOP	17 (*R*)
	PhCH_2OH	(+)-DIOP	19 (*R*)
α-NpPhSiH_2[c]	(−)-Menthol	(−)-DIOP	31 (*R*)
	(−)-Menthol	(+)-DIOP	49 (*R*)
	(−)-Ephedrine	(−)-DIOP	42 (*R*)
	(−)-Ephedrine	(+)-DIOP	46 (*R*)
	(+)-Ephedrine	(−)-DIOP	44 (*S*)
	(+)-Ephedrine	(+)-DIOP	40 (*S*)
	(−)-Menthol	NMDPP[d]	22 (*R*)
	(−)-Menthol	MDPP[d]	44 (*R*)
SiH_2 (cyclic)	(−)-Menthol	(+)-DIOP	56 (*S*)

[a] $(Ph_3P)_3RhCl$ (7); (DIOP)RhCl(S) (6), S = solvent.
[b] Estimated after conversion into appropriate silanes, $R^1R^2R^3Si^*H$.
[c] Double asymmetric synthesis.
[d] Catalyst of the type $L_2^*Rh(S)Cl$; L* = NMDPP or MDPP, S = Solvent.

7.2. Asymmetric synthesis of alkoxyhydrosilanes by dehydrogenative alcoholysis

In connection with the catalytic hydrosilylation, asymmetric alcoholysis of dihydrosilanes in the presence of chiral rhodium complexes as catalysts deserves to be described.

It has been known for some years that the group VIII metals and metal salts catalyze the dehydrogenative solvolysis of hydrosilanes with amines, alcohols and carboxylic acids [69]. Studies on the scope of these processes and the mechanistic investigations were limited to the use of monohydrosilanes. Recently, it has been found that chlorotris(triphenylphosphine)rhodium(I) (**7**) is exceedingly effective for the selective alcoholysis of hydrosilanes including polyhydrosilanes [33, 70].

Aminolysis and thiolysis of hydrosilanes were also effectively catalyzed by the rhodium(I) complex (7) [71].

Very recently, asymmetric synthesis of optically active alkoxyhydrosilanes has been accomplished by way of this type of reaction. Alcoholysis and treatment of the product with an appropriate Grignard reagent led to the known optically active hydrosilane, equation (31). Results are summarized in Table 23.

$$\begin{array}{c} R^1R^2SiH_2 + R^*OH \xrightarrow{(Ph_3P)_3RhCl} R^1R^2HSi^*OR^* + H_2 \\ \downarrow R^3MgX \\ R^1R^2R^3Si^*H \end{array} \tag{31}$$

Results of simple and double asymmetric alcoholysis of prochiral dihydrosilanes catalyzed by the chiral rhodium(I) complex, (DIOP)Rh(S)Cl (6), are also shown in Table 23. Optical yields of the simple asymmetric alcoholysis using the chiral rhodium(I) complex were rather low, while the double asymmetric one provided good results [72]. The mechanism of the asymmetric induction may be closely similar to that of the asymmetric hydrosilylation.

8. CONCLUSION

Reviewed here is the recent advances in catalytic asymmetric hydrosilylation of olefins, carbonyl compounds including keto esters, and imino compounds using transition metal complexes of chiral phosphine ligands. All data available up to August 1975 are compiled in Tables. The asymmetric addition of hydrosilanes to carbon–oxygen double bonds is well regarded as one of the most efficient catalytic asymmetric reactions, providing a unique method of reduction in organic synthesis.

9. ACKNOWLEDGMENTS

The authors would like to take this opportunity to express their sincere appreciation to Professor Y. Nagai, Gumma University, for his valuable suggestions and to Dr. T. Hayashi, Kyoto University, for his enthusiastic collaboration throughout the studies reviewed here.

We also deeply thank Professors H. B. Kagan, Université de Paris-Sud, and R. J. P. Corriu, Université des Sciences et Techniques du Languedoc, for fruitful discussions.

10. ABBREVIATIONS FOR LIGAND NAMES

Name of ligand	Abbreviation
(*R*, *R*)-(−)-2,3-*O*-Isopropylidene-2,3-dihydroxy-1,4-*bis*(diphenylphosphino)butane	(−)-DIOP
(*R*)-(+)-Benzylmethylphenylphosphine	(*R*)-BMPP
(*R*)-(−)-Methylphenyl-*n*-propylphosphine	(*R*)-MPPP
(−)-Menthyldiphenylphosphine	MDPP
(+)-Neomenthyldiphenylphosphine	NMDPP
(*S*)-α-[(*R*)-2-Diphenylphosphinoferrocenyl]ethyl-dimethylamine	(*S*)-(*R*)-PPFA
(*R*)-α-[(*S*)-2-Dimethylphosphinoferrocenyl]ethyl-dimethylamine	(*R*)-(*S*)-MPFA
(*S*)-α-[(*R*)-1′,2-*Bis*(diphenylphosphino)ferrocenyl]-ethyldimethylamine	(*S*)-(*R*)-BPPFA

11. REFERENCES

[1] For recent pertinent reviews, see:
a) J. D. Morrison and H. S. Mosher, *Asymmetric Organic Reactions*, Prentice-Hall, Inc., Englewood Cliffs, New Jersey, (1971).
b) J. W. Scott and D. Valentine, Jr., *Science*, **184**, 943, (1974).
c) L. Marko and B. Heil, *Cat. Rev.*, **8**, 269, (1973).
d) D. R. Boyd and M. A. McKervey, *Quart. Rev.*, **22**, 95, (1968).
e) J. Mathieu and J. Weill-Raynal, *Bull. Soc. Chim. France*, 1211, (1968).
f) H. J. Schneider and R. Haller, *Pharmazie*, **28**, 417, (1973).
g) T. D. Inch, *Synthesis*, 466, (1970).
h) S. Yamada and K. Koga, *Selective Organic Transformations*, B. S. Thyagarajan, (Ed.), Wiley-Interscience, New York, Vol. 1, p. 1, (1970).
i) E. I. Klabunovskii and E. S. Levitina, *Uspekhi Khim.*, **39**, 2154, (1970).
j) K. Yamamoto, *Kagaku to Kogyo*, **26**, 193, (1973).
k) *Chemistry of Asymmetric Reactions, Kagaku Sosetsu*, Chem. Soc. Japan, (Ed.), No. 4, (1974).
l) J. Hetflejs, *Chem. Listy.*, **68**, 916, (1974).
m) Y. Izumi and A. Tai, *Stereo-differentiation – A New Concept for Asymmetric Reactions*, Kodan-sha, Tokyo, and John Wiley & Sons, New York (in press).

[2] *a*) W. S. Knowles, M. J. Sabacky and B. D. Vineyard, *Chem. Commun.*, 10, (1972); *Chem. Technol.*, **2**, 590, (1972).
b) W. S. Knowles, M. J. Sabacky, B. D. Vineyard and D. J. Weinkauff, *J. Amer. Chem. Soc.*, **97**, 2567, (1975).

[3] H. Kagan and T.-P. Dang, *Chem. Commun.*, 481, (1971); *J. Amer. Chem. Soc.*, **94**, 6429, (1972).

[4] For a review, P. Pino, G. Consiglio, C. Botteghi and C. Salomon, *Homogeneous Catalysis II, Advances in Chemistry Series*, **132**, Chapt. 20, (1974).

[5] For a review, B. Bogdanović, *Angew. Chem.*, **85**, 1013, (1973).

[6] *a*) H. Nozaki, H. Takaya, S. Moriuti and R. Noyori, *Tetrahedron*, **24**, 3655, (1968); *Can. J. Chem.*, **47**, 1242, (1969).
b) Y. Tatsuno, A. Nakamura, A. Konishi and S. Otsuka, *Chem. Commun.*, 588, (1974).
c) T. Aratani, Y. Yoneyoshi and T. Nagase, *Tetrahedron Lett.*, 1707, (1975).

[7] For a review, I. Ojima, *J. Syn. Org. Chem. Japan*, **32**, 687, (1974).

[8] *a*) C. Eaborn and R. W. Bott, *Organometallic Compounds of the Group IV Elements*, A. G. MacDiarmid, (Ed.), Vol. 1, Part 1, Marcel Dekker, New York, (1968).
b) E. Ya. Lukevits and M. G. Voronkov, *Organic Insertion Reactions of Group IV Elements*, Consultants Bureau, New York, (1966).
c) R. N. Meals, *Pure Appl. Chem.*, **13**, 141, (1966).

[9] A. J. Chalk and J. F. Harrod, *J. Amer. Chem. Soc.*, **87**, 16, (1965).

[10] J. A. Osborn, F. H. Jardine, J. F. Young and C. Wilkinson, *J. Chem. Soc. A*, 1711 (1966).

[11] Y. Kiso, K. Tamao and M. Kumada, *J. Organometal. Chem.*, **76**, 95, (1974), and references cited therein.

[12] *a*) S. Takahashi, T. Shibano and N. Hagihara, *Chem. Commun.*, 161, (1969).
b) J. Tsuji, M. Hara and K. Ohno, *Tetrahedron*, **30**, 2143, (1974).

[13] *a*) K. Yamamoto, T. Hayashi and M. Kumada, *J. Organometal. Chem.*, **28**, C37, (1971).
b) W. Fink, *Helv. Chem. Acta*, **54**, 1304, 2186, (1971).

[14] *a*) R. N. Haszeldine, R. V. Parish and D. J. Parry, *J. Chem. Soc. A*, 683, (1969).
b) F. de Charentenay, J. A. Osborn and G. Wilkinson, *J. Chem. Soc. A*, 787, (1968).

[15] F. A. Cotton and G. Wilkinson, *Advanced Inorganic Chemistry*, 3rd Ed., Chapt. 24, Wiley-Interscience, New York, (1972).

[16] A. J. Chalk, *Trans. N.Y. Acad. Sci., II*, **32**, 481, (1970).

[17] *a*) J. F. Harrod and C. A. Smith, *J. Amer. Chem. Soc.*, **92**, 2699, (1970).
b) J. F. Harrod, C. A. Smith and K. A. Than, *J. Amer. Chem. Soc.*, **94**, 8321, (1972).

[18] C. Eaborn, D. J. Tune and D. R. M. Walton, *J. Chem. Soc. Dalton*, 2255, (1973), and references cited therein.

[19] L. H. Sommer, L. E. Lyons and H. Fujimoto, *J. Amer. Chem. Soc.*, **91**, 7051, (1969).

[20] K. Yamamoto, T. Hayashi and M. Kumada, *J. Amer. Chem. Soc.*, **93**, 5301, (1971); K. Yamamoto, T. Hayashi, M. Zembayashi and M. Kamada, *J. Organometal. Chem.*, **118**, 161, (1976).

[21] K. Yamamoto, Y. Uramoto and M. Kumada, *J. Organometal. Chem.*, **31**, C9, (1971); K. Yamamoto, T. Hayashi, Y. Uramoto, R. Ito and M. Kumada, *J. Organometal. Chem.*, **118**, 331, (1976).

[22] Y. Kiso, K. Yamamoto, K. Tamao and M. Kumada, *J. Amer. Chem. Soc.*, **94**, 4373, (1972).

[23] *a*) J. D. Morrison, R. E. Burnett, A. M. Aguiar, C. J. Morrow and C. Philips, *J. Amer. Chem. Soc.*, **93**, 1301, (1971).
b) J. D. Morrison and W. F. Masler, *J. Org. Chem.*, **39**, 270, (1974).

[24] T. Hayashi, Ph.D. thesis, Kyoto Univ., (1975).

[25] For reviews,
a) B. R. James, *Homogeneous Hydrogenation*, Wiley-Interscience, New York, (1973).
b) R. E. Harmon, S. K. Gupta and D. J. Brown, *Chem. Revs.*, **73**, 21, (1973).

[26] R. R. Schrock and J. A. Osborn, *Chem. Commun.*, 567 (1970); *J. Amer. Chem. Soc.*, **94**, 2397, (1971).

[27] *a*) P. Bonvicini, A. Levi, G. Modena and G. Scorrano, *Chem. Commun.*, 1188, (1972); *Chem. Commun.*, 6, (1975).
b) M. Tanaka, Y. Watanabe, T. Mitsudo, H. Iwane and Y. Takegami, *Chem. Lett.*, **239**, (1973).
c) C. J. Sih, J. B. Heather, G. P. Peruzzotti, P. Price, R. Sood and C. F. H. Lee, *J. Amer. Chem. Soc.*, **95**, 1676, (1973); C. J. Sih, J. B. Heather, R. Sood, P. Price, G. Peruzzotti, L. F. H. Lee and S. S. Lee, *J. Amer. Chem. Soc.*, **97**, 865, (1975).

[28] *a*) Y. Ohgo, S. Takeuchi and J. Yoshimura, *Bull. Chem. Soc. Japan*, **44**, 583, (1971).
b) Y. Ohgo, Y. Natori, S. Takeuchi and J. Yoshimura, *Chem. Lett.*, 33, (1974).

[29] R. Calas, E. Frainnet and J. Bonastre, *Compt. Rend.*, **251**, 2987 (1960).

[30] E. Frainnet, *Pure Appl. Chem.*, **19**, 489, (1969).

[31] S. I. Sadykh-Zade and A. D. Petrov, *Zh. Obshch. Khim.*, **29**, 3194, (1959); *Doklady Acad. Nauk, SSSR*, **121**, 119, (1958).

[32] *a*) I. Ojima, M. Nihonyanagi and Y. Nagai, *Chem. Commun.*, 938, (1972).
b) I. Ojima, T. Kogure, M. Nihonyanagi and Y. Nagai, *Bull. Chem. Soc. Japan*, **45**, 3506, (1972).

c) I. Ojima, M. Nihonyanagi, T. Kogure, M. Kumagai, S. Horiuchi, K. Nakatsugawa and Y. Nagai, *J. Organometal. Chem.*, **94**, 449, (1975).

[33] R. J. P. Corriu and J. J. E. Moreau, *Chem. Commun.*, 38, (1973); *J. Organometal. Chem.*, **114**, 135, (1976).

[34] C. Eaborn, K. Odell and A. Pidcock, *J. Organometal. Chem.*, **63**, 93, (1974).

[35] *a*) I. Ojima, T. Kogure and Y. Nagai, *Tetrahedron Lett.*, 5035, (1972).
b) T. Kogure, I. Ojima and Y. Nagai, *Abstracts of the 21st Symposium on Organometallic Chemistry*, Japan, 114, (1973).

[36] I. Ojima, M. Nihonyanagi and Y. Nagai, *Bull. Chem. Soc. Japan*, **45**, 3722, (1972).

[37] H. C. Brown and V. Varma, *J. Amer. Chem. Soc.*, 88, 2871, (1966).

[38] For a review, E. C. Ashby and J. T. Laemmie, *Chem. Revs.*, **75**, 521, (1975).

[39] K. Yamamoto, T. Hayashi and M. Kumada, *J. Organometal. Chem.*, **46**, C65, (1972); *J. Organometal. Chem.*, **113**, 127, (1976).

[40] *a*) I. Ojima, T. Kogure and Y. Nagai, *Chem. Lett.*, 541, (1973).
b) I. Ojima and Y. Nagai, *Chem. Lett.*, 223, (1974).

[41] W. Dumont, J. C. Poulin, T. P. Dang and H. B. Kagan, *J. Amer. Chem. Soc.*, **95**, 8295, (1973); *Compt. Rend.*, **277**, C41, (1973).

[42] K. Yamamoto, T. Hayashi and M. Kumada, *J. Organometal. Chem.*, **54**, C45, (1973).

[43] *a*) T. Hayashi, K. Yamamoto and M. Kumada, *Tetrahedron Lett.*, 4405, (1974).
b) D. Marquarding, H. Klusacek, G. Gokel, P. Hoffmann and I. Ugi, *J. Amer. Chem. Soc.*, **92**, 5389, (1970).
c) T. Hayashi, T. Mise, S. Mitachi, K. Yamamoto and M. Kumada, *Tetrahedron Lett.*, 1133, (1976).

[44] T. Hayashi, H. Ohmizu, S. Baba, K. Shinohara, K. Kasuga, K. Yamamoto and K. Kumada, *Abstracts of the 22nd Symposium on Organometallic Chemistry*, Japan, 308A, (1974).

[45] R. J. P. Corriu and J. J. E. Moreau, *J. Organometal. Chem.*, **64**, C51, (1974); *J. Organometal. Chem.*, **85**, 19, (1975).

[46] P. Meakin, J. P. Jesson and C. A. Tolman, *J. Amer. Chem. Soc.*, **94**, 3240, (1972); C. A. Tolman, P. Z. Meakin, D. L. Lindler and J. P. Jesson, *J. Amer. Chem. Soc.*, **96**, 2762, (1974).

[47] K. W. Muir and J. A. Ibers, *Inorg. Chem.*, **9**, 440, (1970).

[48] R. S. Cahn, C. K. Ingold and V. Prelog, *Angew. Chem. Internat. Edit.*, **5**, 385, (1966).

[49] R. Glaser, *Tetrahedron Lett.*, 2127, (1975).

[50] T. Hayashi, K. Yamamoto and M. Kumada, *Tetrahedron Lett.*, 3, (1975).

[51] I. Ojima, T. Kogure and Y. Nagai, *Chem. Lett.*, 985, (1975); T. Kogure and I. Ojima, *Abstracts of the 24th Symposium on Organometallic Chemistry, Japan,* 115B, (1976).

[52] G. Vavon and A. Antonini, *Compt. Rend.*, **232**, 1120, (1951).

[53] A. Horeau, H. B. Kagan and J. P. Vigneron, *Bull. Soc. Chim. France*, 3795, (1968).

[54] Y. Ohgo, Y. Natori, S. Takeuchi and J. Yoshimura, *Chem. Lett.*, 709, (1974).

[55] *a*) I. Ojima, T. Kogure and Y. Nagai, *Tetrahedron Lett.*, 1899, (1974).
b) I.Ojima and Y. Nagai, *Chem. Lett.*, 191, (1975).
c) I. Ojima, T. Kogure and M. Kumagai, *J. Org. Chem.*, **42**, (1977), in press.

[56] V. Prelog, *Helv. Chim. Acta*, **36**, 308, (1953).

[57] I. Ojima, M. Kumagai and Y. Nagai, *Abstracts of the 23rd Symposium on Organometallic Chemistry*, Japan, 111B, (1975); Also presented at the *7th International Conference on Organometallic Chemistry*, Sept. 1975, Venice (Italy).

[58] I. Ojima, T. Kogure and Y. Nagai, *Tetrahedron Lett.*, 2475, (1973).

[59] E. Frainnet, A. Bazouin and R. Calas, *Compt. Rend.*, **257**, 1304, (1963).

[60] N. Langlois, T.-P. Dang and H. B. Kagan, *Tetrahedron Lett.*, 4865, (1973).

[61] L. H. Sommer, *Stereochemistry, Mechanism, and Silicon*, McGraw-Hill, Inc., New York, N.Y., (1965); *Intra-Science Chem. Rept.*, **7**, No. 4, p. 1, (1973).

[62] R. J. P. Corriu and G. F. Lanneau, *Tetrahedron Lett.*, 2771, (1971); *J. Organometal. Chem.*, **64**, 63, (1974).

[63] A. Holt, A. W. P. Jarvie and G. J. Jervis, *J. Chem. Soc., Perkin II*, 114, (1973).

[64] J. F. Klebe and H. Finkbeiner, *J. Amer. Chem. Soc.*, **90**, 7255, (1968).

[65] T. Hayashi, K. Yamamoto and M. Kumada, *Tetrahedron Lett.*, 331, (1974).
[66] T. Hayashi, K. Yamamoto and M. Kumada, unpublished data.
[67] *a*) J. Tsuji and K. Ohno, *Tetrahedron Lett.*, 2173, (1967).
b) M. C. Baird, C. J. Nyman and G. Wilkinson, *J. Chem. Soc. A*, 348, (1968).
[68] R. J. P. Corriu and J. J. E. Moreau, *J. Organometal. Chem.*, **91**, C27, (1975).
[69] L. H. Sommer and J. D. Citron, *J. Org. Chem.*, **32**, 2470, (1967).
[70] I. Ojima, T. Kogure, M. Nihonyanagi, H. Kono, S. Inaba and Y. Nagai, *Chem. Lett.* 501 (1973).
[71] *a*) I. Ojima, M. Nihonyanagi and Y. Nagai, *J. Organometal. Chem.*, **50**, C26, (1973).
b) H. Kono, I. Ojima, M. Matsumoto and Y. Nagai, *Org. Prep. Proc. Int.*, **5**, 135, (1973).
[72] R. J. P. Corriu and J. J. E. Moreau, *Tetrahedron Lett.*, 4469, (1973); *J. Organometal. Chem.*, **120**, 337, (1976), and references cited therein.

Author index

C

I

J

K

L

M

N

O

P

T

U

V

W

Y

Z

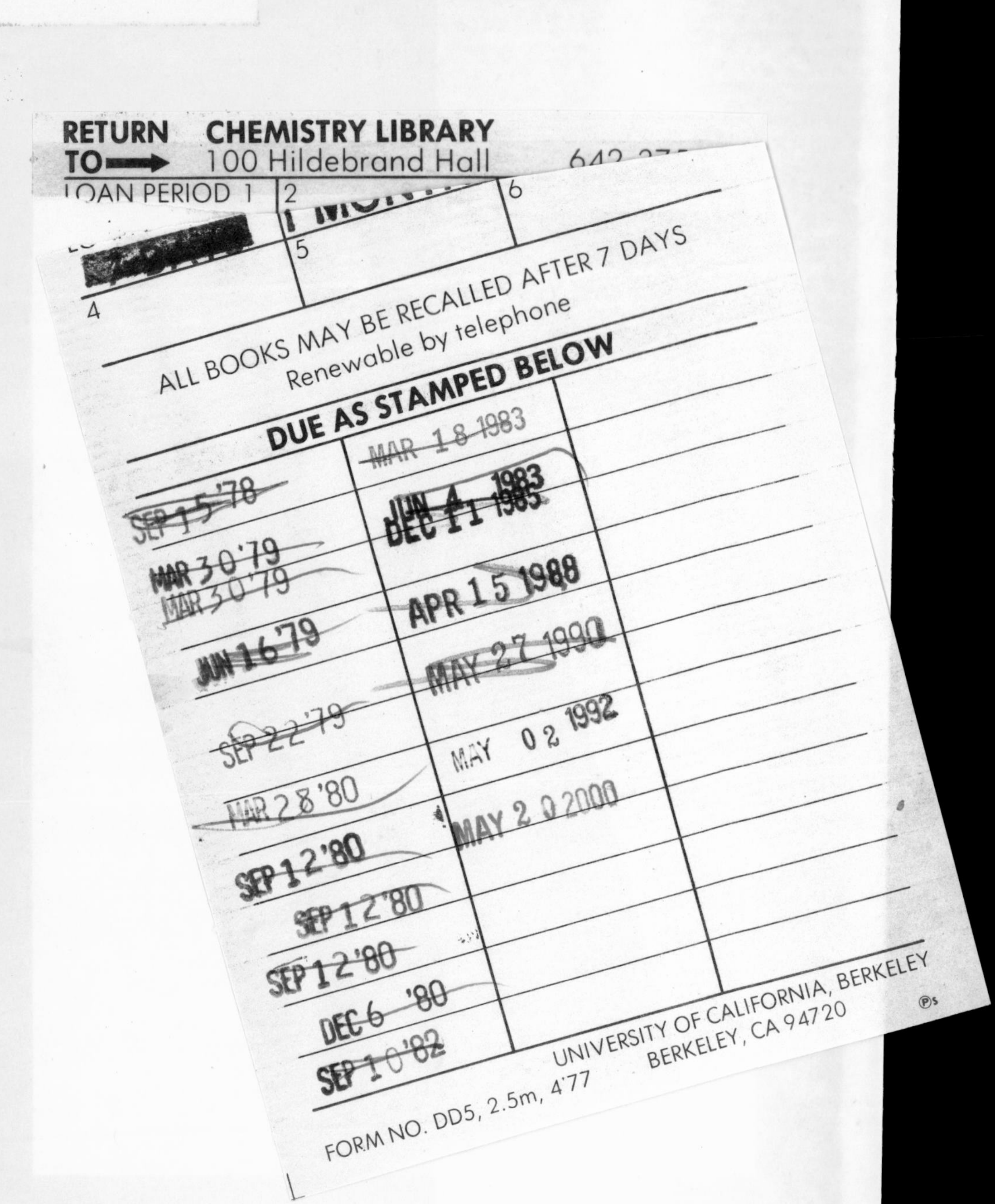